Understanding Voice over IP Technology

NICK WITTENBERG

Australia • Brazil • Japan • Korea • Mexico • Singapore • Spain • United Kingdom • United States

Understanding Voice over IP Technology, First Edition
Nick Wittenberg

Vice President, Career and Professional Editorial: Dave Garza

Director of Learning Solutions: Sandy Clark

Managing Editor: Larry Main

Senior Product Manager: John Fisher

Senior Editorial Assistant: Dawn Daugherty

Vice President, Career and Professional Marketing: Jennifer McAvey

Marketing Director: Deborah S. Yarnell

Marketing Manager: Erin Coffin

Marketing Coordinator: Shanna Gibbs

Production Director: Wendy Troeger

Production Manager: Mark Bernard

Content Project Manager: Christopher Chien

Senior Art Director: David Arsenault

Technology Project Manager: Christopher Catalina

Production Technology Analyst: Thomas Stover

Library of Congress Control Number: 2008938851
ISBN-13: 978-1-4354-2727-3
ISBN-10: 1-4354-2727-0

Delmar
5 Maxwell Drive
Clifton Park, NY 12065-2919
USA

Cengage Learning is a leading provider of customized learning solutions with office locations around the globe, including Singapore, the United Kingdom, Australia, Mexico, Brazil and Japan. Locate your local office at: **international.cengage.com/region**

Cengage Learning products are represented in Canada by Nelson Education, Ltd.

For your lifelong learning solutions, visit **delmar.cengage.com**

Visit our corporate website at **cengage.com**

Printed in the U.S.A.
3 4 5 XX 13

Contents

Preface

The technology of voice over IP provides for making telephone calls over data networks such as the Internet and it has now reached critical mass. I know this because as a technology instructor I spend more and more time teaching Voice over IP or "VoIP." I wrote this book because I couldn't find any material suitable for classroom use or self-study. There is a good selection of *reference* books on VoIP; this is not one of them.

In order to be successful as teaching material, this book needs to meet at least three criteria. First, the material needs to be organized in a logical way, be of appropriate technical level for the reader and progress by building upon material that has been presented in early chapters. Second, feedback in the form of exercises must be provided so that the learner can judge whether he understands the material provided. Third, labs need to be included that provide hands-on experience. Hands-on experience is the most exciting and, dare I say "fun," part of the learning experience. This presents challenges when learning about technology that, until recently, has been confined to the central office of the telephone company. Nevertheless, this challenge has been met by using programs that can be freely downloaded from the Internet and are suitable for both classroom use and an individual using a home computer.

This book is divided into three sections: an overview of VoIP, the basics of TCP/IP, and the technical details of VoIP. The first six chapters provide the introduction to the technology of VoIP: what it is, the different services based upon it, a first look at the protocols and what infrastructure is required to support it. These chapters provide an easy entry for the reader who is anxious to jump in and also for those who want an overview of VoIP with a minimum of technical detail.

A good working knowledge of the TCP/IP protocol is fundamental to understanding how VoIP functions and is a prerequisite for tackling the technical details of voice over IP. In my classes I find that students have experience with TCP/IP ranging from barely recognizing an IP address to full protocol analysis, and everything in between. Chapters 7 to 13 provide the TCP/IP background that the reader needs to understand the technical details of VoIP. Completing these chapters provides assurance that the reader has the prerequisite knowledge. By including this section, I know that this book is a single source for all the material needed and the reader does not have to refer to another textbook or take another course.

The final section on advanced VoIP is found in Chapters 14 through 20. It covers the technical details of VoIP, particularly quality of service, codecs, and the signaling protocols of H.323, SIP, and Megaco. Some practical considerations for deploying VoIP are also covered such as writing a dial plan and troubleshooting VoIP when it passes through network address translation (NAT).

INSTRUCTOR NOTE

The modular nature of this book allows a course to be designed for different audiences. I frequently encounter an audience that needs to understand the basics of VoIP but is not interested in the deep technical details. The audience might include managers, salespeople, computer help desk, or those tasked with recommending a VoIP system. For this audience, the first section of the book provides a nice one-day course. I also encounter training institutions that provide a dedicated TCP/IP course as part of the curriculum. In this situation, the TCP/IP section of this book can be assigned as review and not covered in classroom instruction. In my experience, however, most students need this book's material from cover to cover. Therefore, customize the course as you see fit; you know your audience best. A complete set of slides in a PowerPoint presentation is included on the Instructor's Resources CD-ROM, ISBN 1435427289.

LAB REQUIREMENTS

The hands-on labs in this book were designed to accommodate as wide a variety of hardware situations as possible. Microsoft Windows XP is the base operating system, but Windows Vista can be used as well. All the software used in the exercises can be downloaded from the Internet and is free to use. The utilities used in the labs can also be found on the included CD-ROM. This makes the exercises accessible to those who are doing self-study in the home environment as well as a classroom. Naturally, the solitary reader will not be able to phone another person when called to do so in a lab. The only other hardware required is a headset with microphone. I often use VoIP telephones in the classroom, but these are neither assumed nor included in any exercises.

Understanding Voice over IP Technology is professionally designed course material that gives the student the in-depth knowledge of voice over IP and the

TCP/IP protocol that it is based on. Voice over IP technology is making deep inroads in all types organizations, even residential telephony, such that it will eventually become the standard telephone technology. I hope that you enjoy this book and that it contributes to your future success in this exciting industry.

ACKNOWLEDGMENTS

The publisher and author would like to thank and acknowledge the many professionals who reviewed the manuscript to help us publish this text. A special acknowledgment is due to the following instructors who reviewed the chapters in detail:

Warren Koontz, Rochester Institute of Technology, Rochester, NY
Eliazar Martinez, El Centro College, Dallas, TX
Mike Murphy, Foothill College, Los Altos Hills, CA

Nick Wittenberg

SECTION

1

Voice over IP Overview

CHAPTER

1

Voice over IP Overview

Objectives

Upon completion of this section, the reader will:

- *Understand that voice over IP (VoIP) is a technology that transmits telephone calls over an IP-based data network.*
- *Appreciate that telephone calls can be made over the Internet with a substantial savings in cost but that this technology has problems with voice quality.*
- *Know that organizations are slowly embracing VoIP to replace their traditional analog telephone systems and decrease their long distance telephone charges.*

INTRODUCTION

Voice over IP (VoIP) is the technology used to describe telephone services over a transmission control protocol / Internet protocol (TCP/IP) network. The applications for this technology vary, including making telephone calls over the Internet as well as intra-organizational telephony. However, nothing is simple—VoIP requires modifications to network infrastructure and protocols; it requires new hardware, such as IP telephones; it requires new software. This lesson looks at the broad picture of sending voice over a data network.

1.1 WHAT IS VoIP?

VoIP is an application that uses a combination of hardware and software to provide telephony services across an IP-based network.

Hardware and software

At a minimum, the hardware required for VoIP includes network hardware such as network adapters, routers, and switches. Now add gateways and IP PBXs. If a personal computer is being used as a soft phone, the PC also needs a sound card, microphone, and headphone. In a business environment, IP telephone handsets will most likely be used. Software may need to be added to the personal computer, and software in routers, gateways, and PBXs will need to be modified.

Telephony services

A VoIP system provides simple telephone conversations between two persons. This is as expected. It can also provide the additional extended services of a modern telephone system such as conference calling, call forwarding, and call display. Some additional services that can be provided by VoIP may be surprising, however. One such is moving the telephone to another location and having the call follow automatically. Employees can be reached by just dialing their extension, even if they are working from home that day.

IP-based network

The Internet protocol (IP) is a component of the TCP/IP suite of protocols used to transmit data in a reliable fashion over wide area networks (WANs) and local area networks (LANs). Although there are many protocols, IP is the only protocol allowed on the Internet and therefore it has become the overwhelming favorite to move data on modern systems. IP was not designed with voice traffic in mind; however, modifications to the protocol allow it to provide excellent voice quality. These will be examined in upcoming sections of this book.

People can use the Internet as the transmission medium for telephone calls. For users who have free or fixed-price Internet access, Internet telephony software essentially provides free telephone calls anywhere in the world. To date, however, Internet telephony does not offer the same quality of telephone service as direct telephone connections. When organizations require the highest voice quality, they may avoid the public Internet, but still use VoIP technology within their company and to their branch offices.

1.2 RESIDENTIAL / SMALL-BUSINESS VoIP

Voice over the Internet

VoIP lets you make long distance calls over the Internet right from your computer. The driving force behind the development of these software tools is the money users can save on long distance charges. The Internet phone was particularly alluring when long distance telephone charges were high. Since the advent of serious price competition in this field, the charges for long distance in North America have dropped extremely low so this exercise may not be worth the effort if your long distance needs are minimal. However, if you often make telephone calls to places outside North America, for business or family, you may find some serious savings in Internet telephone.

In this section we will explore the different forms that Internet telephone takes—computer to computer and computer to person via an ITSP (Internet telephony service provider).

Computer to computer

You can use computer-to-computer Internet telephone if the computers at both ends of the conversation are similarly equipped with a sound card, microphone and speakers, a headset, or a USB telephone. Then, all you need is the right software and you can start talking.

This is a relatively new field in software. The first product came on the market in 1994. Pioneered by Vocal Tec, Internet Phone is still a leader in its field. Others soon followed, such as Web Phone and Microsoft's NetMeeting. Currently the market leader with the most mind share is Skype. Its website claims 173 million downloads as of this writing.

Internet telephone software may be free or cost only a modest amount, and certainly the avoidance of long distance charges is compelling. So what is the catch?

Computer-to-computer Internet telephone is not without its wrinkles

- Both computers have to be similarly equipped with sound cards, speakers, and microphones. So obviously you can't phone just anyone. You need to know if the other user is equipped to receive your call.
- Both computers have to be running the same Internet telephone software.

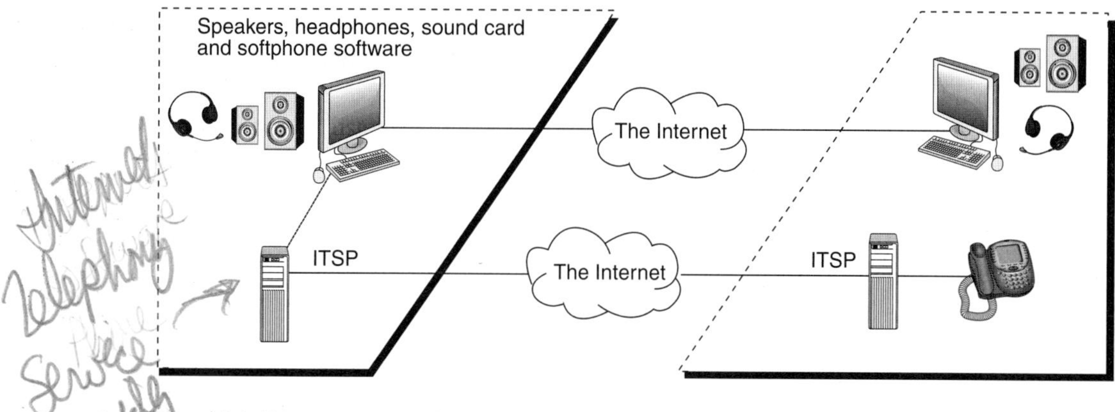

Figure 1-1: Voice over the Internet

- Both machines have to be up and running plus logged onto the Internet; otherwise, how will the receiving machine "ring" when the call is coming in?
- The sound quality can range from good to very poor. After all, it is traveling over the Internet and sound quality cannot be guaranteed.

Despite these problems, this type of telephone service does have its ardent followers. More users will be attracted once they learn that some of the Internet telephone programs have been enhanced with video. This is a natural fit for these services since once you are already talking, why not add video and see each other as well? As attractive as this sounds, a reality check is in order here; with poor connections or limited bandwidth, the video will be close to useless.

Figure 1-1 illustrates the concept of the computer-to-computer telephone call and the equipment needed to make it. The diagram also illustrates the role of an Internet telephony service provider.

Internet Telephony Service Provider (ITSP)

After looking at the way the computer-to-computer telephone service works, you can see one huge flaw. What if the other party doesn't have a computer, let alone the hardware and software needed to make the telephone service work? The alternative is to enlist the aid of an ITSP.

An ITSP will take your Internet telephone call and place it onto the national telephone system of the person you are calling. As the originator of the call, your computer must be set up just as is required for computer-to-computer telephone; namely you will need the sound card, speakers, and microphone. However, the recipient will receive a real telephone call and can talk to you using his regular telephone handset so you have no worries about computer hardware and software capabilities at the other end.

The ITSP will take your call either at a local server and send it across a private network to the destination country, or send your call across the Internet to the ITSP's node in the remote country. In either case, the ITSP must have a server in the remote country that transfers the call to the national telephone system. It is important to know this limitation if you are trying to call someone in a country that is not well served by the Internet. Naturally, the ITSP will expect to be paid for this little service.

Residential VoIP

Residential VoIP makes a regular telephone service available over a broadband Internet connection. The Internet connection can be either DSL, cable, or fiber to the home. The providers of this type of service offer telephone service at very low rates in comparison to the legacy telephone companies. Figure 1-2 illustrates how the residential VoIP service and a family area computer network can share a high-speed Internet connection.

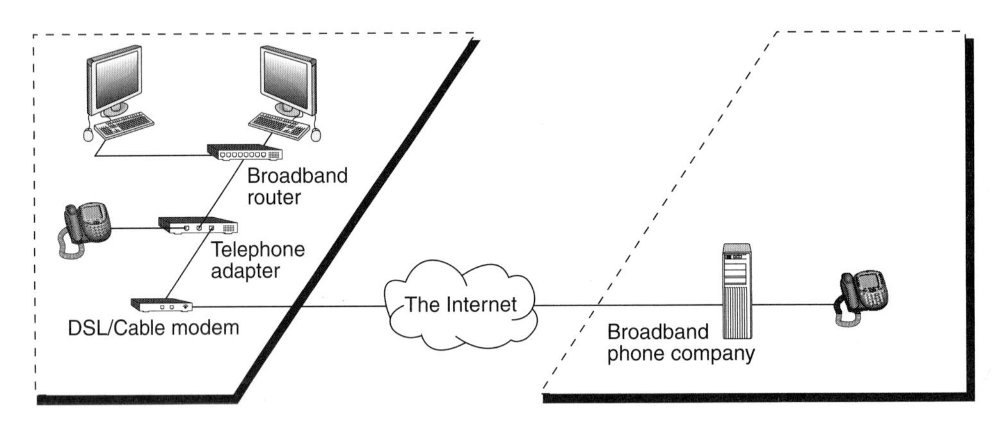

Figure 1-2: Residential VoIP

Standard telephone features are included with the service, including voicemail, call waiting, and call forwarding. There are two surprising features that cannot be duplicated by the traditional telephone companies.

- Virtual telephone numbers—this option will assign an additional telephone number to you from another area code. Anyone calling you from that area code will dial a local number, thereby avoiding all long distance charges.
- Telephone portability—this means that you can take your telephone adapter with you if you travel, and anyone calling your local number will reach you wherever you are. Naturally, you will need to find a high-speed Internet connection at your destination. But since many hotels now offer a high-speed service, this is a practical option.

1.3 CORPORATE VoIP

Corporate VoIP

Corporate VoIP falls into three distinct types of systems:

1. Analog at the ends, digital IP in the middle
2. Digital IP at the ends, analog in the middle
3. Digital IP end to end

Naturally, the permutations and combinations of digital and analog are endless.

VoIP on managed IP backbones

This is system type one, analog at the ends and digital in the middle. Converting voice to data and sending it across a wide area network to avoid long distance charges (toll bypass) is not a new idea. It is a common ploy used by large organizations to provide telephone services between offices. The current implementation of this idea uses leased lines and technology such as T1 lines, frame relay, or ATM. Figure 1-3 illustrates this technology. Without a leased data line, a call must enter the public system telephone network (PSTN) and long distance charges apply. With a leased line, voice and data can be multiplexed and sent over long distance while avoiding long distance charges. Voice must be digitized by the PBX before it can be sent over the data lines. If the long-haul network is replaced by an IP network, the architecture of the combined voice and data network changes to that illustrated in Figure 1-4.

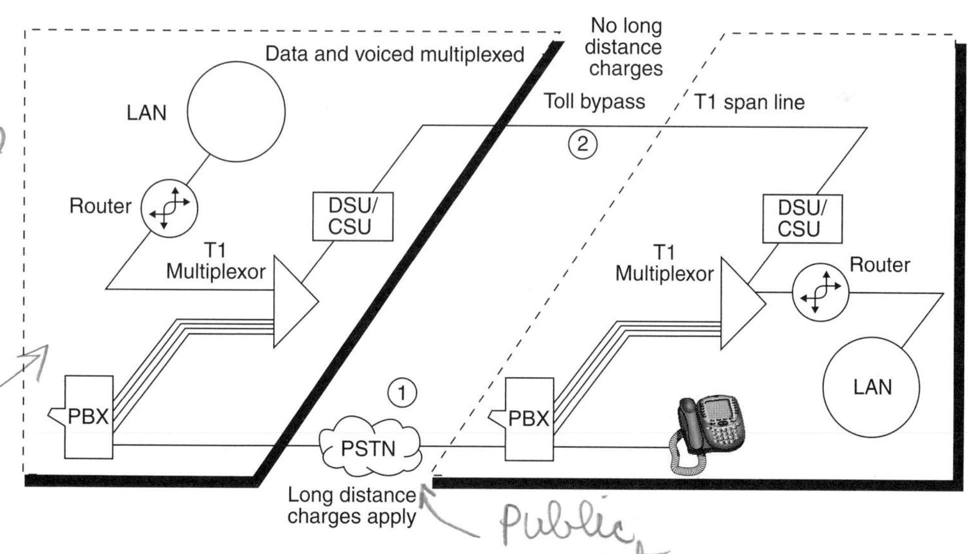

Figure 1-3: Traditional long distance—PSTN or toll bypass

The major change is that VoIP packetizes the digital voice data before transmitting it across the wide area network. This is a requirement if the voice traffic is to mingle with the computer data as it traverses the IP backbone network.

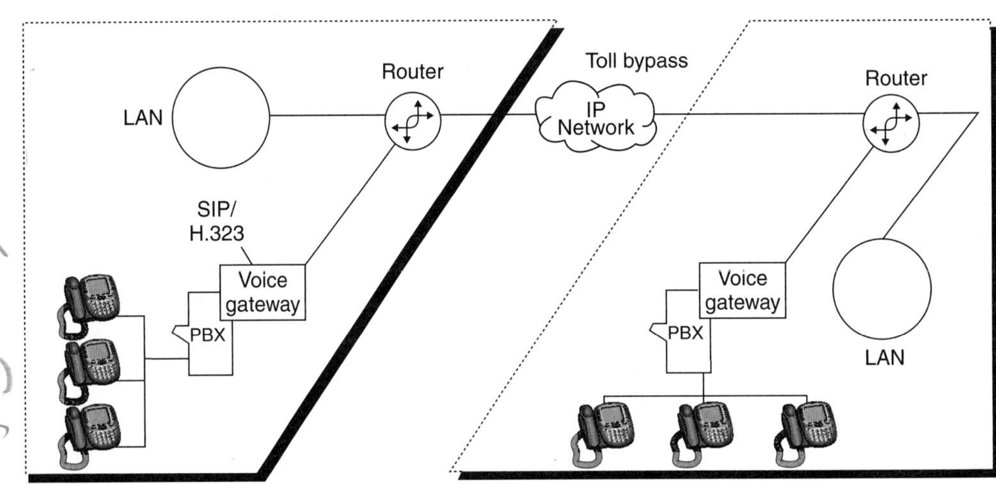

Figure 1-4: Voice on a managed IP backbone

Because a traditional voice system is used inside the organization, all the telephone lines are routed to a PBX. To packetize the voice, a VoIP gateway is used. This may be a separate box or it could be a component of the PBX. The digitized voice is then routed over the WAN via routers. The gateway must match the control protocols used by the VoIP system. The most common control protocols used by VoIP are SIP and H.323. The gateway has many functions, just one of which is to look up the address of the person being called and make a connection. A gateway is a machine which connects two different types of systems. Since the machine connects an analog and digital system, gateway is the correct term for it. The relationship of these components is illustrated in Figure 1-4.

VoIP takes over the company telephone system

The second scenario involves an IP-based telephone system replacing the analog telephone system inside the organization. Earlier, this was characterized as the digital at the ends, analog in the middle system. This scenario uses analog in the middle because telephone calls must still funnel into the PSTN to reach outside parties or branch offices that aren't reached by managed IP backbones. See Figure 1-5.

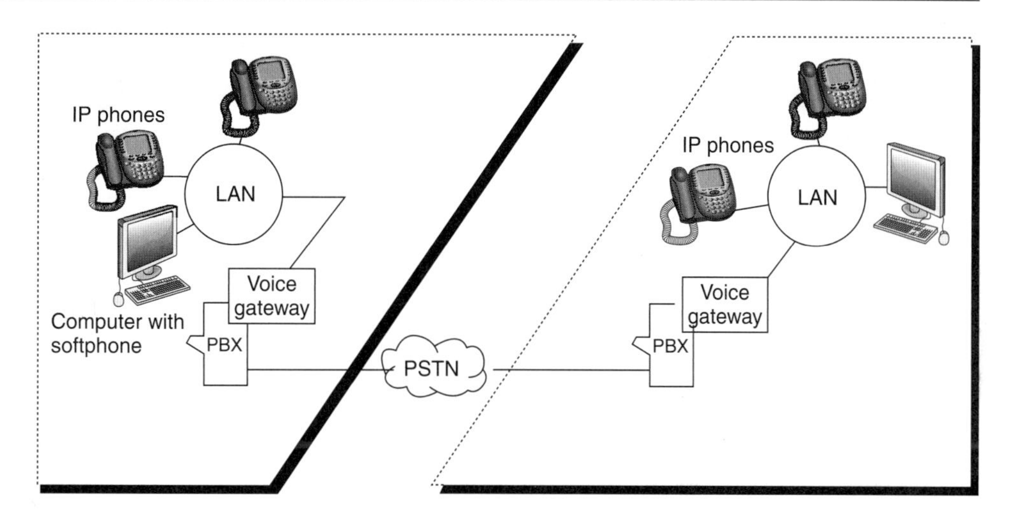

Figure 1-5: VoIP takes over the company telephone system

Installing an IP-based telephone system may be attractive if the company wants to manage a single unified communication system. For example, only one wire has to be routed to each workstation instead of separate LAN and telephone cables. It is easier to make moves, adds, and changes on an IP-based telephone than a traditional PBX system. The drivers for this technology are explored in greater depth in an upcoming section. In this system, telephones are IP based or the user uses the communication software of her computer. Note that the telephones are connected to the LAN and that therefore, the telephone system is dependent on the data network being configured and functioning properly. Since data networks don't traditionally approach telephone networks in reliability, this could prove a challenge.

VoIP end-to-end

VoIP end-to-end is a combination of the previous two systems and is characterized as being digital at the ends as well as digital in the middle. Not only is the telephone system digital but the network that carries voice to the other offices of an organization is also digital.

Figure 1-6 illustrates the system. Note that the devices that connect different parts of the pure VoIP system are called VoIP gatekeepers, not gateways. The gatekeeper is used to coordinate voice communications across the IP backbone. Since all portions of the system are digital VoIP, a gatekeeper is used, not a gateway. A gateway is used when connecting different types of systems, such as a digital VoIP system to an analog voice system. Note that even users of a pure VoIP system still need to dial outside the organization and this connection to the public telephone system requires a VoIP gateway.

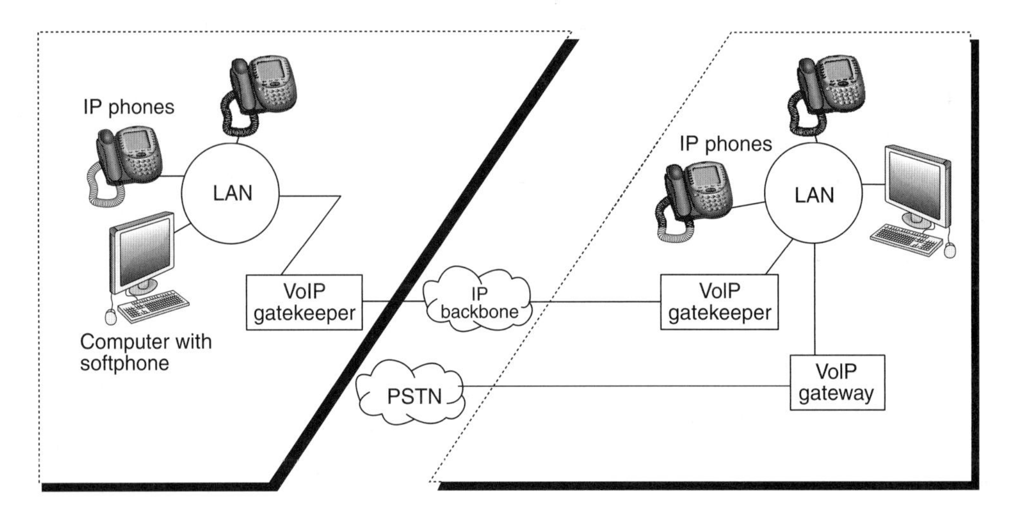

Figure 1-6: VoIP End-to-End

1.4 VoIP FOR TELEPHONE COMPANIES AND SERVICE PROVIDERS

Telephone companies and VoIP

For telephone companies, VoIP is a major disruptive influence. It requires re-engineering their voice networks and it allows competitors into their market.

The telephone companies' voice systems are based on fixed lines with circuit switching. Although voice traffic is already digitized in order to travel over T3 and ATM circuits, these are not packetized. The packet switching services they do offer, X.25 and frame relay, are losing their installed base.

Switching over to an IP-based network fabric will cause disruption and require an expensive upgrade. On the other hand, an IP-based network can provide some efficiencies. Because VoIP can compress telephone conversations, the telephone companies can make better use of their lines.

Realistically, the telephone companies will change over to an IP-based infrastructure more because they have to than because they want to. Corporate customers will demand VoIP services, residential customers will be seduced into thinking they should have it, and competitors will be offering it.

Converting carriers to VoIP

A superclass softswitch is used to overlay or add VoIP capabilities to the networks of carriers. The superclass softswitch supports multiple core applications on a single platform, including local, long distance, and tandem. It provides these services with the reliability that carriers are expecting. As Figure 1-7 illustrates, the following

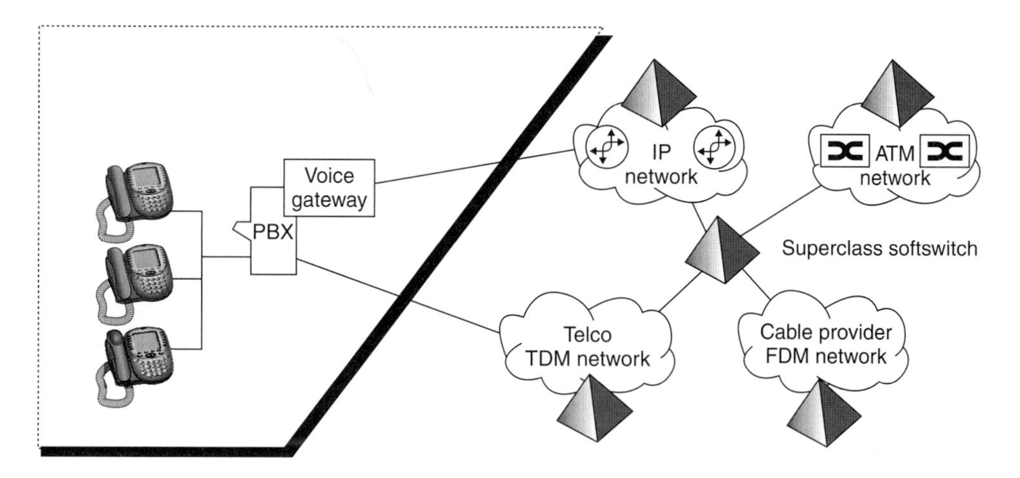

Figure 1-7: Converting Carriers to VoIP

carriers or service providers can use a superclass softswitch to convert their core networks to VoIP:

- Carriers that currently provide voice services, the traditional telephone companies, typically use a TDM (time domain multiplexing) core infrastructure based on SONET and the DS/T services. Their telephone switches are the DMS series from Nortel, the 5ESS switch from Lucent, and the EWSD from Siemens. These systems can be overlaid with VoIP services by adding a superclass softswitch at which point they are known as "hybrids."
- Cable companies use a technology known as FDM (frequency division multiplexing). VoIP services can be added to these systems, giving cable companies a comprehensive package of services to offer to their customers.
- Service providers that already run an IP or ATM network can also integrate VoIP services by deploying softswitch technology.

SUMMARY

This module provides the basic introduction to voice over the IP protocol. Unless you know how traditional analog telephone systems work and how data networks work, VoIP is hard to understand.

Section 1.1: What Is VoIP?

This section defines VoIP as a hardware and software system that provides telephony services over an Internet protocol system.

Section 1.2: Residential/Small-Business VoIP

This section first looks at telephone services over the Internet. Both computer-to-computer and computer-to-ITSP are examined. Since the quality of voice over the Internet is not up to corporate standards, corporations are pursuing VoIP services over leased lines.

Section 1.3: Corporate VoIP

Organizations see VoIP in three areas: toll bypass, replacing the corporate telephone system with an IP-based telephone system, and pure VoIP end to end. This section examines these three areas.

Section 1.4: VoIP for Telephone Companies and Service Providers

VoIP is a double-edged sword for the telephone companies. It is a disruptive technology that will require them to spend a lot of money to overhaul their systems and it will allow new competitors to enter their business. On the other hand, they will be able to run their systems in a more efficient manor. By deploying a superclass softswitch, their networks can be made VoIP ready.

Review Questions
Voice over IP Overview

1. In order to relay a VoIP call from within a company to the public telephone system, __________ is required.

    a) a PBX
    b) a gateway
    c) a router
    d) a firewall

2. Which one of the following reasons may prevent computer-to-computer telephone calls?

    a) Each computer may have different telephony software.
    b) One of the computers may not have an IP telephone attached to it.
    c) Both of the computers have sound cards, microphones, and speakers.
    d) One of the users hasn't set up an account with an ITSP yet.

3. In order to achieve cost savings with toll bypass, a company needs

    a) an account with an ISP giving it access to the Internet.
    b) a PBX at both head office and the branch office.
    c) a special long distance plan.
    d) a leased line to its branch office that is located in a different area code.

4. In order for a telephone to be connected directly to the local area network,

    a) the PBX must have an IP interface.
    b) there must be a jack on the back of the computer to plug the telephone into.
    c) it must be an IP telephone.
    d) it can be an analog telephone with an IP address.

5. In order for a VoIP telephone call to go from one data network to another, it must be routed through

    a) a gatekeeper.
    b) a digital PBX.
    c) a router.
    d) a gateway.

CHAPTER

2

Why VoIP?

Objectives

Upon completion of this section, the reader will:

- *Appreciate that data and voice networks have been coming together for some time now.*
- *Will be able to answer if the final network of the future will be a switched TCP/IP network.*
- *Understand the benefits of voice over IP as well as some of the challenges involved.*
- *Know who and what factors are driving the adoption of VoIP.*

INTRODUCTION

Voice over data networks and data over voice networks are definitely merging. This chapter explores this issue as well as why you might want to implement VoIP technology. To be fair, the downsides of VoIP must also be explored. Finally, who and what are driving this technology forward? These are the issues that are covered in this chapter.

2.1 CONVERGING NETWORKS

Data over voice networks

With the introduction of modems in the early 1960s, data could be transmitted over the public telephone system.

Voice over data networks

AT&T introduced the T services in 1962 as a means of increasing the capacity of its central inter office trunk lines. Originally designed to multiplex 24 voice circuits over a 4-wire cable, it now transmits any combination of voice, data, and video. It wasn't until the early 1980s that T1 circuits became available for corporate clients.

ISDN

Integrated services digital network (ISDN), developed in the late 1980s, was a successful attempt to provide services for voice, data, and video traffic to small businesses as well as large corporations. ISDN was attractive to small businesses because the service included two connections (called 2B+D) over one telephone wire. You could have a telephone conversation, fax a document, and cruise the Internet, as long as you used two out of the three, at the same time. The transmission rate was also attractive, 64 kbps at a time when modems were speedy at 24 kbps. ISDN's time has now passed because of the emergence of high-speed access to the Internet.

The Internet

The Internet, and its precursor, ARPANET, languished for most of its history. From its inception in 1969 to the early 1990s, the Internet was mostly of interest to academics and the military. The introduction of the World Wide Web made the Internet of interest to everyone and since then, the growth of the Internet has put it on par with the public system telephone network.

Now we are in the situation of having two parallel networks. Merging the two together makes some sense as long as all types of data can be serviced equitably. The analog voice network is actually a hybrid with a digital core, and analog services only to the end user. On this basis, it makes sense to convert the last analog portions to digital. In addition, the preferred method of moving data is through packet switching networks instead of point-to-point connections.

The only portion of the telephone system that is still analog, at least in advanced economies, is the "last mile" from the customer's premises to the telephone company's central office. Only when these links have been converted to digital will the two networks have truly converged.

2.2 BENEFITS OF VoIP

Voice over IP can be a complex and expensive technology to install. What benefits can organizations see that would be an incentive to make a change from their well-established telephone system?

Reduced long distance costs

Reducing long distance costs is always attractive to any organization, but only realistic if certain conditions are met. First, those calls are being made to other locations of the same organization. Furthermore, the organization must have fixed-rate leased lines to those locations. If a long distance call enters the public telephone system, savings on that call cannot be realized.

Traditionally, leased lines to other locations were installed if the organization needed to move computer data between sites. By converting voice to data, voice could join the other data and be transmitted to other locations. As long as the voice did not increase the need for more bandwidth, and hence increase the leased line charges, long distance charges could be avoided. This concept is known as toll bypass.

VoIP provides the toll bypass benefit and, in fact, may use the same leased lines. It is natural to ask the question, "What is the difference between the two systems?" The answer is the packet. In order for IP to work, the information that is being transmitted is divided into sections, and control information is added to each section in the form of a "header." The resulting combination is called a "packet." An example of control information is the address of the destination. Why go to all this trouble? Because each packet is independent and when transmitted on a packet switching network will find its way to the destination. In other words, a fixed link does not have to be established between one computer and the other computers that it needs to communicate with.

Data networks are naturally packet based and by converting voice to packets, they can be managed along with the other data on the system.

If you already have toll bypass using leased lines, switching to VoIP on a managed IP backbone may not save a lot of money. However, better management and the increased services described later in this chapter may still make it a worthwhile exercise.

More calls with less bandwidth

Traditional voice digitization techniques require 64 kbps (thousands of bits per second) to function properly. This rate is a function of the way voice is digitized using a technique known as pulse code modulation (PCM). PCM provides pristine voice quality and is the hallmark against which other voice digitizing methods are measured. The 64 kbps rate is responsible for the way T1 lines operate. Each T1 line has 24 channels of 64 kbps each (which are individually referred to as DS0) and was originally devised to transmit 24 long distance calls simultaneously.

Today, state-of-the-art toll quality voice can be achieved at as low as 2 kbps, but is more typically provisioned at 8 kbps. Methods of generating digitized voice below 64 kbps PCM can use several techniques. One is data compression. Silence suppression is also a key component of many of these techniques. In other words, when there is a lull in the conversation, no packets are generated.

If each voice stream requires 8 kbps, then one DS0 channel can carry 8 conversations instead of one, a clear savings.

More and better enhanced services

No one will replace a system with one that has fewer features. Therefore the VoIP system has to at least equal the traditional PBX when it comes to voice services. These include the familiar ones such as call forwarding, call waiting, third-party calling (charging the call to a telephone other than the caller or the recipient), collect calling (charge reversal), caller identification, and so on. Table 2-1 lists and describes many services available with VoIP, any of which might be assigned to a consumer depending on the plan you select when you sign up with a VoIP service provider.

VoIP makes it easy to interact with services that are also data based such as e-mail and unified messaging as well as call center applications.

These last examples fall into the category of computer telephony integration (CTI). VoIP will further enhance CTI. This is especially true when the Internet and the World Wide Web are mixed in.

One intriguing new service applies to home workers. You can reach a worker who is at home by her extension number. When that worker is back in the office, you can still reach her with the same extension number. She made a simple configuration change and now her office telephone rings instead of her home phone.

Administration and maintenance savings

Administration and maintenance savings fall into several categories:

- One cable system to maintain. Since the telephone system and the LAN use the same cable system, UTP cat5, only one system needs to be maintained.
- Administration can be centralized, probably within the IT group. Head count can be decreased.
- Moves, adds, and changes are easier to make. If an employee moves to a different floor, just relocate his telephone. The DHCP based IP phone will instantly put the employee back on the telephone system. His extension will stay the same.

Three-Way Calling	Talk with two parties at the same time.
Anonymous Call Blocking	Block calls that do not have an originating telephone number.
Area Code Selection	Receive a telephone number with a different area code than the area that you are physically located in. This feature is used when a VoIP provider offers you a second telephone number and depends upon the areas that the service provider has a point of presence.
Call Blocking	Blocks calls coming from a specific number. Useful if you want to block calls from a pesky telemarketer or an ex-boyfriend or girlfriend.
Call Forwarding	Calls are forwarded to another telephone number or go straight through to voice mail.
Caller ID	Displays the number or name of the person calling you. Helpful if you want to screen your calls.
Caller ID Blocking	Hide your call ID when calling others.
Call Transfer	Transfer a call that you have already answered to another telephone.
Call Waiting	If you are in the middle of a conversation, a tone will alert you to another call coming in. You can put the first party on hold while you answer the second call or let it go through to voice mail.
Change Phone Number	Some VoIP service providers allow you to change your telephone number yourself online or by calling customer service.
Do Not Disturb	An incoming call will not ring your telephone but will go directly to voicemail instead.
Distinctive Ring	Use a different ring tone for your business and home lines or to different callers.
Enhanced Voicemail	Pick up your voicemail from any phone, not just your home phone. Have your voicemail messages forwarded as email attachments.
Extra Virtual Numbers	Additional telephone numbers with one service. Combine this with a different area code and you have a local presence in a long distance location.
Fax	Attach an analog fax machine to an analog telephone adapter and fax over your VoIP line.
eFax	eFax is a service of a VoIP service provider such that when a fax is sent to you, it is forwarded by the provider as an email attachment.
International Blocking	Block international calls by a business that wants to control its costs or a family that has small children that play with the phone.
Last Number Redial	Call the last number that you dialed with the push of a button.
Local Number Portability	The ability to keep your telephone number even when you change telephone service providers.
Line Unavailable Forward	Forward a call from an unavailable line, say your high-speed connection is down, to an available telephone, say an analog line or cell phone.
Multi-ring	Ring several telephones at once or in sequence (a hunt group), for example business, home, and cell phones. Whichever telephone you answer, you use.
PC to Phone Option	Use the softphone software loaded on your computer to make telephone calls. Particularly attractive to someone traveling with her laptop.

Personalize Contact List	Assign a friendlier name to Caller ID with Name. Instead of "Last Name" showing up on your display, "Mom" will be used.
Repeat Dial if Busy	Redial automatically if you receive a busy signal.
Telemarketer Blocking	The automatic dialing system of some telemarketers produces a 2–3 second pause before the agent gets on the telephone. This feature detects the pause and disconnects the call.
Travel Globally	When traveling from home, you can still receive your calls if you have access to a high-speed Internet connection and brought your VoIP telephone, ATA adapter, or have softphone software on your laptop.
Unlimited to US/Can	Unlimited calling within North America without long distance charges. A common plan offered in North America.
Voicemail	The ability to receive and save a message from a caller even when you are not available to pick up the phone.

Table 2-1: VoIP telephone service features

Notes: 1) Most of these features are also available with the analog telephone service.
2) Some of these features require support by the VoIP provider or are available only if the equipment supports it.

2.3 ISSUES WITH VoIP

Organizations may be hesitant to implement VoIP for several reasons. They may wish to test out and play with the technology before committing to a full scale implementation.

One cable system

One of the benefits of VoIP, a single cable system to install and manage, can also be a weakness. If the network is down both your data and telephone systems are out of action.

IT is not telephony

Telephone systems normally have five 9s (99.999%) uptime. Data networks cannot say the same. To put this in perspective, five 9s means that the system can be down only 5.26 minutes per year. IT personnel will need to take extra telephony training and be cautious around their networks. No more playing around with the server in the middle of the day. Data networks are also notorious for weak security and viruses. These will have a negative impact on the telephone service.

Upgrade the system

Besides the telephone equipment itself, the network infrastructure will probably need to be upgraded. This may include upgrading the network cabling and devices such as hubs and switches to handle fast Ethernet. Routers will need to be upgraded to handle quality of service.

Power can't be interrupted

Analog telephone sets receive power over the telephone wire. This provides dial tone even when there is a power outage. In order to provide this same service, servers and other network devices need uninterruptible power supplies (UPS). IP telephone sets will need to rely on the building's protection from power outages or embrace Power over Ethernet.

911 emergency service

Emergency services depend on being able to find someone who dials 911. This is not a problem with the PSTN since telephone numbers, including extension numbers in office buildings, are mapped to physical locations. However, IP telephones can easily be moved from location to location. When plugged in, they receive their IP configuration from a DHCP (dynamic host configuration protocol) server and they are back in service again. Some method must be used to map IP addresses to physical locations in order for emergency services to respond quickly to the right location.

2.4 DRIVING VoIP

VoIP technology will not be adopted unless there is a measurable benefit to stakeholders. The stakeholders are as follows.

Users

Users, either as individuals or organizations, need to see some benefits of VoIP. The benefits were discussed in an earlier section and include savings in telephone charges, savings in administration and maintenance costs, and greater capabilities. What end users want is relatively well understood: high-quality voice, "always-on" dial tone, easy-to-use features, and new applications.

Service providers

Service providers fall into two categories, those who provide only data services including access to the Internet (Internet service providers—ISPs) and those who provide analog voice service as well as data (the telephone companies, telcos). The first group has a clear desire for the success of VoIP since it will increase the volume of data, hence revenue, and compete for business with the traditional telephone companies.

The second group is ambivalent about VoIP. On the one hand, as providers of data services, they will benefit from the increase in data traffic. On the other hand, as suppliers of the analog service, their traditional revenue will decrease as VoIP becomes more popular.

The traditional telephone companies know that VoIP will become the standard technology for the telephone system; it is just a question of when. They all have plans to implement it and are trying to work it into their business plans.

Manufacturers

Manufacturers are looking for an opportunity to generate revenue. Whenever a new technology field opens up, there is an opportunity for startup companies to provide cutting edge products. This is the case with VoIP where startups were the first out with innovative products. Now that customers have shown an interest, established companies have followed with their own products. Companies that exist in the IP space, particularly Cisco, see VoIP as a brand new marketing opportunity. The companies that have traditionally supplied the telcos with their voice gear have a need to protect their installed base. Nortel, Lucent, and Alcatel are as interested in upgrading PBXs and central office equipment to VoIP as they are in end-to-end VoIP. In some ways this last group is ambivalent about VoIP since their sales of traditional voice gear have plummeted while their sales of VoIP equipment have not increased enough to fill the vacuum.

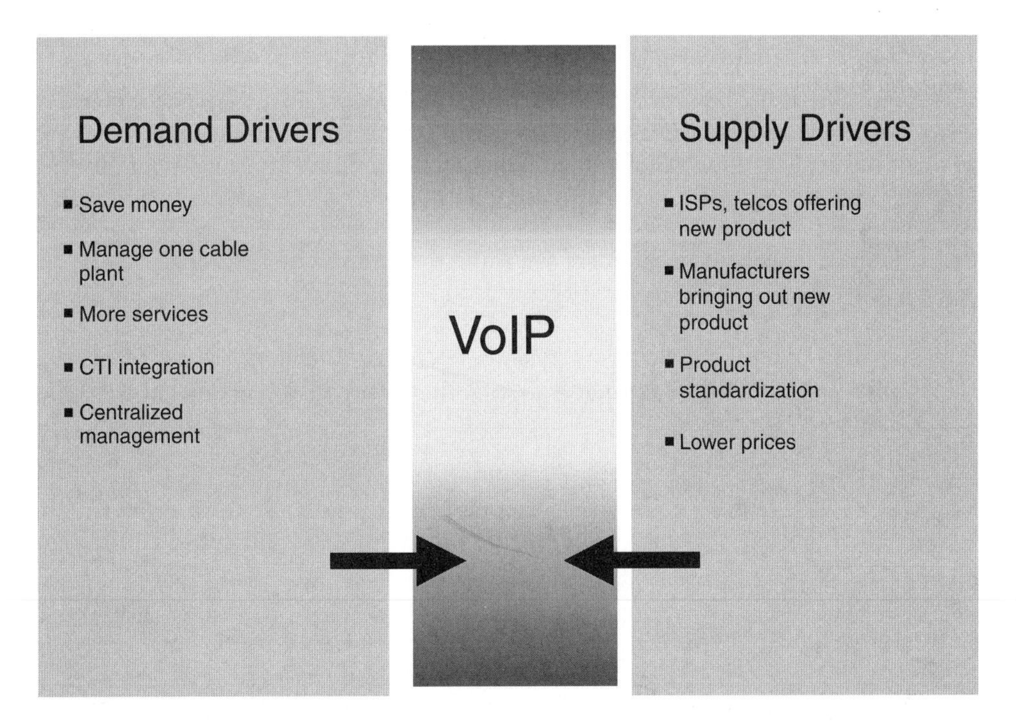

Figure 2-1: Drivers for VoIP technology

SUMMARY

This chapter looked at the reasons why VoIP is gaining in popularity and some of the issues involved in deploying it.

Section 2.1: Converging Networks

Voice and data networks traditionally have had separate characteristics because they moved different kinds of data. Now voice is typically converted into digital form for transmission and the two networks have for all purposes become one. Being able to manage a single network instead of two separate ones, excites many network administrators.

Section 2.2: Benefits of VoIP

This section examined some of the benefits touted for voice over IP, including reduced long-distance costs, more calls with less bandwidth, more and better enhanced services, and administration and maintenance savings.

Section 2.3: Issues with VoIP

This section looked at some of the issues with a voice over IP deployment. These include being dependent on one single cabling system and the need to train IT people in telephony technology. The organization may also need to upgrade the data infrastructure for bandwidth and switches and routers for quality of service features. Uninterruptible power is also a requirement because without power going to VoIP servers and the telephones, there will be no dial tone. Finally, 911 service is a concern if emergency responders can't find the location of an emergency.

Section 2.4: Driving VoIP

This section looked at what is driving this new technology, including organizations and manufacturers.

Review Questions Why VoIP?

1. Why does VoIP have a problem with the 911 emergency service?
   a) The 911 operator keeps the caller on the line which interferes with other people making telephone calls.
   b) IP telephone handsets are identified by IP address, which is not mapped to a physical location.
   c) IP cannot guarantee that a call to 911 will get through.
   d) The emergency services in most cities have not upgraded to VoIP yet.
2. Without compression, what is the normal bandwidth required for one telephone conversation?
   a) 2 kbps
   b) 32 kbps
   c) 64 kbps
   d) 8 kbps
3. Why haven't data networks been considered suitable for telephone service in the past?
   a) They use more expensive cable than telephone networks.
   b) Administrators of data networks are over-qualified to run telephone networks.
   c) They have five 9s uptime.
   d) They are prone to be infected by viruses.
4. Which is not considered a benefit of VoIP?
   a) Increase in the number of IT jobs
   b) Unified management of data and telephone infrastructures
   c) More calls with less bandwidth
   d) Closer integration of computer and telephony applications
5. In order to maintain dial tone on a VoIP system,
   a) you can implement Power over Ethernet.
   b) you can use rechargeable batteries in the IP phone.
   c) you can put a UPS on the telephone.
   d) you can plug the telephone into the computer.

CHAPTER

3

How Does VoIP Work?

Objectives

Upon completion of this section, the reader will:

- *Appreciate that making a telephone call requires two phases, setting up the call (call control) and the actual transmission of the conversation. The PSTN and VoIP require the same two phases.*
- *Have been exposed to the control function of a telephone call and the functions that this performs. In addition, the reader will understand how the control function is implemented on a PSTN using SS7 and on a voice over IP network using H.323 or SIP.*
- *Understand the four steps that are required to put voice on an IP network: sampling, quantizing, encoding, and packetizing.*

INTRODUCTION

Voice over IP has parallels to the traditional public switched telephone network. Both systems need to identify the end points of a telephone call and route conversations between them. Both systems also need to be able to physically put the conversation on the wire, keep it coherent as it travels across the system, and be made understandable when it reaches the destination. This chapter examines how these functions are accomplished by a voice over IP system.

3.1 HOW DOES VoIP WORK?

Voice over IP is a complex operation that encompasses two phases: call control and voice transmission.

Call control

Call control is the mechanics of setting up and ending a telephone call. This is a function of the PSTN and includes such tasks as setting up a route to the telephone dialed, ringing the bell, billing the customer for long distance charges, and ending the call. For VoIP to provide telephone service, then it too must provide these basic calling functions. The call control function is independent of the voice transmission function.

Voice transmission

Voice transmission actually sends the sounds across the network. For VoIP this includes digitizing the voice, compressing and encoding it, placing it in IP packets, and routing it to the destination telephone.

As you will shortly see, these two functions require entirely different technologies.

3.2 CALL CONTROL ON THE PSTN

Before a call can take place, the first major issue that needs to be addressed is: "How can the caller's phone find the called user's phone and establish contact with it?" This requires some form of signaling that is understood by the calling and called users' clients. It also requires a suitable naming and addressing scheme. This scheme will have to deal with traditional telephone numbers and IP addresses. Ideally the signaling messages in the VoIP system will be based upon some common format to enable a wide range of terminals, and hence users, to communicate.

SS7

Signaling system 7 (SS7) architecture performs out-of-band signaling in support of the call-establishment, billing, routing, and information-exchange functions of the

PSTN. It identifies functions to be performed by a signaling network and a protocol to enable their performance.

Whenever you dial a telephone number, you are using SS7. Whether the party you are trying to reach is in the same city or around the world, you are using SS7.

SS7 is responsible for the following services:

- Basic call setup, management, and tear down
- Wireless services such as personal communications services (PCS), wireless roaming, and mobile subscriber authentication
- Local number portability (LNP)
- Toll-free (800/888) and toll (900) wire line services
- Enhanced call features such as call forwarding, calling party name/number display, and three-way calling

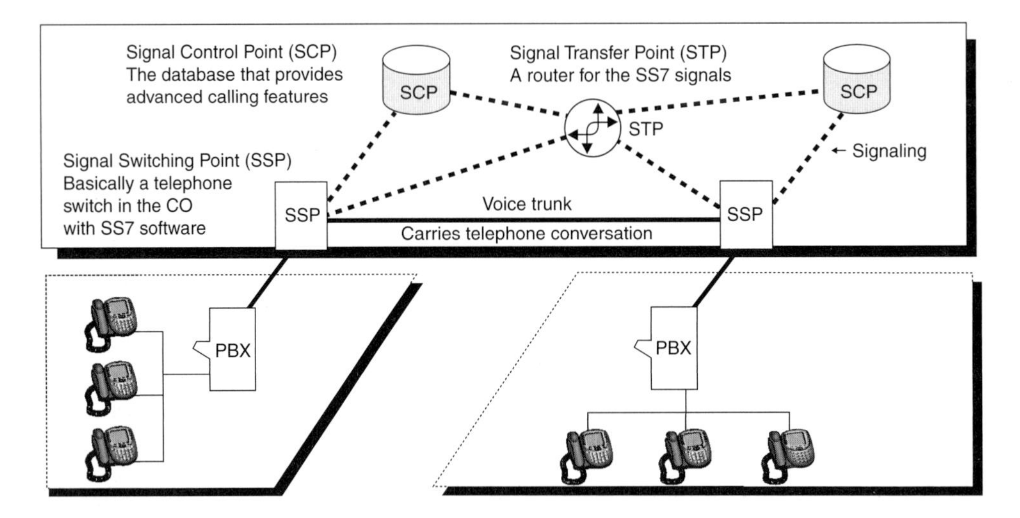

Figure 3-1: Signaling System 7

SS7 provides services

SS7 is connected to databases that allow it to provide extra services. 800/888/900 numbers couldn't exist without these. Caller ID, call forwarding, and calling cards are other good examples of services that need these databases.

SS7 entities

Figure 3-1 illustrates an SS7 system and its relationship with the voice network. A brief description of the nodes found in the SS7 architecture follows.

Signal Switching Points (SSP) are hybrid devices that form part of the voice network as well as have an interface to the SS7 network. These are the central office switches that look up the route of a call and run SS7 software. They refer a request for call routing to the SCP.

Signal control points (SCP) maintain databases of routes, local or long distance, for the voice traffic to take. Therefore when the SSP needs to do a lookup on how to route the voice call, the SCP can provide the answer. In addition, the SCP's databases maintain records of 800/888/900 numbers, and name and number pairs for call display and so on.

Signal transfer points (STP) are basically the routers used by the SS7 system. They forward requests from the SSPs to the SCPs. STPs make the SS7 network a packet switching network.

Out of band

SS7 uses an entirely separate network from the voice network and its technology is different as well. It uses a packet switching network to connect the different signaling entities.

Out of band just means that the signals used to provide these services do not travel over the same lines as the voice call itself. To see the difference between in-band and out-of-band signaling, you can examine the two different phases of setting up a telephone call. Setting up the telephone call proceeds differently from the user's telephone to the telephone company's central office (CO) and from one CO to another CO. When you pick up a residential analog telephone you alert your local CO that you want service because your phone goes to an off-hook condition. Your telephone accomplishes this by closing a loop allowing current to flow. The CO makes a −48-V DC current available on your line that doesn't flow when you are on-hook. When you pick up the handset, your telephone closes the circuit through a resistor and current flows. The CO detects the current flow and provides dial tone, which is the sound you normally hear indicating that you can now dial your number. As you dial numbers, they are sent over the same line using the dual tone multifrequency (DTMF) scheme to the CO. Because these supervisory signals are traveling over the same line as the voice conversation, they are in-band. SS7 signaling is out-of-band because it uses different lines to set up the call.

The benefits of out-of-band signaling

- Signaling does not use up a voice channel. A voice channel is dedicated to one conversation while it is in progress and represents revenue for a telephone company.
- Because the signaling is digital and packetized, control over many thousands of calls can use a single line. Signaling is also high speed and gives better performance. Call setup is normally 3–5 seconds instead of 10–15 with older systems.

- Better security. Previous to SS7, signaling went over the voice lines and it used audible tones. When telephone hackers learned how to manipulate these tones, they were able to avoid long distances charges.

Telephone hackers, also called "phone phreaks," learned how different audible frequencies represented different control signals to the telephone system. Since making these sounds proved difficult for people without perfect pitch, a mini-industry developed for making a device, often called a "blue box," which would emit the correct frequencies to fool the telephone system and allow the user to avoid paying long distance charges. The implementation of the SS7 system allowed the telephone companies to put a stop to this practice.

3.3 CALL CONTROL WITH VoIP

The call manager on an IP telephone system provides the management of the telephone calls, similar to the function that SS7 does for the PSTN. Although in theory two IP telephones could communicate directly, this would require that one user know the IP address of the other user's telephone, an unlikely occurrence. The call manager is referred to as a gatekeeper or proxy server. The call manager provides all of the following services.

Call establishment

Establishing a call requires finding the recipient's telephone. On the PSTN the telephone is identified by the familiar telephone number. With VoIP, the telephone is identified by an IP address. Since the IP address of a telephone is not easily found by the user making the call, other types of identification may also be used. Some forms of identification that are used by VoIP include a name, extension number, PSTN telephone number, or an address similar to an e-mail address. Because IP cannot use these forms of identification, the call manager must look up the identification used and find the IP address of the destination telephone. This process is called address resolution. On the other hand, if the recipient is on an external telephone system, the call manager must forward the request to the PSTN access device, the VoIP gateway.

Call initiation

The call manager then proceeds to initiate the call, by using H.323, SIP, or proprietary signaling to contact the called party. Upon a successful connection, the gatekeeper "hands off" the call to the two connected phones or phone-gateway pair. If the call cannot go through, because the called phone was in use or is otherwise unavailable, the gatekeeper will then inform the caller with a busy signal or redirect the call to an automated attendant or voicemail system.

Admissions control

The call manager can control access to the telephone system to authorized and registered endpoints. Devices unknown to the administrator will be disallowed.

The list of administrative functions that the call manager is responsible for is lengthy. Other responsibilities include managing moves, adds, and changes to extensions, updating dial plans, maintaining logs, and troubleshooting.

Bandwidth control

Connections can be disallowed if the system has no more bandwidth. A percentage of the bandwidth of the network may be reserved for data or some critical usage. The call manager may also restrict the number of people participating in a video/audio conference. Having sufficient bandwidth to support all of the simultaneous calls that users want to make plus having enough for the data applications of the network can be a challenge. This issue will be explored later in this book.

Zone management

The call manager will manage its zone, defined as the endpoint devices that register with it. This last includes maintaining a real-time list of calls in progress in order to provide a busy signal as required.

Additional services

The call manager provides some of the services that a traditional PBX provides such as call hold, call transfer, call forwarding, call waiting, three-way calling, and voicemail. Refer back to Table 2-1 for a more complete list of these services.

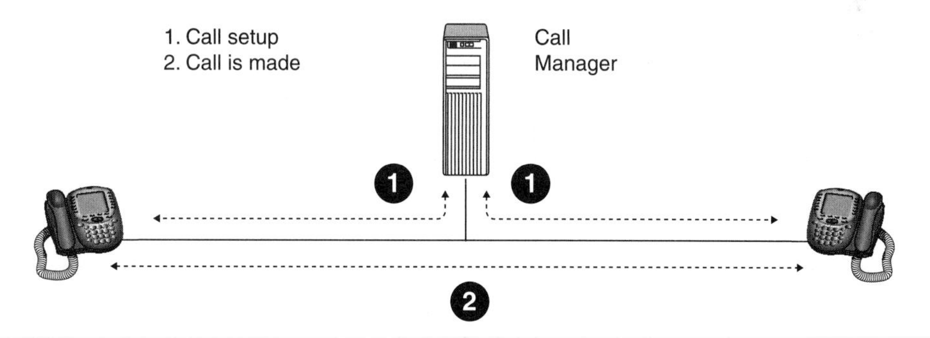

Figure 3-2: Setting up a call

The call manager can be a dedicated box, be integrated into a network device such as a router, or even be software sitting on a server or a user's computer.

The call control functions used by the call manager are H.323 and SIP plus some proprietary protocols such as Cisco's SKINNY protocol or the IAX2 protocol. These protocols will be discussed shortly.

As Figure 3-2 illustrates, when the phone at one end is picked up a connection is made to the call manager, which in turn contacts the second VoIP device. Only when the call setup is completed do the two endpoint devices communicate with each other on a peer-to-peer basis. Because the transmission of the conversation has different requirements than setting up the call, it uses the real-time transport protocol (RTP) over the user datagram protocol (UDP).

3.4 CALL SETUP SERVICES

This section looks at the call setup services provided by the H.323 and SIP protocols. These are open standards and widely supported. Proprietary protocols are occasionally found, including SCCP (Skinny Call Control Protocol) owned by Cisco and IAX2 (Inter-Asterisk eXchange V2), which was developed by the programmers of the Asterisk open source IP-PBX. Asterisk's IAX2 has been submitted to the IETF for ratification and may well become an open standard. Nevertheless, these last two do not have wide industry support, are not often encountered, and will not be discussed further in this book.

H.323

H.323 is a set of protocols used by VoIP to establish and manage telephone calls. H.323 is controlled by the International Telecommunication Union (ITU). The ITU is a United Nations Agency whose mandate includes telecommunication standards. Because many telephone companies are associate members of ITU and because the ITU promotes standardization and interoperability, the recommendations of the ITU are widely followed in the telecommunications industry.

Not only voice

H.323 is a very broad set of standards that provides for video, audio, and data conferencing. The networks that H.323 was designed to work with do not provide quality of service. Ethernet is the primary example of this type of network. H.323 is not specific to Ethernet or any network; it will work with all of them.

An umbrella for many protocols

H.323 is actually a framework for multimedia conferencing and therefore uses other protocols to handle many functions. See Figure 3-3.

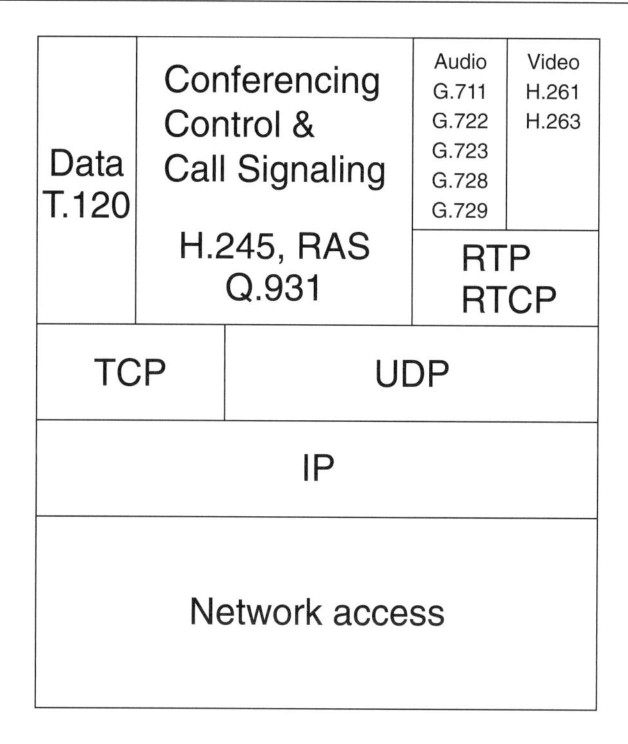

Figure 3-3: The H.323 protocol suite

Control and call signals

H.225/Q.931 — Used for call signaling and call setup
H.225/RAS — Registration/Admission/Status is a protocol used to communicate with a gatekeeper
H.245 — Used to negotiate channel usage and capabilities
Audio codecs — G.711, G.722, G.723, G.728, G.729
Video codecs — H.261, H.263
RTP/RTCP — Used for sequencing audio and video packets

The details of these standards will be covered in later chapters: H.323 (H.225, H.245, Q.931, RAS) in Chapter 17, "H.323," audio codecs in Chapter 15, "Voice to Digital," and RTP in Chapter 14, "How IP Handles Voice."

H.323 components

H.323 defines four major components for a network-based communications system: terminals, gateways, gatekeepers, and multipoint control units. Figure 3-4 illustrates a typical H.323 zone with its components.

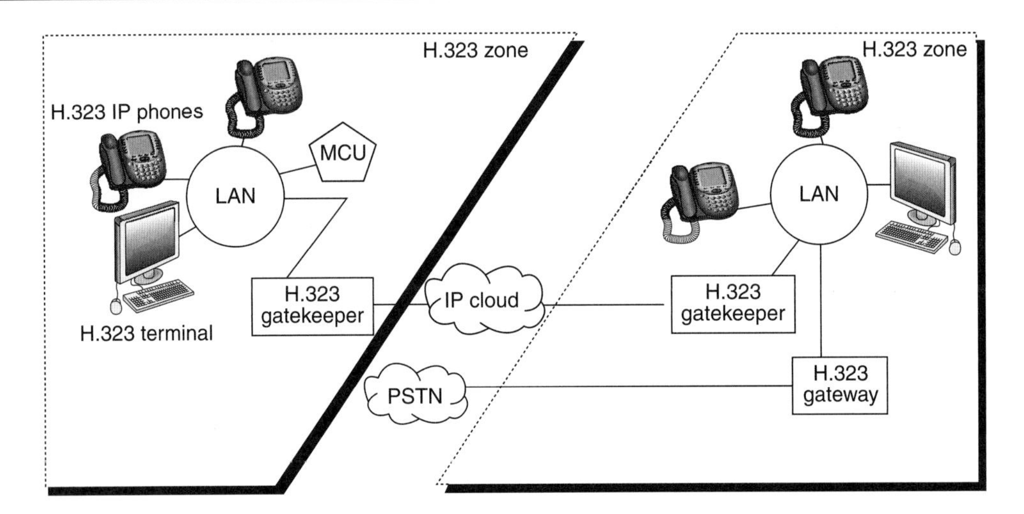

Figure 3-4: H.323 zones

Terminals on an H.323 VoIP system are computers running softphone software and VoIP telephones using the H.323 protocol.

Gateways are the bridge to the PSTN. They are also called "media" gateways highlighting the fact that they interface to the media or cable of the telephone company. The choice of the media gateway depends on the service that the telephone company provides, for example analog, T1, or PRI circuits.

Gatekeepers act as the central point for all calls within their zone and provide call control services to registered endpoints. In a previous discussion, the H.323 gatekeeper was called the call manager. To recap, the functions of the gatekeeper include address resolution and mapping, call initiation and establishment, admissions control, bandwidth control, and zone management.

Multipoint control unit (**MCU**) supports conferences between three or more endpoints. Conferencing can be audio, video, or data.

H.323 is losing popularity

H.323 is a flexible and comprehensive framework for multimedia conferencing, including VoIP, which has been implemented on millions of end devices since 1996 (v1) and 1998 (v2). It is known to work well and is reliable. However, there is room for an alternative call management service because of the following features of H.323.

- H.323 is not dedicated to VoIP and therefore is complicated because of the extra baggage it carries around for video conferencing.
- H.323 is not very flexible. New features can be added but they must be backward compatible.

- H.323 is controlled by the ITU, which moves slowly with new technology, and is slanted toward the traditional PSTN. In the fast-paced world of the Internet, H.323 is evolving slowly. New features must be integrated with, but not break, legacy systems.

SIP is the newer call control protocol and was designed with the Internet in mind. We examine it in the next section.

Session initiation protocol

SIP, the session initiation protocol, is a signaling protocol for Internet conferencing, telephony, presence, events notification, and instant messaging. SIP was developed by the Internet Engineering Task Force (IETF). The IETF is responsible for defining technical standards of the Internet, including IP, TCP, UDP, and many others. The IETF is examined more closely in Chapter 7, "Introduction to TCP/IP."

SIP is used for setting up, controlling, and tearing down sessions on an IP network. Sessions include, but are not limited to, Internet telephone calls and multimedia conferences. SIP is also used for instant messaging and presence. Note that SIP is designed for managing sessions (connections) whereas H.323 was designed for multimedia conferencing. VoIP falls within the scope of both and yet neither was designed explicitly for it.

Two other Internet protocols closely resemble SIP because they are request-response protocols: HTTP and SMTP (the protocols that power the World Wide Web and e-mail). Consequently, SIP sits comfortably alongside Internet applications. Using SIP, telephony becomes another Web application and integrates easily into other Internet services.

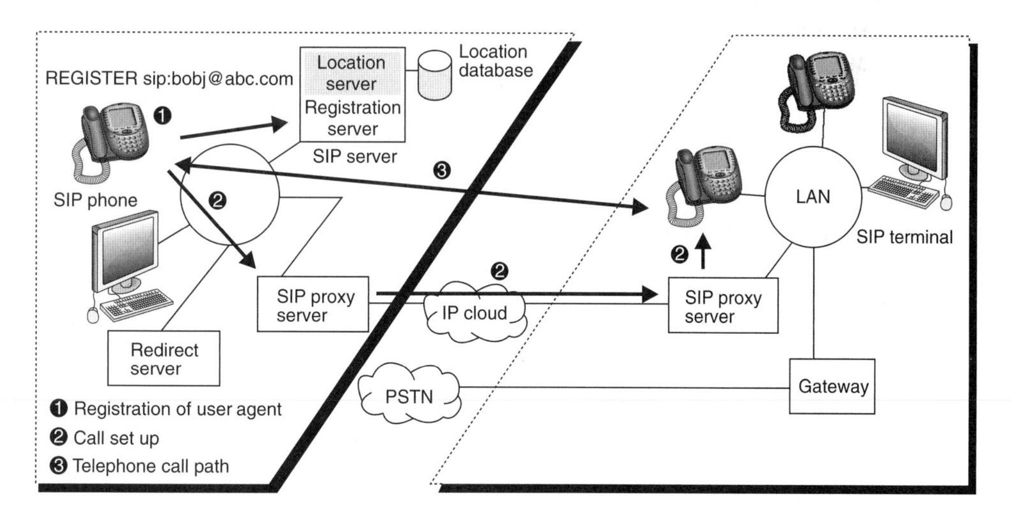

Figure 3-5: SIP architecture

SIP architecture

SIP uses the following components: SIP URI, registration server, location database, proxy server, and redirect server. Although they are separate functions, they can also be combined on a single physical machine. Refer to Figure 3-5 as you follow the discussion.

SIP URI — Every endpoint on the VoIP system has a SIP URI for identification. The URI for Bob Jones at ABC.com might be:

sip:bobj@abc.com

Notice how much this URI resembles an e-mail address; this is not an accident. If placed on a Web page, clicking the URI will instigate a phone call to that endpoint. URI stands for universal resource identifier. It is very similar to the URL, or universal resource locator, that you use in the address field of your Web browser. The URI doesn't locate the SIP endpoint device; instead, that job is left to the location server.

If calling a telephone number on the PSTN, the URI will look like: SIP:5551212@gateway, where gateway is the name of the machine that acts as the gateway to the PSTN.

Registration server — The registration server authenticates the user, and adds the mapping between URI and network address to the location server's database. When the user agent starts up, the first message it sends is a REGISTRATION request.

Location database — The location database maintains the database of name to location (usually an IP address) mappings. The information in the database is usually acquired from the user agent registrations, but may be acquired in other ways as well, such as DNS. The database may be queried in various ways, although LDAP is the most common.

When a user agent wants to connect to a remote SIP endpoint, it queries the location database in the location server for the contact information.

Proxy server — Proxy servers, as their name suggests, act on behalf of user agents, routing SIP messages to correct destinations.

Redirect server — A redirect server differs from a proxy server in that it does not forward messages but simply does a location lookup and returns one (or more) addresses for the destination and leaves it up to the original user agent to contact the destination at these addresses directly.

3.5 TRANSMITTING VOICE

Once a call has been set up between two or more devices, the caller starts speaking. At this point the voice signal has to be converted into a digital signal, formatted for TCP/IP transmission and sent along the network to the destination, where all of the preceding steps have to be reversed.

Digitizing voice

The steps involved are:

1. Converting the voice into a digital signal
   a) Sampling
   b) Quantizing
2. Making the voice data smaller
   a) Silence suppression
   b) Compression

Sampling and quantizing

The first step in converting analog voice signals into digital is called sampling. The voice signal is sampled 8,000 times per second using a technique call pulse amplitude modulation (PAM). This sampling frequency is derived from the telephone system that restricts the sound frequency to a range of 4 KHz. This many samples are sufficient to reproduce the original sound accurately because it is twice the frequency of the original sound. Each sample is encoded in 8 bits, which produces a data stream of 64,000 bits per second (kbps) using a technique called pulse code modulation (PCM). The process of converting one sample into 8 bits is called "quantizing" because the infinite possible values of a voice sample must fit into one of 256 discrete values available for the digital octet ($2^8 = 256$). An "octet" is also referred to as a "byte" as long as it contains 8 bits. The device that produces a digital signal from an analog one is called a codec, which is an abbreviation of code/decode. It also includes a component known as the ADC or analog/digital converter. Normally a codec is embedded in a microchip called a digital signal processor (DSP).

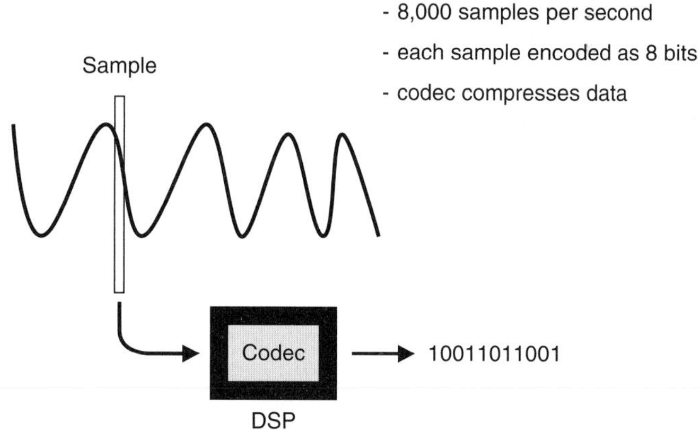

DSP = Digital Signal Processing

Codec = Code/decode

Figure 3-6: Codec converts sound to digital

PCM produces a 64-kbps stream of data with excellent voice quality. This process allows long distance calls to be placed on the T1 lines of the telephone company for transmission. One voice call takes up one channel, not a very efficient scheme. With VoIP, we want to cram as much voice into as little digital signal as possible. And instead of diverting our digital voice signal directly onto a T1 line, we need to packetize it and send it over an IP network.

Silence suppression and compression

It has been estimated that as much as 60 percent of a voice conversation is silence. Deleting these empty bits decreases the amount of data needed for the voice transmission. Taking all of these empty bits out of the transmission produces an eerie, otherworldly quality to the conversation. In fact, voice engineers compensate by putting some background "comfort noise" back into the conversation, which also helps to determine if the other party is still on the line.

In addition to silence suppression, the digital data that represent the voice can be compressed with modern compression techniques, similar to that used for computer data.

The net effect of these techniques is to reduce the bandwidth required for a voice conversation down from 64 kbps to 32 kbps, 16 kbps, 8 kbps, or even less. Eight voice conversations at 8 kbps can take place over the same circuit as a single conversation at 64 kbps. A slightly lower quality of voice may be a reasonable tradeoff for the increase in capacity of the system.

Encoding and compression techniques are published as standards by the ITU. Expect to see these when looking at specifications for VoIP equipment. The original PCM at 64 kbps is G.711 and is always supported by VoIP equipment. Other important encoding and compression standards are listed in Table 3-1.

Codec	G.711 PCM	G.726 ADPCM	G.728 LD-CELP	G.729A CS-CELP	G.723.1 ACELP/ MP-MLQ
Kbps	64	32	16	8	5.3/6.3

Table 3-1: Common codecs used with VoIP

On a VoIP system the voice packets may arrive at slightly different intervals degrading quality, a condition known as jitter. Jitter buffers are memory areas used to store voice packets as they arrive. The steady release of the voice samples to the listener from the jitter buffer is called playout. The playout is steady and constant, and as long as the jitter buffer receives an ample supply of voice packets, the system appears to have a fixed delay leading to better voice quality.

Packetizing voice

Once the voice data has been digitized, compressed, and the silence suppressed, it has to be divided into sections for placing into IP packets.

VoIP is inefficient for small voice packets, while large voice packets lead to long delays. The VoIP packet will have overhead in the form of headers. The headers for IP, UDP, and RTP add up to 40 octets. If the data was as small as 40 octets, the packet would only be 50 percent efficient, a very poor situation. The largest size packet that can exist on an Ethernet system is 1,500 octets. Take away the 40 octets for the header and you still have 1,460 octets available. That translates into 1,460 samples of uncompressed voice, or about one-fifth of a second (182 ms). If the voice data is also compressed, then about 1.5 seconds of a conversation could be carried by a single packet. If a packet with this much voice is lost or arrives out of turn, the conversation will be severely disrupted.

Typically, 10 ms to 30 ms (average 20 ms) of voice is placed inside one packet. A 20 ms of uncompressed voice takes 160 octets. Compressed at 4 to 1, 20 ms would take 40 octets.

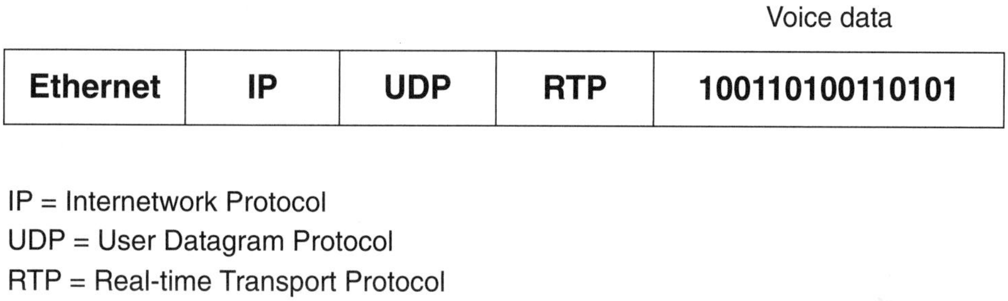

Voice data

Ethernet	IP	UDP	RTP	100110100110101

IP = Internetwork Protocol
UDP = User Datagram Protocol
RTP = Real-time Transport Protocol

Figure 3-7: The voice packet

Transmission of voice by IP

Figure 3-7 illustrates how the voice data is placed inside an IP packet and then transported over an Ethernet network. The three protocols of the TCP/IP protocol suite used by the voice data are real-time transport protocol (RTP), user datagram protocol (UDP), and the Internet protocol (IP).

- RTP is used to reorder packets that have arrived out of turn.
- UDP is used to identify which service is the target of the data in the packet.
- IP identifies the destination network and host and routes the packets.

Why UDP and not TCP?

TCP is used for call control and setup but UDP is used for the voice transmission itself. TCP is the protocol used when guaranteed delivery of a packet is required. If a packet is lost, TCP provides a retransmission mechanism that continues to transmit the same data until it is finally received. This same mechanism makes TCP unsuitable for voice transmission. Retransmission of a lost packet will introduce a gap in the conversation. It is better that one packet, which represents typically 20 ms of conversation, stays lost than that the conversation is interrupted. UDP does not provide a retransmission facility and is therefore the protocol of choice for voice transmission.

SUMMARY

This lesson looked at the mechanics of transmitting voice over a packet switching IP network.

Section 3.1: How Does VoIP Work?

Voice over IP is a complex operation that encompasses two phases: call control and voice transmission.

Section 3.2: Call Control on the PSTN

This section introduced the concept of call control and the functions required to set up and manage a call on the PSTN. SS7 is the system used to manage calls on the public system telephone network and it is examined first to see which functions will need to be emulated on a VoIP system. The three components of the SS7 system are the SSP, which does a lookup to route a call, the SCP, which maintains a database of routes, and the STP, which is the router for the SS7 system.

Section 3.3: Call Control with VoIP

This section examines how the call manager provides the call control functions on a VoIP system. H.323 and SIP are two common call control protocols and they are described in this section. Call control on a VoIP system requires the following functions: address resolution, call initiation, admissions control, bandwidth control, zone management, and additional services.

Section 3.4: Call Setup Services

This section looks at the call setup services provided by the H.323 and the SIP protocols. These are open standards and widely supported.

Section 3.5: Transmitting voice

This section looks at how voice is transmitted over a VoIP system. The four steps involved are sampling, quantizing, encoding, and packetizing. The codec is responsible for the conversion of the voice to data that can be sent over the IP network, and its selection determines bandwidth and voice quality.

Review Questions How Does VoIP Work?

1. How does VoIP use SS7?

   a) It maps the name to the IP address in the call database.
   b) It provides the out-of-band call setup for VoIP.
   c) It provides the SSP, SCP, and STP for VoIP.
   d) Because it provides call control for the PSTN, SS7 is not used by VoIP directly.

2. Which one of the following functions does the call manager not provide to a VoIP system?

   a) Encoding and compression
   b) Admissions control
   c) Zone management
   d) Bandwidth control

3. Why can SIP integrate easily into Internet services?

   a) It is controlled by the IETF.
   b) It uses a request-response protocol similar to HTTP.
   c) It uses the TCP/IP protocol.
   d) It supports audio, video, and instant messaging.

4. An H.323 zone is defined as

   a) all the IP telephones and terminals on a subnet.
   b) all the IP telephones and terminals in a company.
   c) all the IP telephones and terminals that register with the H.323 gatekeeper.
   d) all the IP telephones and terminals on a network.

5. Which one of the following encoding schemes would allow you to carry the most telephone conversations for a fixed bandwidth?

   a) G.723.1
   b) G.711
   c) G.729A
   d) G.726

CHAPTER

4

Quality of Service

Objectives

Upon completion of this section, the reader will:

- *Understand what factors influence quality of service (QoS) and how this affects voice quality.*
- *Appreciate that two approaches can be taken when lack of bandwidth affects QoS: restrict the number of sessions to the available bandwidth or share what bandwidth there is but give priority to services that need it.*
- *Know some of the techniques used to provide QoS to VoIP. These include 802.1p, TOS, DiffServ, RSVP, and MPLS.*

INTRODUCTION

When you pick up a telephone, you expect to converse in a clear and understandable way. The ever-present public system telephone network provides this service. A voice over IP system needs to provide the same. Anything less would provide an excuse to avoid the technology. The public Internet is notoriously haphazard. Watching a Web page stall as it loads is a fine example of the pitfalls of this complex technology. Quality of service is an issue with voice over IP and is usually listed as one factor in the slow adoption of the technology. In the case of VoIP, QoS means that the telephone conversation sounds as close to real life as possible, or at least similar to a conversation over the public telephone network.

This chapter looks at the factors that affect quality of service for VoIP and then examines the techniques that may be applied to a VoIP system that can give the users a true high-quality telephone experience. Chapter 16, "Implementing QoS," looks at these same factors but from a technical and packet level viewpoint.

4.1 FACTORS THAT AFFECT QUALITY OF SERVICE

Bandwidth

Bandwidth refers to the amount of data that can pass over a network in a time frame, usually measured in bits per second (bps). If many services are using the network, then bandwidth must be shared among them. If there is not enough bandwidth, then "something's gotta give." Either packets can't get onto the network in a reasonable time or excess packets are discarded. Bandwidth is listed as a contributing cause of quality of service problems because it affects the other factors discussed here: delay, jitter, and packet loss.

Delay

Delay refers to the promptness with which packets containing the voice data show up such that there are no gaps in the conversation or degradation in the voice quality.

The typical delay for a signal going end to end on the PSTN is 30 to 50 ms. A millisecond is abbreviated to ms and this represents 1/1000 of a second. Consider this the best-case scenario. At the other end, delays of 250 ms or longer make carrying on a conversation extremely difficult. This is the typical delay when making telephone calls using satellites in geosynchronous orbit. At this delay, a caller doesn't receive responses fast enough and speaks over the other caller. Table 4-1 gives general guidelines for delay and quality.

0–50 ms	Excellent quality, goal to strive for
50–150 ms	Acceptable voice quality
150–250 ms	Voice has noticeable degradation
> 250 ms	Unacceptable

Table 4-1: End-to-end signal delay and voice quality

Delay can be caused by many factors in a VoIP system, including the efficiency of the codec, processing of the packets, router delays, number of routers in the path, and congestion caused by insufficient bandwidth. Table 4-2 illustrates the effect on delay that common codecs produce. The delay is mostly due to the processing required for compression. The efficiency of a codec can vary depending on the proficiency of the programmer who coded it; therefore take the values in the table as a guideline only.

Codec		Throughput	Typical delay
G.726	ADPCM (adaptive differential pulsecode modulation)	16, 24, 32, 40 kbps	0.125 ms
G.728	LD-CELP (Low-delay-code excited linear prediction)	16 kbps	2.5 ms
G.729	CS-ACELP (Conjugate-structure algebraic-code-excited linear prediction)	8 kbps	10 ms
G.723.1	Multi-rate coder	5.3, 6.3 kbps	30 ms

Table 4-2: Typical delay introduced by common codecs

Jitter

Jitter is defined as variable delay. In other words, the packets are showing up with different intervals between them. Jitter degrades the voice quality, which becomes noticeable to the listener. Jitter is caused by differences in packet processing, router forwarding, and congestion of the network.

Lost packets

Lost packets are a fact of life on any network and an IP network is considered "best efforts." TCP compensates for lost packets with a retransmission function. However, with real-time applications such as voice and video, retransmissions cause unacceptable delay. Therefore VoIP uses UDP instead of TCP. UDP does not retransmit and therefore lost packets stay lost. Some codecs do not tolerate lost packets well; even a few lost packets will degrade voice quality. In general, the higher the compression that a codec produces, the less tolerance for lost packets there is. As a guide, the

G.723.1 codec can only tolerate a packet loss of less than 1% before voice quality degrades, whereas G.711 can tolerate a packet loss in the 7–10% range.

Echo

Echo in a telephone circuit refers to the speaker's voice bouncing back from certain disjunctions of the circuit such that the speaker can hear parts of his conversation. Echo occurs even in a traditional switched telephone circuit, but since the round-trip time is less than 50 ms, the effect is masked and not noticeable. When the round trip is longer than 50 ms, such as in a long distance call, echo canceling techniques need to be used.

With VoIP, round-trip times are always greater than 50 ms and therefore echo cancellation needs to be employed.

Table 4-3 follows and lists different applications that run over a network, their typical bandwidth, and their sensitivity to delay, jitter, and loss. Although VoIP doesn't take much bandwidth, it is very sensitive to problems on a network. This table also highlights some other issues that cause problems on modern networks. All video applications require a high bandwidth, which can crowd out other applications causing poor performance for all. This can clearly be seen for a residential VoIP service. If someone is downloading a video file or using streaming video, another person may have quality problems when making a telephone call.

Computer data moving across a network is very sensitive to the loss of data. If even one bit of a file transfer is lost, a data file will become corrupt. TCP is used to transmit computer data across a network because it provides guaranteed delivery. TCP depends on a regular contact between the two computers to maintain a connection. This can be disrupted by VoIP because voice packets are often given priority over computer data packets. When this happens, TCP is forced to retransmit, thereby causing additional congestion on the network.

Application	Bandwidth	Sensitivity to:		
		Delay	Jitter	Loss
VoIP	Low	High	High	Medium
Video Conferencing	High	High	High	Medium
Streaming Video on Demand	High	Med	Med	Medium
Streaming Audio	Low	Med	Med	Medium
Client/Server Transactions	Med	Med	Low	High
E-mail	Low	Low	Low	High
File transfer	Med	Med	Low	High

Table 4-3: Applications and Quality of Service demands

4.2 BANDWIDTH: SHARE OR RESTRICT

With unlimited bandwidth, any number of telephone calls as well as audio, video, and data can make their way down a cable. When the bandwidth is limited, some hard decisions have to be made.

In the previous section, bandwidth was discussed as a factor in quality of service. There are two approaches to rationing bandwidth. In the restrict approach, access to the bandwidth is parceled out until all of it has been taken. Then no more sessions are allowed to initiate until a "slot" becomes available again. In the sharing approach, all sessions are allowed to initiate; however, certain traffic is given priority and transmitted promptly through intermediate points, such as routers. Nonpriority traffic is held back at the routers and switches until the high-priority traffic has been released. At times of congestion, nonpriority traffic can move very slowly and can even time out, at which point the session is lost and retransmission is required.

Ethernet or wireless LANs do not have a natural mechanism for restricting access. Integrated services and the RSVP protocol were devised to do just that. Alternatively, differentiated services were devised to provide priority services when sharing bandwidth between VoIP and other data.

Restricting access with integrated services (IntServ)

IntServ is the IETF's answer to providing QoS on an IP network by restricting access to available bandwidth. IntServ was first described by the IETF in 1994 (RFC1633). RSVP is the primary mechanism used by IntServ to request QoS parameters for a data flow over an IP network. RSVP was described by the IETF in 1997 in RFC2210.

Resource reservation protocol (RSVP)

An RSVP host will request specific QoS parameters from the network, for example 16 kbps, 100 ms stable delay, etc. RSVP routers will provide the parameters when they set up the session. A session, or end-to-end connection is called a data flow in RSVP speak. A particular bandwidth can support only so many data flows. For example, a 1.544 Mbps link can support 24 64-kbps data flows and no more. Once the bandwidth is used up, RSVP will decline any further requests until a slot is freed up. This type of system requires that both end-point hosts as well as intermediate routers support RSVP.

RSVP places a heavy burden on routers since they must track each data flow to make sure that service guarantees are met. This then is the Achilles heel of RSVP. It cannot scale to the extremely large networks we can expect to see with VoIP in the future.

Differentiated services (DiffServ)

DiffServ was defined by the IETF in 1998 in RFC2475. DiffServ provides an alternative to IntServ for providing QoS on a network. It is designed to answer the weaknesses of IntServ.

DiffServ flags individual IP packets with a priority level using the type of service (TOS) field in the IP header. Therefore routers can handle packets according to their class of service and need not track them as part of a data flow. In other words, routers can handle packets individually as they normally do. In addition, the end-point hosts don't necessarily need to understand DiffServ; intermediate routers can flag packets appropriately instead. This approach scales well as systems grow larger.

Queuing

Once a packet that has been flagged with a class of service reaches a router or switch, how is it supposed to be handled? This is a relevant question since packets of different priorities may reach the router or switch at the same time and multiple packets of the same priority may also reach it. In order to provide priority handling, packets must be placed in separate queues as they are received according to their class and then released ahead of packets with a lower priority. Many queuing methods have been devised of which the following is a sample. Figure 4-1 illustrates the behavior of priority queues.

- Class-Based Queuing (CBQ) — CBQ divides all user traffic into categories and assigns bandwidth to each class.
- Per-VC Queuing — Per-VC queuing uses a separate output buffer for each virtual circuit. Each buffer can be given a priority, so voice virtual circuits (for example) can be given precedence over other data-carrying virtual circuits.
- Random Early Discard (RED) — This method relies on rules based on probability to instruct a router to begin dropping packets when established queuing thresholds are crossed.
- Weighted Fair Queuing (WFQ) — Each packet stream that WFQ is applied to is buffered separately and receives bandwidth on a variable, weighted basis.
- Weighted Random Early Discard (WRED) — WRED tries to identify the low-priority traffic and randomly discard those packets when congestion occurs.

Why so many queuing algorithms? Well, any queuing system can handle priority packets efficiently. It is the low-priority traffic that causes problems when packets are held back because the associated TCP connections time out. Connections that time out create duplicate traffic, which saturates the link giving poor performance for the entire system. Although the different queuing systems use different algorithms, their common goal is to avoid choking the nonpriority traffic. In addition, some of the queuing algorithms identify packets that they can discard in order to relieve congestion. Typically, these are management packets whose loss won't affect the transfer of voice or computer data.

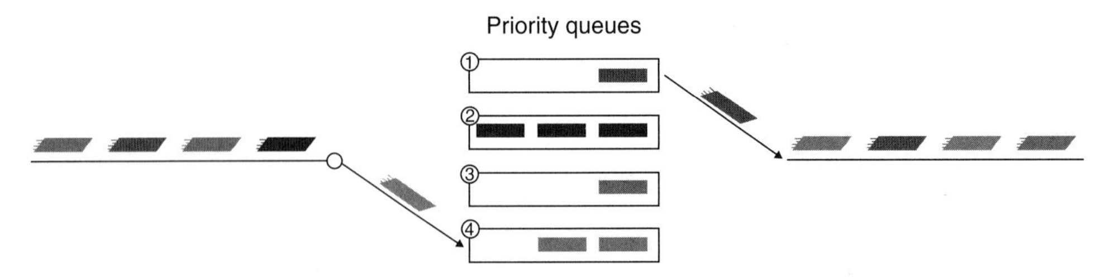

Figure 4-1: Using queues to give priority

4.3 PROVIDING QoS

Over-provisioning bandwidth

Since lack of bandwidth leads to many of the QoS problems we have discussed, it makes sense to increase bandwidth. It is difficult to analyze the requirements of VoIP and integrate it with the other data needs of the network. Buying more bandwidth than you need is a sensible reaction to this situation. On the company LAN this is realistic and cost effective since the cost of Ethernet components is at an all-time low. If the LAN uses 10 Mbps Ethernet, upgrade to 100 Mbps. If the LAN is already 100 Mbps, upgrade the backbone or switch/router links to gigabit Ethernet.

In contrast, buying more WAN bandwidth can be a serious proposition for an organization because there will be increased monthly expenses that may be difficult to justify. In this situation, a serious analysis of the company's communication needs should be undertaken in an effort to minimize potential WAN costs.

Dealing with delay

Delay is a complex topic since it can be introduced at so many places along the path that the voice travels end to end.

Algorithmic (codec) delay

Voice has to be captured and the codec has to digitize, quantize, compress, and silence suppress it before passing it onto the protocol stack. Codecs introduce different delays depending on their processing needs. In general, the more processing required because of compression, the more time delay is introduced. A trade-off between compression efficiency and time delay needs to be made and if the delay is leading to unacceptable voice quality then a different codec needs to be chosen.

Protocol fine-tuning

The IP protocol has to make a trade-off between large, efficient packets and the time delay created by waiting for additional voice data to be created to add to the packet.

Certain parameters, such as buffer size and timing, can be fine-tuned. However, only the control packets use TCP. Voice is carried by UDP, which has no parameters that can be fine-tuned.

Processing power

The voice packet has to be processed by many machines, including the end-point telephones or user PCs, call managers, gatekeepers, gateways, and routers. The power of these machines affects the delays introduced during transmission. Machines may need to be upgraded to produce acceptable voice quality. The most likely upgrade is to the microprocessor in servers, switches, and routers.

Because a VoIP telephone uses a dedicated digital signal processing (DSP) chip in contrast to a personal computer with a VoIP softphone, which must use its general purpose CPU for processing, a noticeable increase in quality may result from changing over to a VoIP phone.

Route complexity

At each router along the path, the voice packet has to be processed, introducing some delay. Simplifying the route would help with the delay problem. However, once the packet enters the service provider's network, a corporation has very little control over it.

Network induced delay

On a congested network, the voice packets are mingled with data packets. This can mean they are delayed or discarded. The easy way to deal with the problem is to over-provision the LAN with more bandwidth than needed. Since this may not be feasible or cost effective with a WAN, other techniques are needed. The next section delves into some of the techniques used to give voice a higher priority than data on the network.

Handling jitter

Since jitter is the variance in the delay of packets arriving, it can be compensated for by using a playout buffer. The playout buffer receives the packets in the uneven intervals that they arrive in and then plays them out to the receiver at a steady pace.

Suppressing echo

All VoIP circuits use echo cancellers. The echo canceller compares the voice data received from the packet network with voice data being transmitted to the packet network. The echo from the telephone network hybrid is removed by a digital filter on the transmit path into the packet network.

4.4 PACKET LEVEL TECHNIQUES FOR PROVIDING QoS

When voice and computer data share a network, the VoIP traffic needs to be given priority if the quality of the conversation is to be maintained. The following is a list of some techniques used to give VoIP packets precedence over data packets: 802.1p, IP TOS (type of service), differentiated services (DiffServ), and multiprotocol label switching (MPLS). These techniques are grouped in this section because they make a change to the structure of the packet carrying the voice data, either by modifying the meaning of some data or by adding extra information to the packet. In either case, the packet is marked for special handling. To round out the discussion, the resource reservation protocol (RSVP), which doesn't modify packets but uses different kinds of packets, is also described.

Despite the availability of these techniques, the quality of a voice conversation may not meet acceptable standards for the following reasons:

- These techniques are not universally implemented. In order to be effective the VoIP systems of the two end organizations must use the same technique, plus all the service providers in between must also use it.
- For peak efficiency, the VoIP phone or softphone must mark the packets with QoS tags. If an intermediate device, such as a router does the marking, then a portion of the trip has been made without the benefit of the QoS feature.
- Even though a packet is marked with a certain QoS value, the service providers that forward the packet may treat that value in different ways. There is no standard treatment of packets based on their QoS value.

The following is a short description and introduction to these techniques. A detailed and more technical examination is found in Chapter 16.

802.1p

Ethernet does not have a native ability to give packets a priority level. However, the 802.1p specification does just that. The 802.1p specification can be described only along with the 802.1Q specification because the two are always found together. 802.1Q is used by virtual LANs (VLANs). When multiple VLANs span remote sites, there is a need to identify the VLAN that a packet belongs to. 802.1Q inserts 4 bytes between the Ethernet header and the payload and this is used for VLAN identification. 802.1p uses 3 bits of the 4 bytes to assign seven priority levels to LAN frames. The service levels can be mapped to IP type of service (TOS) levels or supported in routers with a number of other mechanisms. Figure 4-2 illustrates the 4 bytes inserted in the Ethernet packet and the relationship between 802.1Q and 802.1p.

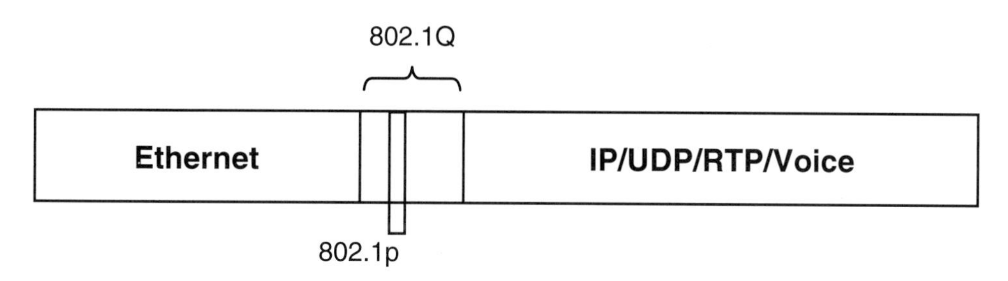

Figure 4-2: The 802.1p Protocol

IP precedence

The IPv4 protocol has a field in its header called type of service (TOS) that can be used to give the packet a precedence level. Three bits in this field are available for precedence, giving eight levels numbered from 0 to 7. Seven is the highest level. Traditionally, this field has been ignored by TCP/IP stack programmers. In order to use this field, all IP devices along a route must honor this field. IP precedence is rarely encountered because it has been superseded by DiffServ, which is described next. The usage of the TOS field is illustrated in Figure 4-3.

DiffServ

Differentiated services has been described earlier in this chapter as the preferred method to mark an IP packet with a precedence level. It makes use of the same TOS field in the IPv4 header as IP precedence does, which can lead to confusion. A system can use either DiffServer or IP precedence, but not both. DiffServ uses 6 of the 8 bits in the TOS field to define 64 levels of service. The Internet Engineering Task Force (IETF) defines the TCP/IP protocol, and when they designed DiffServ they also made sure that it was backward compatible with IP precedence. This is crucial since the two definitions of the same field will co-exist on the Internet for the foreseeable future. Whichever definition is used, the fact remains that all of the devices in the system must adhere to it if it is to be useful. Figure 4-3 illustrates the two interpretations of the TOS field in the IP header.

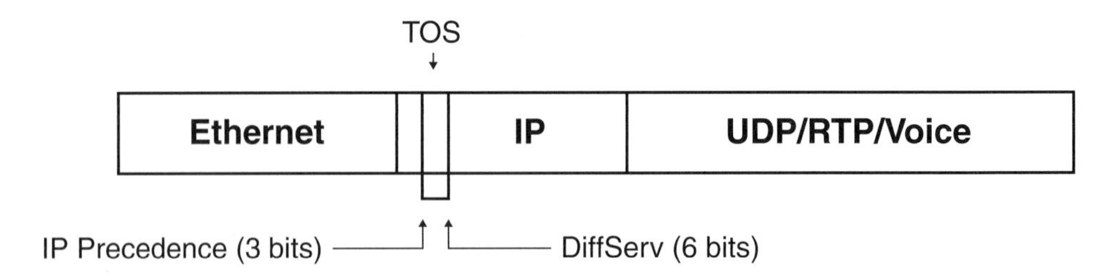

Figure 4-3: IP Precedence and DiffServ

RSVP

The resource reservation protocol (RSVP) takes a different approach from the previous three methods. In 802.1p, IP precedence, and DiffServ, a bit pattern in the packet signaled the handling that was desired for it by routers, gateways, and endpoints. These are non-stateful techniques because the devices don't have to track the packets, they just handle them individually as they show up.

RSVP in contrast doesn't deal with individual packets, but instead it controls the flow of packets for a connection. In short, a device using the RSVP protocol will set up a connection to the endpoint and request a certain class for service from all the intermediate devices along the way. This is a stateful service since all of the intermediate devices will have to track the persistent information about all of the connections through it. This places a heavy burden on these devices. In addition, cooperating service providers will have to agree on the definitions of the service levels. Because of this restriction, RSVP will be most useful on single-provider VoIP systems.

MPLS

Multiprotocol label switching (MPLS) is a technique used to forward packets in an efficient and fast way through a network by adding a simple label to each packet at an edge router (i.e. the first router that supports MPLS that the packet reaches). The label includes forwarding information in a form that switches can use. Since most large routers are also switches, MPLS allows the switching function of the router switch to forward the packet instead of doing a lookup in the router table for the next hop destination of the packet. This increased efficiency decreases the latency of the packets as they move across the network. MPLS supports QoS by integrating with both IntServ and DiffServ.

The label is a field that identifies the destination and class of service. In the case of Ethernet, the label is a 4-byte field, called a shim, inserted between the Ethernet and IP headers. While Ethernet is restricted to a LAN, MPLS can easily travel over a WAN if ATM or frame relay is used. MPLS is popular with service providers because it provides good control and management of their networks and allows them to prove quality of service and virtual private networks. Figure 4-4 illustrates the MPLS label.

Ethernet	Label 1	Label x	IP	UDP/RTP/Voice

Figure 4-4: MPLS

SUMMARY

This module looked at the issue of quality of service insofar as voice over IP is concerned.

Section 4.1: Factors that Affect Quality of Service

This section looks at the major factors that affect QoS, particularly bandwidth, delay, jitter, lost packets, and echo. The relationship between delay and compression is examined.

Section 4.2: Bandwidth: Share or Restrict

Bandwidth is a major factor in QoS simply because it affects the other factors. Two approaches can be taken to a lack of bandwidth. The first approach simply restricts connections when bandwidth runs out. This approach is used by IntServ. The other approach is to let all traffic on but give priority to traffic that is delay sensitive. This is the approach taken by DiffServ.

Section 4.3: Providing QoS

This section identifies some of the techniques used to provide QoS. Over-provisioning is a simple answer to bandwidth woes but is not practical for expensive WAN links. A careful analysis of bandwidth needs and intelligent network design can provide good WAN service at an affordable cost. Different techniques are used to tackle delay, jitter, and lost packets and are discussed in this section.

Section 4.4: Packet Level Techniques for Providing QoS

This section looks at the changes made to packets by various QoS techniques. 802.1Q/p adds an extra field to the packet as does MPLS on Ethernet. Precedence and DiffServ make use of the TOS field already present in the IP header.

Review Questions Quality of Service

1. Compressing voice data may deteriorate the quality of the signal by introducing

   a) jitter.
   b) echo.
   c) noise.
   d) delay.

2. Which of the following techniques restricts access to the data channel when there is not enough bandwidth?

   a) RSVP
   b) TOS
   c) DiffServ
   d) 802.1p

3. IP precedence and DiffServ are mutually exclusive because

   a) both use the same queues in the router switch.
   b) IP precedence is a proprietary protocol of Cisco whereas DiffServ is controlled by the IETF.
   c) they both use the same TOS field in the IP header.
   d) routers are configured for DiffServ whereas switches are configured for IP precedence.

4. Which one of the following does not make use of information in the voice packet or change the voice packet structure in order to implement QoS?

   a) 802.1p
   b) RSVP
   c) MPLS
   d) DiffServ

5. If a single packet in a voice conversation were lost, how would the user notice the effect?

   a) He wouldn't notice.
   b) There would be a hiss on the line.
   c) There would be a gap in the conversation.
   d) The person on the other end would repeat what she said.

CHAPTER

5

Is Your Network Ready for VoIP?

Objectives

Upon completion of this section, the reader will:

- *Understand the technology and standards (802.3af) for providing power over the Ethernet cable to VoIP telephones.*
- *Appreciate that the wiring closet may need to be provisioned with extra power, uninterruptible power supplies, and air conditioning.*
- *Gain insight into bandwidth issues when voice is added to an existing data network.*
- *Have a checklist of minimum infrastructure requirements before VoIP can be implemented.*

INTRODUCTION

You can decide if VoIP is right for your enterprise after analyzing its merits. Next you will need to consider the total cost of ownership. You must appreciate that the cost of a VoIP system is not restricted to new telephone handsets, IP-PBXs, call manager boxes, and gateways. Your complete data network may need to be re-engineered. This chapter looks at some of these issues such as providing power over Ethernet, upgrading the wiring closet, examining bandwidth needs, upgrading the data infrastructure, and considering the organizational issue of who will look after the telephone system.

5.1 VoIP CONSIDERATIONS

Is your system ready for voice over IP? Are you? Before implementing VoIP, the following factors need to be considered:

- Power to VoIP telephones. If power can be delivered over the network cable, you can decrease desktop clutter.
- Power in the wiring closet. If power to the telephone handsets is to be delivered over the network cable, an increased supply of power will need to be available for the network switches.
- Uninterruptible power supplies (UPS). Disruption of power is as unacceptable for the telephone system as it is for data systems.
- Bandwidth. Is there enough bandwidth on the network to accommodate both data transmission and telephony?
- Quality of service. Do the switches and routers of your existing network need to be upgraded for quality of service?
- Do the administrators of your network have the telephony frame of mind, meaning that they will try to maintain 99.999% uptime even if network maintenance needs to be done in the middle of the night?

5.2 POWER OVER ETHERNET

The concept of power over Ethernet (PoE) is simple—run a single cable to a device and provide both data communications and the power needed for the device's operation. The need for this arrangement is obvious for certain kinds of devices, such as wireless access points placed in the ceiling or a monitoring webcam put in an awkward location where no AC outlet is installed; but what are the benefits of PoE to IP telephones? Some benefits include the following:

- Uninterruptible power can be provided over the LAN cable instead of the building's wiring system or individual UPSs for each telephone.
- Telephones can be placed in locations that have no AC power source available.

- A telephone handset can be manufactured without regard to the country it will be used in if the AC power doesn't have to be considered.

Architecture

The specification for PoE (IEEE 802.3af) defines two types of power source equipment: end-span and mid-span.

- End-span refers to an Ethernet switch with embedded power over Ethernet technology.
- Mid-span devices resemble patch panels and typically have between six and 24 channels. They are placed between legacy switches and the powered devices. Each of the mid-span ports has an RJ-45 data input and data/power RJ-45 output connector. See Figure 5-1 for a graphic depicting these two devices.

For new deployments, you'd typically buy an end-span Ethernet switch. Mid-span devices make sense if you want to add PoE to a network and don't want to replace the switches.

However, it is probably wise to consider replacing your old devices with new end-span switches because you can amortize it over a long life of servicing IP phones, wireless LAN access points, and webcams.

With either type of power-sourcing equipment, you can safely mix legacy Ethernet devices and new LAN-powered devices. A safety feature built into the PoE specification requires the power provider to test the end equipment for its ability to accept power over the UTP cable.

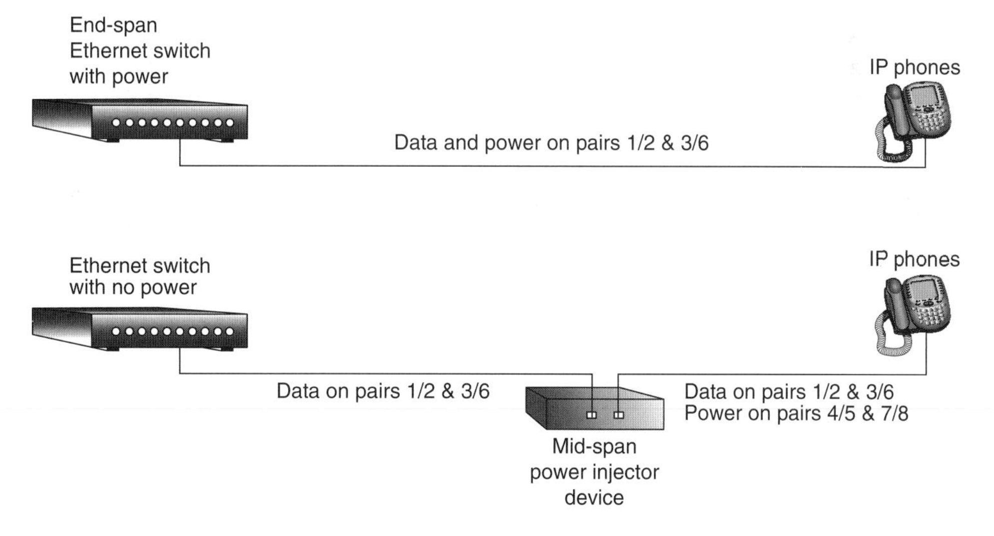

Figure 5-1: Power over Ethernet

Power

The 802.3af specification calls for delivering 48 volts DC down the cable. The maximum power to any device is 12.9 watts. Since most IP telephones are in the 3- to 10-watt range (typically 6/7 watts), the delivered power is adequate.

PoE works with existing cable plant (including Category 3, 5, 5e, or 6), horizontal and patch cable, patch-panels, outlets, and connecting hardware, without requiring modification.

Which cable pairs?

An unusual difference between end-span and mid-span devices is the cable pairs they use to deliver the power to the end device. Typically, only two pairs of the four pairs in an unshielded twisted pair (UTP) bundle are used to transmit data. This is the case for both 10BaseT and 100BaseTX running over Cat5/5e/6 cable. The only exception to this is 100Base4.

Mid-span devices inject the power onto the unused cable pairs in the bundle whereas end-span devices used the same cable pairs as are used for the data. This flexibility is built into the 802.3af specification so that manufacturers have some leeway in designing their equipment. This has no impact on your purchase decisions. If the VoIP telephone is 802.3af compliant, it will work.

5.3 THE WIRING CLOSET

Power

Plan for a sufficient power service in the wiring closet. A typical wiring closet service is 110-V AC at 15 amps. This provides a total of 1,650 watts of power. Note that a 110-V AC supply cannot provide more than 30-amp service. If your needs are higher than this, you will need to provision your wiring closet with a 220-V 20-amp service. Your power budget needs to include PoE switches and uninterruptible power supplies as well as any other machines located in the closet such as routers.

The power required for the switches, which provide PoE, vary with the number of VoIP telephones that are serviced from that closet.

When calculating your power budget, keep these points in mind:

- PoE 802.3af specification provides for a maximum power draw for each circuit of 15.4 W at the provider end (the switch). The power loss over 100 M gives a net 12.9 W at the power user end (the IP telephone).
- VoIP telephones are much more efficient than the theoretical maximum. Most IP telephones are rated at 5 W during use and draw no power when they are idle. Cisco notes that their IP phones draw 7 W only when they are ringing.

In order to make an educated calculation or guess for your power budget you need to know the following:

- How many VoIP telephones you will have
- How much power each telephone draws when it rings
- How much power each telephone draws when it is in use during a conversation
- How many telephones are ringing at any one time
- How many telephones are in use for conversations at one time
- How much power the switches draw bare, that is before they supply power to the telephones and how many switches there are

In order to calculate the power budget, assumptions will need to be made. After the calculation is made, you will also need to factor in a safety margin for peak times.

Air conditioning

Although the PoE switches generate heat when providing power to the remote equipment, the heat will not be concentrated in the wiring closet. It will also dissipate along the wires and at the telephones. Nevertheless, the temperature tolerances of the switches will need to be adhered to and air conditioning may need to be considered.

Uninterruptible power supplies

Providing uninterruptible power to the switches is a basic requirement. When deciding on how much power to provision, take into account the power requirements of the telephones and how long you want them to stay functioning if there is a power interruption.

5.4 BANDWIDTH

Having sufficient bandwidth for the organization's telephone calls can minimize a host of VoIP issues.

How much is enough? Obviously factors such as the number of users, the type of voice and messaging applications, and how the network is used on the data side are all considerations. Unless a thorough analysis of current and future traffic patterns can be carried out, the easiest plan is to over-provision your LAN. Unfortunately, over-provisioning a WAN involves hard dollars and may not be so easy to justify. In that case, a proper analysis of your telephony needs may be required.

The amount of bandwidth required to carry voice over an IP network is dependent upon a number of factors. Among the most important are:

- Codec (coder/decoder) and sample period
- IP header
- Transmission medium
- Silence suppression

The codec determines the actual amount of bandwidth that the voice data will occupy. The IP/UDP/RTP header can generally be thought of as a fixed overhead of 40 octets per packet. The transmission medium, such as Ethernet, will add its own headers and checksums to the packet. Finally, some codecs employ silence suppression, which can reduce the required bandwidth by as much as 50 percent.

Finally, the voice traffic does not have the cable all to itself; it is incremental to the data traffic that already exists on the network before you install VoIP. Add the bandwidth that the other data requires to your calculations of "enough" bandwidth.

5.5 INFRASTRUCTURE UPGRADES

If your current network infrastructure is suitable for voice over IP, upgrades will be minimal. If major upgrades are required, you will need to anticipate them, budget for them, and schedule them before embarking on the VoIP implementation. Look at the following list to determine the upgrade needs.

Cabling

Your cable plant needs to run at least 100 Mbps Ethernet. If you already have cable that meets the Cat5 specification, no further upgrade is required. If your cable is older UTP or coaxial cable, an upgrade is required.

Cable installations today meet at least the Cat5e specification. However, you should consider Cat6 to future-proof your system. Cat6 cabling anticipates 1 Gbps Ethernet. In any case, consider the following:

- Your cabling plant should meet the Cat5 specification all the way through. There should be no sections still using older cable.
- If in doubt, call in a cabling contractor and have him run verification tests on your system.

100 Mbps Ethernet

Your Ethernet system should be at least 100 Mbps (100BaseTX) throughput. If by chance you are still using 10Base2 or 100Base4 Ethernet, upgrade. If you are still using token ring, you will also want to upgrade.

Switches, not hubs

Hubs are not suitable for a VoIP network since collisions on the Ethernet system decrease throughput too much. Switches decrease collisions and are required. In addition, switches that have quality of service features are recommended for a VoIP system.

Quality of service

QoS features, such as 802.1Q/p are implemented in switches and routers. DiffServ and MPLS are also implemented here. You may already have switches and routers without these features. You will either need to upgrade these pieces of equipment, or if this is not feasible, replace them.

Virtual LANS

Virtual LANs allow similar devices to be grouped together such that they can communicate easily with each other, but not with devices belonging to other VLANs. This allows the traffic from devices that generate voice data to be segregated from the traffic of devices that generate computer data. VLANs make marking VoIP traffic for QoS purposes easier. Creating VLANs require Ethernet switches that support this feature.

5.6 TELEPHONY STATE OF MIND

If you lose your email, employees will fume, but if you lose dial tone, the world will come down about your ears.

Five 9s reliability

Everyone has been conditioned by over a hundred years of service to expect the telephone system to work. Although we can be conditioned otherwise (look at cellular telephone systems), we expect the telephone to work even during power outages. If the enterprise wants to replace the traditional PBX with a VoIP system, it needs to provide the same level of service.

Who is in charge—Telecom or Data?

Does the VoIP system belong to the telecom group or the data group? Most enterprises don't actually have a telecom group since management of the telephone system is often outsourced to the telephone company. Will VoIP be added to the data group's duties or will it be contracted out? If the data group does take it over, do they have the technical skill set or can they train to acquire it?

Telephony mind set

Maybe even more important than the technical skill set is the telephony mind-set. Can the support group avoid downtime at all costs? Will they avoid rebooting servers and routers except in the middle of the night?

Security

Because VoIP is a data service as well as a telephony service, it is subject to the evils of the data world. This includes viruses and hackers. The telephone system makes a tempting target for mischief and the enterprise will have to be on guard against it. In addition, the enterprise will have to accommodate an endless patching cycle for the software that runs the VoIP system.

SUMMARY

Before you can implement VoIP, your network needs to be ready. This chapter looked at the various factors that you need to consider.

Section 5.1: VoIP Considerations

Is your system ready for voice over IP? Before implementing VoIP many factors need to be considered.

Section 5.2: Power over Ethernet

The technology of PoE allows you to run power to IP telephones, thus avoiding the need to plug them into the building power supply. Assuming that the Ethernet switches are attached to UPSs, users can continue to make telephone calls even when there is an interruption in the power supply.

Section 5.3: The Wiring Closet

Examine your wiring closet and decide if you need to upgrade the power service and provide air conditioning and uninterruptible power supplies.

Section 5.4: Bandwidth

Analyze your LAN system and decide whether adding voice to your data network will require an upgrade to your Ethernet system.

Section 5.5: Infrastructure Upgrades

You may need to move from 10 Mbps Ethernet to 100 Mbps Ethernet or add gigabit links between your switches. If you have any Ethernet hubs still in place, they will need to be replaced by switches. You may also need to upgrade your cabling system to Cat5 if it isn't already there.

Section 5.6: Telephony State of Mind

Consider the changes you may need to make to your network administration, particularly in regard to merging the data and telephone groups. With telephone traffic on your data network, special attention will also need to be paid to reliability and security.

Review Questions: Is Your Network Ready for VoIP?

1. If you would like to add PoE to your network infrastructure, but your Ethernet switches do not support this feature, what do you do?

   a) Replace your switches with Ethernet hubs that have the PoE feature.
   b) Add uninterruptible power to your building power supply so that you will never have a blackout.
   c) Purchase a mid-span power injection device for your network.
   d) Add a UPS to your Ethernet switch so that it will not fail during a blackout.

2. What is the specified power that a PoE circuit can provide to an IP phone at the maximum cable length?

   a) 12.9 W
   b) 7 W
   c) 15.4 W
   d) 9.4 W

3. What is the easiest way to provide sufficient bandwidth for VoIP plus data on your LAN?

   a) Over-provision the network.
   b) Use gigabit Ethernet.
   c) Use fiber optic cable.
   d) Use a high-compression codec.

4. If your network can't provide QoS, which of the following devices might you need to upgrade?

   a) routers, switches, and hubs
   b) switches and hubs
   c) hubs and routers
   d) routers and switches

5. What is not a concern of the administrator of a VoIP system?

   a) viruses
   b) users talking too long
   c) rebooting the call manager in the middle of the day
   d) hackers

CHAPTER

6

Components of the Voice over IP System

Objectives

Upon completion of this section, the reader will:

- *Appreciate the different types of VoIP products available, including call manager software, hardware platforms, media gateways, IP telephones, and ancillary services.*
- *Know who the major players are and into which camp they belong—traditional telecommunications supplier or enterprise data supplier.*
- *Understand how the manufacturers have segmented the VoIP market by customer size and developed products that target these markets.*
- *Be familiar with the major VoIP platforms of the important VoIP manufacturers.*

INTRODUCTION

Whether installing a new VoIP system or upgrading an existing telephone system with VoIP capabilities, familiarity with the components is a fundamental requirement.

Telephone systems are a new field for some VoIP manufacturers, the main business of others. The former see VoIP as a new business into which they can expand. The latter see VoIP as a threat as well as an opportunity and want to service their installed base with new products. The products they produce, however, mostly overlap. This chapter looks at how the different vendors approach VoIP systems and the products they offer. Special attention is paid to the market segment that the equipment is designed for. Large, medium, and small organizations need different capabilities and capacities and the manufacturers have designed their systems to appeal to these different customers.

6.1 MAJOR PLAYERS

VoIP has brought two reasonably separate sets of companies into collision with each other—those serving wireline and wireless telecommunication companies and service providers, such as Nortel, Avaya, and Alcatel-Lucent and those focused primarily on enterprise data networks, dominated by Cisco. Of course, the two do overlap already. Cisco built a major communications business around its Internet/IP data technology and is also a leader in enterprise voice over IP, while Nortel, Avaya, and Alcatel-Lucent have enterprise network businesses.

But the convergence of voice and data systems around IP has greatly increased the overlap. Nortel, Avaya, and the others are pushing their voice technologies into the data center, while Cisco is eyeing service providers that are increasingly dependent on data rather than voice for new revenue and increased margins.

The traditional voice switch vendors have products based on circuit-switched technologies and particularly time division multiplexing (TDM). Traditional products center around the voice PBX but they also have products that adapt these traditional networks to IP. Cisco, meanwhile, sells pure IP products to handle voice, treating it as just another data stream. A number of VoIP vendors fall into the traditional voice switch camp, including Nortel, Avaya, and Alcatel-Lucent, while Cisco's approach is emulated only by smaller players like 3Com.

The customers of the PBX vendors can't justify the cost of ripping out their current working systems for a wholesale replacement with VoIP gear (rip and replace). Not only do they have an investment in PBXs but in telephones as well. These clients need VoIP features added to their PBX systems. All of the PBX manufacturers offer an upgrade path for these clients. Those enterprises that don't have an investment in a PBX system or are putting in a Greenfield (new) site have the widest possible choice between the vendors and their many systems.

6.2 VoIP PRODUCTS

The components that make up a VoIP system have been introduced in previous chapters of this book; they include the call manager, IP telephones, and the media gateway.

Call manager software

Call manager software is the heart of any VoIP system. It is responsible for call control and session management. Each vendor also enhances this software with as many bells and whistles as it can get away with. Vendors take different approaches to the call manager. Some embed it in their proprietary platforms while others make versions available for industry standard platforms such as Intel processors and standard operating systems such as Windows or Linux.

The platform

The platform for VoIP is a computer that runs the call manager software. All vendors sell a proprietary solution for their platform. Note that sometimes the platform is a card or special chip that is installed in some other kind of device, such as a router or PBX. Alternatively, all of the components needed for the system are often integrated into a single box for convenience, ease of management, and lower cost. This type of computer is known as a VoIP appliance, giving the illusion that it simply works by plugging it in. In addition, some vendors allow their software to run on industry standard computers running industry standard operating systems, such as Windows or Linux, in order to decrease the cost of a system.

The telephones

All of the vendors manufacture and sell IP telephone sets. Their features can range from very basic to very sophisticated; advanced features include multi-line or colored screens, soft-keys, and support for PoE. Softphones are also supported. This is a software application that adds telephone capabilities to a computer equipped with a microphone and headset and often includes video features as well, but the computer needs a Web camera to take advantage of these.

Media gateways

Media gateways bridge the gap between the VoIP system and the public switched telephone network. When selecting a media gateway, the selection of the interface to the PSTN is particularly important because it must match the communications link that the service provider makes available. Common interfaces include analog, T1/PRI, BRI, and cable.

Ancillary services and servers

All of the vendors want you to buy a "complete solution" from them. Therefore expect their systems to connect to and support services such as multimedia, unified messaging, and call center services.

Table 6-1 lists the major vendors with their products divided by product category. This is only a snapshot at the time this book was published. Because the

	Call Manager Software	Platform	Media Gateway	Other Services
Cisco	Cisco Unified Communications Manager Cisco Unified Communications Manager Business Edition Cisco Unified Communications Manager Express	Media Convergence Servers (also IBM, HP servers) Cisco MCS 7828-H3 Unified Communications Manager Appliance Integrated Services Routers	Cisco Media Gateway Controller	Cisco Unity (unified messaging) Cisco Telepresence
Nortel	Succession Call Server Succession Signaling Server	BCM Communication Server 1000S/M/E, 2100 Superclass Softswitch	BCM Succession Media Gateway MCS5100	CallPilot (unified messaging) Symposium (call center)
Avaya	Avaya Communication Manager	IP Office - Small Office Edition IP Office ECLIPS media servers (S8100, S8300, S8500, S8700)	Avaya G150, G250, G350, G650, G700	Avaya Unified Messenger
Alcatel-Lucent	embedded	OmniPCX Enterprise OmniPCX Office 5020 Softswitch	Alcatel 7505 Media Gateway	
Mitel	embedded	3300 ICP SX-200	embedded	Unified messaging
3Com	3Com NBX system software 3Com VCX system software	NBX V3000 VCX Connect 100,200 VCX 7000	VCX digital and analog media gateways	Unified messaging

Table 6-1: Major VoIP vendors and their products

vendors change their product lines constantly, contact the vendors for up-to-date information.

PRODUCT LINE BY CUSTOMER SIZE

The variety of products available, even from a single vendor, makes it difficult to understand the VoIP marketplace. The easiest concept to grasp is that vendors design and produce equipment according to the customer's needs and ability to pay, and invariably this relates to their size. With each vendor having its strengths and weaknesses, it should come as no surprise that their ability to service clients of different sizes should also vary. One approach is to produce different platforms for each target market. An alternative strategy is to stretch the same platform over different size clients. Some vendors don't have the necessary resources and restrict themselves to a particular target market. Table 6-2 assigns VoIP equipment into the following four customer categories.

Small- and medium-sized business (SMB)

The small- and medium-sized business market is very sensitive to price. Their needs are simple but they value advanced features such as remote extensions for employees working at home, voice mail, auto attendant, and advanced call handling. The basic telephone service and advanced features must be handled in one complete package along with installation, configuration, and support. Because of the large number of businesses in this category, the SMB market is typically serviced by system integrators who install the VoIP system at the customer premises and connect it to the service provider. Prior to the availability of VoIP, SMBs normally used analog key telephone systems (KTS).

Medium-sized organization

A characteristic of this market segment is the deployment of a PBX for telephony services. The medium-sized organization has more sophisticated needs than the SMBs and needs a larger telephone system. Because a PBX is complicated, maintenance is often outsourced. As an alternative to an in-house PBX, the organization may use the Centrex service of its local telephone company. When upgrading to VoIP, a rip and replace strategy for the PBX may not be attractive. By installing VoIP modules on the PBX, VoIP functionality can be added while still preserving the investment in the PBX equipment. The PBX manufacturers offer this option for their equipment.

Enterprise size

An enterprise can have tens of thousands of users and have a distributed geography. An important concern is managing the complexity this entails. Centralized management of

	Small	Medium	Large	Carrier Class
Cisco				
Media Convergence Server		Yes	Yes	
Call Manager Express (embedded in small Cisco routers)	Yes 1–240 users			
Cisco Unified Communications Manager Business Edition/ MCS 7828		Yes 1–500 users		
Cisco Unified Communications Manager			Yes Up to 30,000 users	
Nortel				
BCM	Yes 3–200 users			
Meridien Option 11C		Yes	Yes	
Communication Server 1000,2100		Yes	Yes	
Superclass Softswitch CS2000				Yes
Avaya				
ECLIPS			Yes	
IP Office	Yes	Yes		
Alcatel-Lucent				
OmniPCX Enterprise		Yes	Yes	
OmniPCX Office	Yes			
5020 Softswitch				Yes
Mitel				
SX-200	Yes	Yes		
3300 ICP		Yes	Yes	
3Com				
VCX 7000		Yes	Yes	
VCX Connect	Yes	Yes		
NBX V3000	Yes	Yes		

Table 6-2: VoIP platforms by customer size

distributed systems is a valuable feature of enterprise telephony. Unlike the smaller organizations, the enterprise may be able to support an in-house group with responsibility for the telephone system.

Carrier class

The challenge for carriers, which include wireline telephone companies, cellular providers, and cable companies, is how to VoIP-enable their systems without disrupting their current business model. For those carriers with an investment in

TDM telephone switches, the addition of superclass softswitch software creates a "hybrid," thus allowing them to offer VoIP services to their clientele.

The traditional telephone systems of KTS, PBX, and Centrex have been introduced in the previous paragraphs and an expanded description is presented next.

Key systems

Key telephone systems (KTS) are used by small organizations with a few lines and multiple handsets. They are engineered to be simple, have fewer features, but most importantly, be affordable for small organizations. Key systems do not require dialing a number to gain an outside line and dial tone is provided by the telephone company central office.

PBX (Private branch exchange)

A PBX is a small telephone switch owned by an organization. It provides telephony services to a large number of corporate users while making use of a limited number of outside lines. Typically, the PBX is connected to the telephone company central office with a T1 or PRI trunk. This works because internal users call each other more often than they call outside. All calls to the outside flow through the PBX and a prefix (usually 9) must be dialed in order to access an outside number. Dial tone is provided by the PBX.

Centrex

Centrex is a service provided by the telephone companies that duplicates the functions of a PBX, except that the equipment is located at the telephone company premises. Think of moving the PBX from the client premises to the telephone company. It alleviates the need for the organization to maintain its own equipment and expertise.

Table 6-2 attempts to show how the vendors tackle different-sized clients.

6.3 VENDOR DISCUSSION

CISCO

Architecture for Voice, Video, and Integrated Data (AVVID)

AVVID is Cisco's grand scheme for integrating all the data types (voice, video, and data) under one unified platform.

The AVVID architecture is sprinkled throughout Cisco's product lineup in the sense that many products offer the services that VoIP needs. For example, the

Catalyst Ethernet switches offer the option of Power over Ethernet. In addition, the switches and routers offer quality-of-service features.

For VoIP, the centerpiece is the Call Manager software, which performs most of the VoIP duties.

AVVID product lineup

The following products are the key VoIP pieces in Cisco's AVVID product lineup.

Unified Communications Manager: This is the call manager component of the Cisco VoIP solution. In fact, it was originally called Unified Call Manager. It provides voice, video, mobility, and presence services. In particular, it provides PBX-like functionality, such as call control, hold, call transfer, and other basic and advanced office telephone system features. The Cisco Unified Communications Manager can support up to 30,000 users; however, two variations, the Business Edition and Express, were designed for smaller organizations. Cisco is flexible when providing the software. It is deployed on the 7800 series Media Convergence Servers from Cisco but can also run on servers from IBM and HP.

Media Convergence Server: The 7800 series of Media Convergence Servers are Cisco's proprietary hardware platform for running the Unified Communications Manager. The lineup includes many models designed with various capacities and capabilities. The units are rack mountable and feature generous amounts of memory and hard drive space; some also offer fault tolerance features such as dual power supplies.

Cisco 7900 series IP phones: Cisco makes a variety of IP telephony client devices, such as speakerphones and XML-enabled phones that can act as Web browsing thin clients. When deciding on a Cisco IP telephone, always check the protocols that it supports. Most support the SIP protocol while others support the SCCP (also called "SKINNY") protocol. It should be noted that Linksys, a division of Cisco, also makes a line of IP telephones.

In-line (PoE) power switch modules: Blades are available for Cisco's Catalyst 4000 and 6000 switch lines that can provide AC power to IP telephones, similar to the way PBXs supply power to circuit-switched phones.

Cisco Unity: Cisco Unity is Cisco's unified voice and e-mail server, which provides voicemail for Unified Communications Manager servers and can work with Microsoft Exchange servers to provide end users with one mailbox for voicemail and e-mail.

NORTEL

Nortel has all of the VoIP bases covered. The BCM platform supports the smaller enterprise, Meridien the medium, and Comunication Server for the large. In addition, its Superclass Softswitch can add VoIP capabilities to the TDM networks of the telephone companies and other service providers.

Small- and medium-sized business

Business Communications Manage (BCM) is Nortel's VoIP platform for small to medium size businesses. Supporting a converged data/voice solution, BMC can support either a pure IP or an IP-enabled telephony system for business. Three hardware platforms are available: the BCM50 is designed for 3 to 50 users, the BCM200 is designed for 10 to 20 users, and the BCM400 can be used by 30 to 200 users.

All of the platforms offer support for digital as well as IP telephones, interactive voice response, auto attendant, unified messaging, and more.

Medium-sized enterprise

The Meridian series of PBXs has been very successful and Nortel has expanded its capabilities to include a full set of VoIP features. Meridian 1 Option 11C is a digital PBX for enterprises, delivering advanced applications and carrier-grade five-9s reliability for traditional voice and voice over IP communications. Expandable to 800 lines, it supports digital and IP telephones, IP gateways, mobility (802.11) communications, voice messaging, call center, PC-based system management, and multimedia applications.

The enterprise

For medium to large size business, Nortel offers the Communication Server line of VoIP appliances. The line includes the CS1000S (150–1,000 users), CS1000M (1,000–10,000 users), 1000E (1,000–15,000 users) and the 2100 (2,000–200,000 users).

The Communication Server line is full featured, highly scalable and fault tolerant. There is a model to fill the need of any enterprise situation.

Service provider

Service providers include traditional wired telephone companies, wireless/cellular telephone companies, cable operators, and other service providers. Nortel offers Superclass Softswitches to upgrade these networks to full VoIP capability. The Communications Server 2000 is built on the XA-Core platform, enabling cost-effective migration to packet networks for DMS customers.

Nortel and Microsoft

Nortel and Microsoft formed an alliance in 2006 to aggressively pursue the unified communications market. Their collaboration is called the Innovative Communications Alliance (ICA) and basically combines Nortel's telephony experience and Microsoft's unified communications expertise, which is found in Windows Exchange server.

Nortel and open source

What does a billion dollar telecom manufacturer do when small business customers can't or won't buy its VoIP solution because of the price? If you are Nortel, you design a software-only IP-PBX based on open source software, you partner with IBM whose sales representatives sell it on IBM servers, and you charge a reasonable price. Nortel's product is called SCS500 and is based on an open source product called sipX which was developed by a company called Pingtel. Pingtel no longer exists; it was purchased by Nortel.

AVAYA VoIP PHONE SYSTEMS

AT&T spun off its equipment division to a separate company called Lucent in 1996. Lucent spun off its enterprise equipment division to a new company called Avaya in 2001. Lucent was purchased by Alcatel in 2007. As can be expected of a company with such a long history, Avaya offers a very broad and deep catalog of products for the modern telephone system.

Enterprise Class Internet Protocol Solutions (ECLIPS)

Where Cisco has AVVID, Avaya has ECLIPS. ECLIPS is Avaya's framework for an enterprise VoIP solution that includes the Avaya Communications Manager running on a variety of servers and including a broad range of media gateways.

IP Office: IP Office is Avaya's solution for small- and medium-sized business with support from 2 to 360 extensions. IP Office has a modular design, which starts with a system control unit (the IP500 or IP402) and then a selection of expansion modules to meet the requirements of the business. IP Office Small Office Edition is a special version designed specifically for a very small office or home office in a very small package.

For small organizations of up to 20 users, Avaya offers an unusual product called one-X Quick Edition, which doesn't use a separate server at all; all of the intelligence is in the telephones themselves.

Avaya Communication Manager: The Avaya Communication Manager platform is designed for medium to large organizations. The Communication Manager is the software that provides the call processing plus the advanced features that organizations desire such as support for mobile users, conferencing, and contact center applications. The Communication Manager is partnered with servers that can support a large variety of users from the model S8300, which can support 450 users, up to the model S8730, which can support up to 36,000 users. A large variety of media gateways round out the system, some of which are suitable for branch offices while others can service a campus environment.

ALCATEL-LUCENT

With the purchase of Lucent, Alcatel is now known as Alcatel-Lucent. The company has valuable relationships with telephone companies in the United States as well as Europe. Its product line has expanded to include a wide range of VoIP equipment.

OmniPCX Enterprise: The major VoIP platform for Alcatel-Lucent is OmniPCX Enterprise. This is an integrated, interactive communications solution for medium-sized businesses and large corporations. The solution combines traditional telephone functions with support for Internet-based telephony and multimedia communication.

The OmniPCX Enterprise provides a suite of unified communication applications, including a Web softphone, along with unified messaging and personal assistant applications. Based on a single software platform, the OmniPCX Enterprise is compatible with multiple operating systems, provides powerful communications servers, and features an independent infrastructure.

OmniPCX Office: The OmniPCX Office is Alcatel-Lucent's solution for the small business with up to 200 users. It is a cost-effective solution that supports all of the common VoIP features in an appliance that is easy to install and manage. The appliance uses the Linux operating system and supports the H.323 control protocol.

6.4 OTHER VENDORS

MITEL

Mitel has successfully serviced the small- and medium-sized marketplace with its well-regarded analog PBXs. Building on that experience, the company manufactures a line of VoIP systems that are well suited to small, medium, and large organizations.

Mitel® 3300 Integrated Communications Platform (ICP) provides enterprises with a highly scalable, feature-rich communications system designed to support businesses from 30 users to 60,000 users. The 3300 ICP provides enterprise IP-PBX capability plus a range of embedded applications, including standard unified messaging, auto attendant, ACD, and wireless gateway.

For smaller organizations, Mitel offers the SX-200 IP Communications Platform for up to 600 users, the Mitel 5000 Communications Platform for up to 250 users, and the Mitel 3000 Communications Platform for up to 52 users.

3COM

3Com is a pioneer in the networking business, having produced the first Ethernet adapter for the IBM PC in 1981. The company has been producing VoIP equipment since 1998, when it introduced one of the first VoIP systems, the now discontinued NBX 100 platform. 3Com offers two platforms: the NBX series and the VCX series.

3Com NBX platform

The NBX IP platform is available to meet the needs of small- and medium-sized businesses with between two and 400 users and is advertised by 3Com as being its economical line. The NBX platform is made up of the V3000 model VoIP appliances, which provide either analog or BRI support.

3Com VCX platform

The VCX platform also covers the small- and medium-sized business but can scale to larger organizations. The line is more robust than the NBX platform, with some models having redundant power supplies and RAID hard disk systems. The models in the VCX series, similar to the NBX series, are self-contained appliances. However, the V7000 series is an exception because it is a software suite that runs on the IBM eServer xSeries models using the Linux operating system. By choosing the appropriate IBM Power platform, the system can scale up to 50,000 users.

3Com Convergence Application suite

3Com has a modular approach to adding features to its VoIP systems. The Convergence Application suite is a series of modules that can be added to both the NBX and VCX platforms and allows an organization to customize its VoIP system by adding just the features that it wants. Available modules include messaging, conferencing, telecommuting, presence, and contact centers.

6.5 MORE VENDORS

It would be tempting to stop our discussion of the manufacturers of VoIP equipment at this point and this would imply that we had covered them all. However, this would be misleading. It is true that the vendors described so far are the largest by market share; nevertheless, competition in the VoIP market is intense with hundreds of other manufacturers, big and small, also providing solutions. A listing and description of these manufacturers is outside the scope of this book; however, you will often comes across vendors such as ShoreTel, Linksys, AdTran, and Fonality.

ASTERISK

Open source is a system in which a group of dedicated and talented software developers work on a project and make it available to anyone at no charge. These projects can create software that is sometimes the equal to their commercial competitors. Examples of these include Apache, the world's most popular Web server and Linux, a sophisticated operating system. An IP-PBX is a worthy challenge to a programmer and indeed, many projects are in varying stages of completion. The sipX project was mentioned in the section on Nortel; however, by far the best known is Asterisk. Asterisk is a free, open source IP-PBX software that runs on the Linux operating system. It can be used by the adventurous to create their own VoIP system, although it is more commonly used as a base by systems integrators for an inexpensive, yet powerful, VoIP system for small business. Asterisk was originally developed by Digium, a company that makes VoIP components. Asterisk is both mature and flexible and very customizable. If you don't want to immerse yourself in the technical details, versions are available that provide easy management through a Web interface. If you want to explore the workings of a VoIP system, playing with Asterisk or one of its variations is highly recommended. A search of the Internet will quickly provide a list of websites that you can download the software from.

SUMMARY

This module looked more closely at the manufacturers of voice over IP telephone systems.

Section 6.1: Major Players

We can distinguish between the vendors who are in the business of supplying traditional telephone products and those who have traditionally supplied enterprise data network products. Although their products mostly overlap, there are some differences, particularly when it comes to supporting the PBX system.

Section 6.2: VoIP Products

In this section, VoIP products such as the call manager, VoIP platforms, media gateways, and IP telephones are looked at. One useful way to distinguish between products is by the market segment they are designed for. In order to "right size" their solutions, vendors design for small and medium (SMB) organizations, large enterprises, or the carrier market.

Section 6.3: Vendor Discussion

This section looked at the product lines of the major vendors in more detail.

Section 6.4: Other Vendors

This section looks at some additional vendors.

Section 6.5: More Vendors

This section talks mostly about open source software for telephony applications.

SECTION

2

TCP/IP, the Platform for VoIP

CHAPTER

7

Introduction to TCP/IP

Objectives

Upon completion of this section, the reader will:

- *Understand why TCP/IP is popular and important.*
- *Appreciate the functions of IP, TCP, and the services available such as Telnet, FTP, SNMP, WWW, and e-mail.*
- *Be familiar with the OSI reference model and how TCP/IP relates and does not relate to it.*

INTRODUCTION

VoIP uses the TCP/IP protocol and an understanding of it is fundamental to the understanding of the workings of VoIP. The next seven chapters look at how TCP/IP functions. When these basics are covered, we will return to VoIP and examine it at the packet level.

TCP/IP is the protocol glue that binds the Internet together. In fact, LANs that use other protocols are slowly converting to TCP/IP as well. Therefore, it is important to understand why it took on its central position and to understand the main components of the TCP/IP protocol suite. As a communications protocol, it is important to see its relationship with the other components of the communications system. These connections are illustrated by introducing the OSI reference model of communications.

7.1 WHY TCP/IP HAS SWEPT ALL BEFORE IT

What is a protocol?

A protocol is a formal set of conventions governing the format and relative timing of message exchange in a communications network. Note that it is crucial that both parties understand and agree to use the same set of rules. There are many protocols in the world; for example, you follow a protocol when you meet the Queen of England, when you sit down at a dinner party, or drive on the correct side of the road. The protocols that we are interested in involve the rules that machines use to communicate.

TCP/IP

TCP/IP stands for transmission control protocol/Internet protocol. TCP/IP is actually a collection of many protocols, called a suite, that is used for communications between computing devices.

TCP/IP has become very popular for several reasons

This section lists several reasons for TCP/IP's popularity.

Robust: TCP/IP handles error correction very well. It can still maintain communications when parts of a network are malfunctioning. You would expect this since TCP/IP was funded by the U.S. military and reliable communications is important during war conditions.

Vendor independent: TCP/IP is supported by all major vendors of network hardware and software. This leads to a tremendous selection of components and prices are constantly being driven down because of competition.

Medium independent (transmission medium, network hardware): As a software protocol, TCP/IP runs on any type of local area network (e.g., Ethernet and token ring) as well as on all Wide Area Network technologies.

Built into UNIX: TCP/IP is the networking protocol built into UNIX. This makes it the de facto standard at all UNIX sites. Microsoft Windows also has built in support for it.

Used by the Internet: TCP/IP is the protocol that the Internet uses. It is the only transport protocol used on the Internet. Every site that connects to the Internet must support TCP/IP.

More WAN than LAN

TCP/IP can be used on local area networks as well as wide area networks. TCP/IP was designed for wide area networks. Local area networks originally used protocols specifically designed for them. These are always easier to manage than TCP/IP. They don't have TCP/IP's necessity for managing workstation network addresses and are therefore appropriate for environments in which nontechnical supervisors manage the system. As TCP/IP increased in importance, it was added to the LANs. At this stage, the LAN's native protocol was used to communicate with the file server and TCP/IP was used to communicate with the Internet. Now that all operating systems support TCP/IP, it is commonly the only protocol found on the network.

Weakness of TCP/IP

- Speed: TCP/IP is slower than LAN protocols because of the guaranteed delivery mechanism built into TCP.
- Management: The need to manage the workstation configuration can be a burden on large systems if editing the workstation configuration is done manually. This burden has been largely lifted on modern systems through the use of the dynamic host configuration protocol (DHCP), which automatically configures the workstation.

7.2 WHO MAKES TCP/IP?

Because TCP/IP affects just about all of the computer systems on our planet, changes in the protocol are important to us all. TCP/IP is an open protocol, which means that it is not under the control of any one company. This is in contrast to Novell's IPX/SPX, Microsoft's NetBEUI, or IBM's SNA. Although research into TCP/IP was initially funded by the U.S. Department of Defense, it has passed into the realm of the public domain long ago and is now controlled by the Internet Engineering Task Force.

Internet Engineering Task Force (IETF)

The IETF is a volunteer organization set up specifically to deal with technical issues on the Internet. It is made up mostly of computer industry representatives as well as technical people who are network designers, operators, and researchers, although any interested person is welcome to join. It is easy to join the IETF—just add yourself to the mailing list of any working group. It is a virtual organization in that all communications is done over the Internet. Nevertheless, the IETF meets four times a year at which time proposals are voted upon and standards are ratified.

The standards process

New proposals are always being made to the IETF. These are published in the form of working papers and are generally known as RFCs (request for comments). Originally, RFCs were exactly what the name implied: ideas tossed out in order to get feedback from others. Now, however, RFCs include many kinds of information documents. Subforms of RFCs include FYIs (for your information) and STDs (documents that describe standards).

The majority of Internet protocol development and standardization activity takes place in the working groups of the IETF. Protocols that are to become standards on the Internet go through a series of states or maturity levels (proposed standard, draft standard, and standard) involving increasing amounts of scrutiny and testing. When a protocol completes this process it is assigned a STD number. At each step, the Internet Engineering Steering Group (IESG) of the IETF must make a recommendation for advancement of the protocol. Proposals which do not enter the standards process are flagged as experimental and end up in the "historical" category. Standards which are retired also end up as "historical". Talk about never throwing anything out, all historical RFCs are still on-line and available for reading. The RFC standards process is illustrated in Figure 7-1.

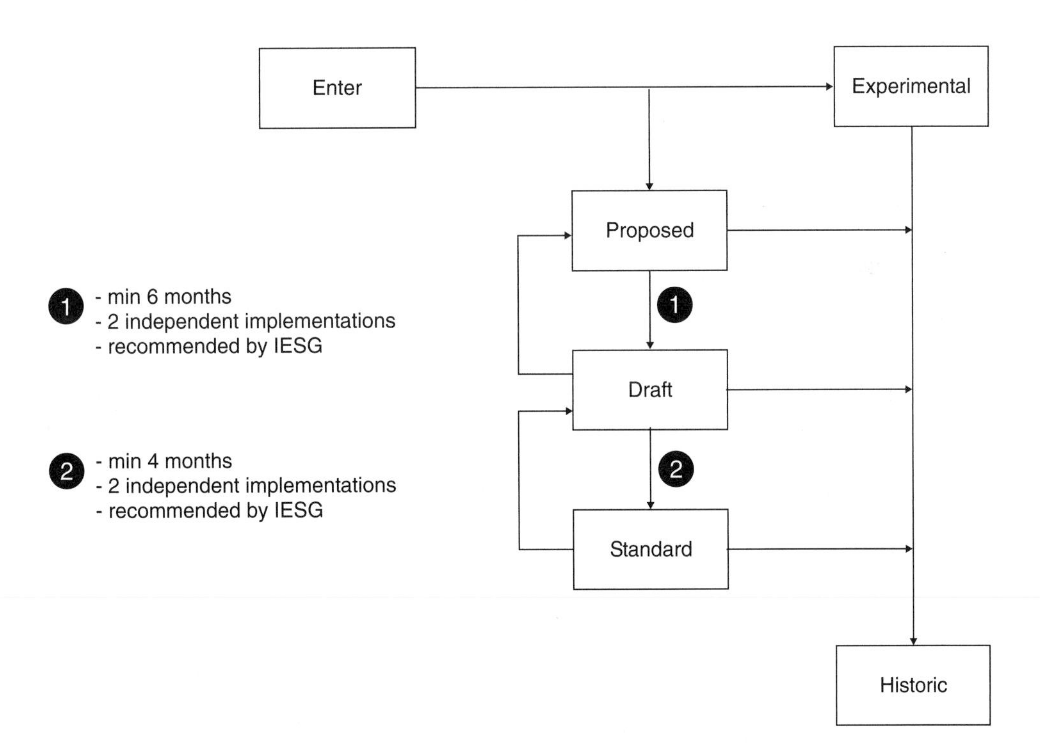

Figure 7-1: The RFC standards process

WHO'S WHO ON THE INTERNET

Internet Engineering Task Force (IETF)

The IETF is the protocol engineering and development arm of the Internet. Though it existed informally for some time, the group was formally established by the IAB in 1986.

Internet Architecture Board (IAB)

The IAB is responsible for defining the overall architecture of the Internet, providing guidance and broad direction to the IETF. The IAB also serves as the technology advisory group to the Internet Society, and oversees a number of critical activities in support of the Internet.

The Internet Engineering Steering Group (IESG)

The IESG is responsible for technical management of IETF activities and the Internet standards process. As part of the ISOC, it administers the process according to the rules and procedures that have been ratified by the ISOC Trustees. The IESG is directly responsible for the actions associated with entry into and movement along the Internet "standards track," including final approval of specifications as Internet Standards.

Internet SOCiety (ISOC)

The Internet SOCiety is a professional membership organization of Internet experts that comments on policies and practices and oversees a number of other boards and task forces dealing with network policy issues.

Internet Corporation for Assigned Names and Numbers (ICANN)

ICANN is contracted to by the U.S. government to supply the IANA (Internet Assigned Numbers Authority) function. The IANA function is responsible for all "unique parameters" on the Internet, including IP (Internet protocol) addresses. Each domain name is associated with a unique IP address, a numerical name consisting of four blocks of up to three decimal digits each, e.g., 204.146.46.8, which systems use to direct information through the network. Also included under the IANA function are protocol numbers and domain names.

World Wide Web Consortium (W3W)

The W3W develops technologies for the World Wide Web, including specifications, guidelines, and tools. The format for Web pages, for example HTML, DHTML, and XML, were developed by the W3W.

The relationship of these governing bodies is illustrated in Figure 7-2.

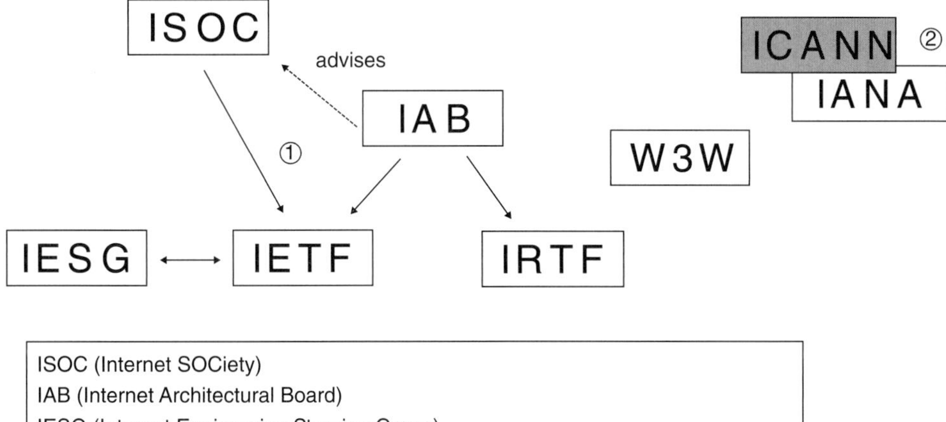

ISOC (Internet SOCiety)
IAB (Internet Architectural Board)
IESG (Internet Engineering Steering Group)
IETF (Internet Engineering Task Force)
IRTF (Internet Research Task Force)
W3W (World Wide Web Consortium)
ICANN (Internet Corporation for Assigned Names and Numbers)
IANA (Internet Assigned Numbers Authority)

1. IETF not incorporated, uses ISOC for corporate functions and funding
2. ICANN under contract to US Dept. of Commerce to provide "IANA" functions

Figure 7-2: Internet governance

7.3 PROTOCOLS AND APPLICATIONS

MAIN COMPONENTS

Internet protocol (IP)

IP is responsible for addressing and routing packets. It provides a connectionless service for end systems to communicate with one or more networks. This means that IP will transmit but has no assurance that the destination machine is actually in operation or that it can accept the transmitted packet.

It defines a 32-bit node address for every machine that is used by the address field of the data packets to identify the destination machine.

Transmission control protocol (TCP)

TCP makes sure data arrives correctly and in order. It provides a reliable end-to-end data transfer service. It also defines a 16-bit port address that identifies an upper-layer protocol.

TCP sets up a session between two machines before it starts transmitting data. It negotiates the size of the data in the packet and how many packets it can send before an acknowledgment is required.

When the packets arrive at the destination, they may be out of order. TCP is also responsible for placing them in the correct order before passing them on to the higher level protocols.

MORE PROTOCOLS

So many protocols

You must realize that TCP/IP is a suite of protocols. While it is true that IP, ICMP, TCP, and UDP are used constantly, there are literally hundreds of extra protocols available to do special jobs. In addition to protocols, TCP/IP also uses many utilities and services. To add to the confusion, sometimes the protocol and the service have the same name.

Examples include the FTP service, which uses the FTP protocol, and the Telnet service, which uses the Telnet protocol.

The following is a small list of protocols.

Internet communications message protocol (ICMP): ICMP is a special-purpose messaging mechanism used to report errors or to provide other information about unexpected circumstances. Because TCP/IP is implemented in software and is independent of the hardware, it cannot depend on the hardware to provide information about failures. ICMP is considered to be the message mechanism of IP itself.

User datagram protocol (UDP): UDP gives application programs direct access to a datagram delivery service. A datagram service does not guarantee delivery or that the data is free from errors. This allows applications to exchange data over networks with a minimum of overhead. UDP is another unreliable protocol, similar to IP.

UDP is a good choice for applications that move only a small amount of data or use a "query/response" model. An example of the former is tiny file transfer protocol (TFTP), of the latter is network file system (NFS).

UDP also cuts down on the overhead of establishing a reliable connection when it is not needed. Therefore it is faster and simpler than TCP although it also exists at the transport layer.

An upcoming chapter explores the characteristics of UDP, which make it the protocol of choice for transporting telephone conversations over an IP network.

Simple mail transfer protocol (SMTP): SMTP is the basic electronic mail protocol. It is used by the e-mail facilities on the Net such as the e-mail programs and e-mail servers. E-mail clients, such as Outlook Express, use SMTP to forward mail to the e-mail server; e-mail servers use SMTP to forward mail among themselves.

Post office protocol (POP): This is the protocol used by the e-mail programs to retrieve e-mail from the e-mail servers.

Simple network management protocol (SNMP): SNMP is a basic troubleshooting and diagnostic protocol. It is used to monitor device status, device performance, connections, configuration, and errors. SNMP has three components: an agent, a database, and a manager.

An agent is software running in the device that is being monitored. It generates information and statistics about the device and stores it. The definition of what types of information the agent can collect is found in a management information base (MIB). The SNMP manager in a management station queries the agent at regular intervals to receive information and take any action required.

SNMP is a simple protocol; therefore the code in the device can remain small and take up minimum amounts of memory.

Table 7-1 lists a more comprehensive list of protocols in a table format.

IP	Internet protocol–provides network identification through the IP address and routing services.
TCP	Transmission control protocol–provides end-to-end guaranteed delivery of packets.
UDP	User datagram protocol–provides best efforts delivery of packets.
ICMP	Internet control message protocol–delivers error messages to IP, provides specialized status and routing messages.
IGMP	Internet group management protocol–used to manage membership in multicast groups.
ARP	Address resolution protocol–resolves IP addresses to MAC addresses.
RARP	Reverse address resolution protocol–used by a client to find its IP address.
SNMP	Simple network management protocol–provides management services, including performance and error statistics and remote control.
Application Services and Their Protocols	
FTP	File transfer protocol–a service used to copy files between machines.
SFTP	A secure version of FTP.
TFTP	Tiny FTP–a small, fast and nonsecure version of FTP.
Telnet	Telnet–provides terminal emulation to host machines.
SMTP, POP3, IMAP4	Simple mail transfer protocol, post office protocol, internet message access protocol–provides e-mail messaging service.
NNTP	Network news transfer protocol–newsgroups.
HTTP	Hypertext transfer protocol–the World Wide Web.
HTTPS	HTTPS–the secure version of HTTP uses encryption.

Table 7-1: The major protocols of the TCP/IP protocol suite

Application Services and Their Protocols	
NTP	Network time protocol–used to synchronize a device's time.
SSH	Secure shell–a secure terminal access, can replace Telnet.
SCP	Secure CoPy is a program to copy files between hosts on a network. It uses SSH for authentication and data transfer.
LDAP	Lightweight directory access protocol–used to access directory listings.
LPR	Line printer remote–provides print spooling service on an IP system.

Table 7-1: *continued*

COMMON APPLICATIONS

TCP/IP is closely associated with applications and utilities. A network is useless if it doesn't allow people to do work and perform tasks. These applications fall within the following three categories.

End user applications

These applications allow end users to do work. The original big three were as follows.

- File transfer protocol (FTP): FTP is the basic file transfer utility of the Internet. Originally, it was run from the command line but now it is available with a graphical user interface. It has been supplanted by the Web browser for downloading files but is still useful for uploading files.
- Telnet: Telnet is a terminal emulation program and is used to access remote hosts. The original machines on the Internet were multi-user hosts such as minicomputers and mainframes and Telnet was the primary tool for accessing them. The importance of Telnet has declined since then but is still important for specialized uses.
- E-mail: E-mail is the third original important application and its importance continues to increase to this day.
- World Wide Web: The World Wide Web is responsible for the popularity of the Internet because of the information that it allows anyone to access. The Web browser is the application that accesses the World Wide Web.

System applications

System applications provide vital network functions, but would not be visible to end users directly. Here are some examples:

- Domain Name System (DNS): DNS resolves domain names to IP address so that servers can be contacted on the Internet.
- Dynamic Host Configuration Protocol (DHCP): DHCP automatically assigns a configuration to an IP host so that the administrator is relieved of manually doing so.
- Network Time Protocol (NTP): NTP allows a host to synchronize its clock with an external reference machine in order to keep the correct time.

Administrator tools

Finally, utilities provide the administrator with management and troubleshooting tools. Examples of these utilities include PING, Nslookup, Ipconfig, ARP, and Netstat. Exercises in this book make use of these utilities.

7.4 OSI REFERENCE MODEL

For a greater understanding of data communications and networking, the open systems interconnect (OSI) seven-layer reference model of communications is usually used. The model was designed by the International Standards Organization (ISO) in order to clarify the relationship of various network components to each other.

The purpose of the model

The OSI model is a model of how communications might work. It was designed by the International Standards Organization with input from many other organizations, including standards committees, manufacturers, and governments. The model includes all of the functions necessary for communications and divides them by functionality into seven divisions, or layers. The purpose of this model includes the following:

- To provide general design guidelines for data communication systems.
- To divide the communication process into well-defined, functional areas (called layers), facilitating the creation of network products and encouraging the interchangeability of network components.
- To promote the goal of communications between different types of systems by encouraging the development of internetworking devices.
- To make sure that all functions needed for communications are addressed and that none are left out.
- To avoid duplication of functions thereby increasing efficiency.
- To design independence for each layer from other layers. This facilitates the interoperability of products from different manufacturers and ensures that the end-user can substitute one vendor's product for another.
- To produce a learning tool that can be used to understand how communications in modern computer systems works.

Some basic rules

The OSI model has some basic rules. They include:

- A layer may communicate with only the layer above and below and may not circumvent intervening layers.
- A layer must be independent of any other layer.

WHAT THE LAYERS DO

It is easier to understand the OSI model if you appreciate that each layer has a job to do. The following is a discussion of each layer and the functions assigned to it. Figure 7-3 illustrates the seven layers in their correct order.

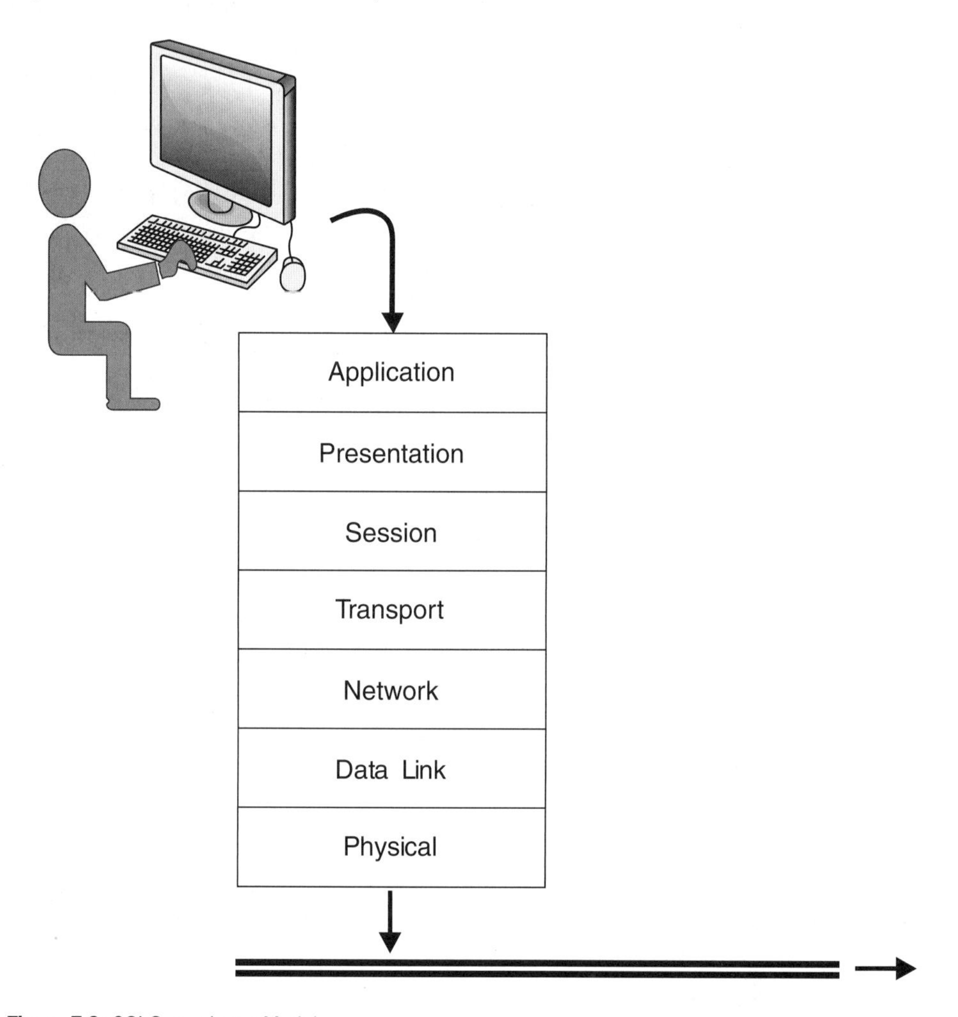

Figure 7-3: OSI Seven-Layer Model

Application layer

The application layer is responsible for providing network services to users and their applications. It is not, however, the user's application or the operating system, such as Windows.

- It provides a user interface—a good example is the type of terminal emulation a host requires, such as VT100 or 3270.
- It provides network services such as file transfer, mail service, or terminal emulation, e.g., FTP, Telnet, SNMP.

Presentation layer

The presentation layer deals with syntax and grammatical rules for presenting data to the application layer (not the user). This includes encoding, decoding, and otherwise converting data. It is responsible for the following.

- Character sets—The code used to represent characters, e.g., the ASCII of personal computers vs. EBCDIC, used by IBM mainframes. A gateway has to convert character sets when a PC communicates with an IBM mainframe. Common character sets are listed in Table 7-2.

ASCII Created by ANSI	7 bits	Used by PCs
ANSI	8 bits	Windows
Unicode Controlled by Uniforum	16 bits	Windows since NT
Shift-JIS Japanese Industrial Standards Committee	8 bits	Japanese character set Katakana
ISO-8859 Latin 1	8 bits	Used in HTML
EBCDIC IBM	8 bits	Used by IBM mainframes

Table 7-2: Character sets

- Compression and decompression of data—Some communication protocols include compression of data. Kermit is one such protocol.
- Encrypts and decrypts data—Moving data over a network safe from prying eyes is the job of encryption. There are many methods available for this, including DES, RSA, and SSL and public/private key schemes.
- Bit order translation—Numbers are sent over the network as binary numbers. The meaning of the numbers is different if the computer starts with the leading bit first or the trailing bit. One is the most significant bit (MSB); the other is the least significant bit (LSB).
- Byte order translation—Just as a byte is made up of bits, a message is made up of bytes. Should the most important byte be transmitted first or the least important? Intel microprocessors transmit the least significant first (called little Endian) while Motorola CPUs (Macintosh) transmit the most important first (called big Endian). Communication between PCs and Macintoshes must take this into account.
- File structure—Another difference between PCs and Macintoshes is their file structure. Macintosh files are comprised of two files, called forks: the resource fork and the data fork. PC files are single entities.

Session layer

A session is an agreement to communicate between two entities, and the session layer controls the setup, termination, and other mechanics of this conversation.

- It establishes and maintains connection, e.g., intent to transmit, was it successful?
- It deals with name recognition (computer name, user name) and login.
- It deals with synchronization of data transmission by placing checkpoints within data stream so that if interrupted, the transmission can take up where it left off.
- It deals with upper-layer errors in the communications process, including problems with memory, storage space, and printing.
- It handles remote procedure calls (RPCs), e.g., running a program on a remote computer.

Transport layer

The transport layer provides extra connection services, including error correction.

- It controls data flow, e.g., slows down transmission if buffers about to overflow.
- It fragments and reassembles data.
- It acknowledges successful transmission.
- It corrects faulty transmission.

Network layer

The network layer is primarily responsible for getting information to the correct computer on the correct network.

- It controls the format of network addresses, for example IP addresses.
- It moves information to the correct address.
- It fragments and reassembles packets.
- It controls the routing function.
- It determines the best path to the destination.

Data link layer

The data link layer creates the entity (the packet or frame), which is put onto the transmission media.

- It controls access to the communications channel.
- It controls the flow of data.
- It organizes data into logical frames.
- It identifies specific computers on the network.
- It detects errors but doesn't correct them.

Physical layer

The physical layer is not the transmission media (cable) itself, but is responsible for the specifications for the media and the electrical signal that goes on it.

- It provides the electrical and physical interface to the network.
- It specifies the standards for media such as the cables and the connectors.
- It specifies type of medium.
- It specifies how signals are transmitted, particularly the encoding schemes.

MAKING A REQUEST

The process of moving data through the OSI model is illustrated in Figure 7-4. It shows a request coming from a user at a workstation to a file server. The data, for example, the request to open a Word file, starts from the user. In fact, it starts with the Word program itself. It is passed to each layer, where it is acted upon until it finally emerges as a packet on the network cable. It is received by the network interface card of the destination machine and then is passed through successive layers until it emerges at the top and is passed to the program running there.

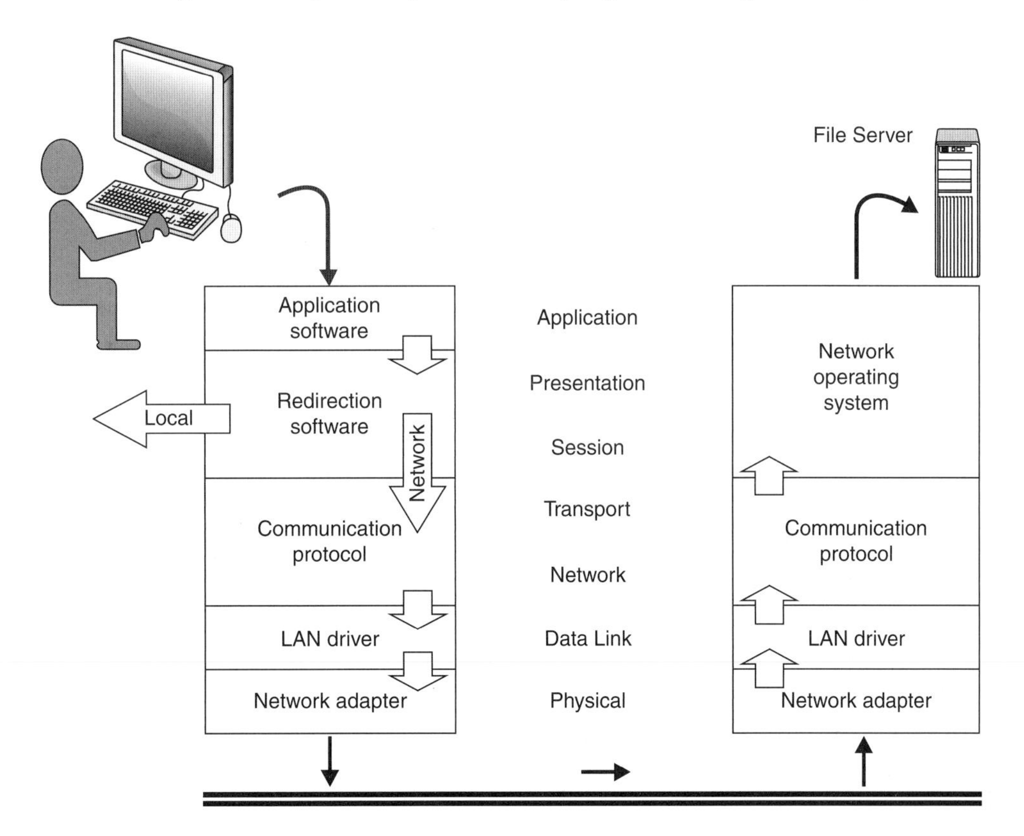

Figure 7-4: Data flow through the OSI model

Notice two things about this process. There may not be a distinct piece of software for each of these layers. The functions of several layers may be combined into a single program. Secondly, layers may be missing entirely if their functions are not needed. OSI is just a model that is useful for learning about data communications and one that developers should try to design for. It does not necessarily represent the real world.

Encapsulation

The process of moving data through successive layers is called encapsulation and is illustrated in Figure 7-5. Each layer takes the frame from the layer above. This becomes the new data, or payload. It then adds information fields to the front of the data, and possibly an error correcting field to the back, and then passes it on to the layer below where the complete frame becomes the new data. Because beginning and ending fields frame the data, this process is often called framing. Because the data is cradled within or enclosed inside, the process is also called encapsulation.

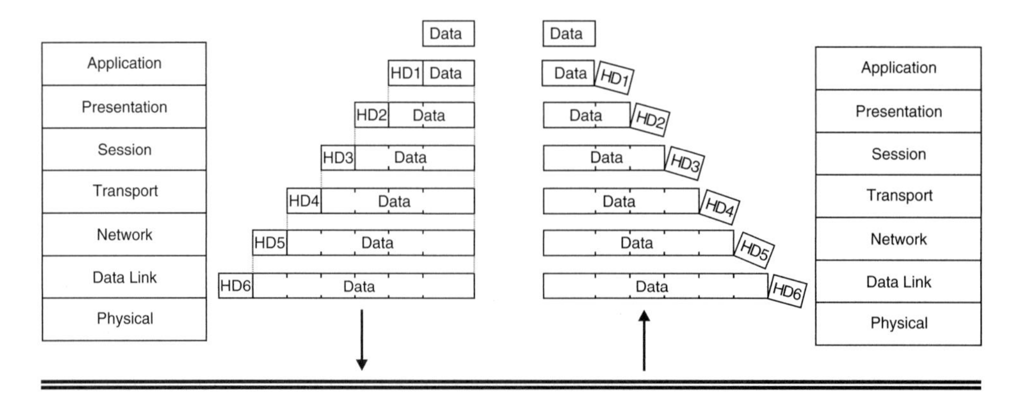

Figure 7-5: Encapsulation

At the receiving end, the process is reversed. At each layer, the framing fields for that layer are stripped off. Error correcting information, if any, is calculated. Then the data is passed up to the next layer and the process is repeated.

Although the process is cumbersome with a great deal of housekeeping involved, performance can still be excellent because it operates at electronic speeds. Improving the speed of communications is a priority for all those involved in this process, including the developers, vendors, and administrators of the systems. An important element of fine tuning, however, is understanding the process just outlined.

TCP/IP IS NOT OSI

Although TCP/IP adheres closely to the OSI model in some respects, it doesn't in others. Keep in mind that TCP/IP was designed and implemented before the final OSI model was published. In fact, it is based on the Department of Defense (DoD) model originally developed for ARPANET, which was the precursor to the Internet. Figure 7-6 illustrates the relationship between the OSI and DoD models and places some of the protocols mentioned in this chapter in context.

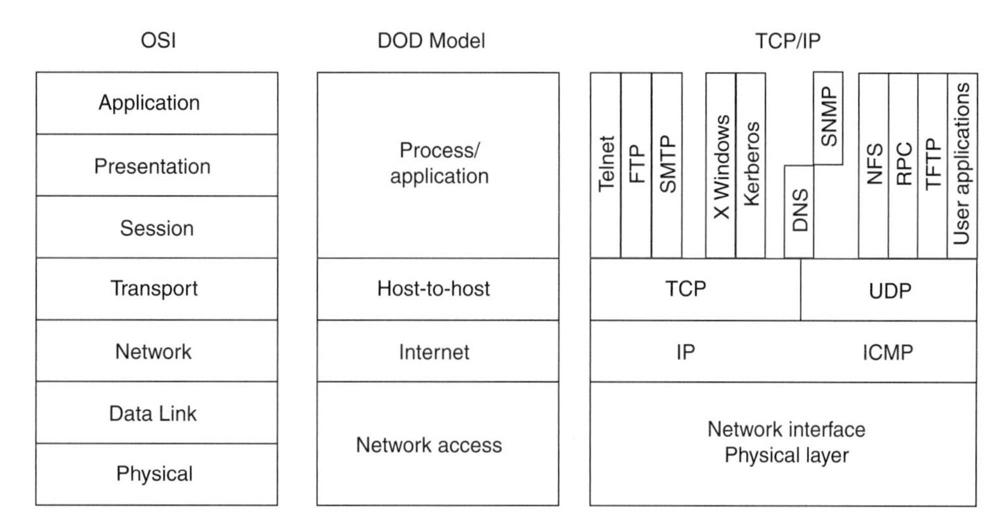

Figure 7-6: TCP/IP is not OSI

In the DoD model, the process/application layer consisted of the applications, such as FTP and Telnet, which are written to use the host-to-host protocols. The host-to-host layer establishes and maintains connections between applications in the host. This could be reliable (TCP) or unreliable (UDP) connections. The Internet layer concerned itself with addressing and routing. The network access layer was responsible for receiving and transmitting the datagrams between hosts.

TCP/IP expects to work with the data link layer, but does not concern itself with any of the details. This makes it independent of the physical aspects of the network. IP is equivalent to the network layer, and TCP and UDP are equivalent to the transport layer.

The TCP/IP services do not neatly divide into the upper three layers of the OSI model. The functions of session, presentation, and application, if they are present, are jumbled within these programs, such as FTP, Telnet, and so on.

SUMMARY

TCP/IP stands for transmission control protocol/Internet protocol. This chapter deals with what TCP/IP is and why it is important. If you understand the material in this chapter, you will understand the following topics.

Section 7.1: Why TCP/IP has Swept All Before It

There are many protocols in the world but TCP/IP has become very popular because it is robust, vendor independent, medium independent, built into UNIX, and used by the Internet. TCP/IP can be used on local area networks as well as on wide area networks. This section also notes the weakness of TCP/IP.

Section 7.2: Who Makes TCP/IP?

This section looks at the different organizations that are involved with TCP/IP and the Internet. The decision-making process of the IETF and the role of RFCs are examined.

Section 7.3: Protocols and Applications

Internet protocol (IP) is responsible for addressing and routing packets. Transmission control protocol (TCP) makes sure data arrives correctly and in order. It provides a reliable end-to-end data transfer service. TCP/IP includes a basic set of services. These include file transfer protocol (FTP), simple mail transfer protocol (SMTP), Telnet, PING, and simple network management protocol (SNMP).

Section 7.4: OSI Reference Model

The open systems interconnect reference model (OSI) was designed by the International Standards Organization (ISO) as a seven-layer model of communications. The process of moving data through successive layers is called encapsulation. TCP/IP fills the functions of two layers of the OSI reference model. IP sits at the network layer, and TCP and UDP sit at the transport layer.

Hands-on Lab 7-1

Examine TCP/IP Configuration under Windows

Lab note: This lab can be performed on any modern version of Windows, including Windows XP, Vista, and Windows Server 2003/2008.

In order to examine the configuration of TCP/IP on Windows, you can use the IPCONFIG command.

1. Start a command window.

   Use **Start** > **All Programs** > **Accessories** > **Command Prompt**

   or

   Use **Start** > **Run** > type in **CMD** > **OK**

2. At the command prompt, run the IPCONFIG command. IPCONFIG runs in the Command window. It does not have a friendly user interface.

   Type in **IPCONFIG**

   Fill in the following information.

   What is the IP address of your machine? ______________________________

   What is the subnet mask? ______________________________

   What is the default gateway? ______________________________

3. A little extra information is provided with the command IPCONFIG/ALL

   Type in **IPCONFIG/ALL**

   Fill in the following information if it is available.

   DHCP server ______________________________

   DNS server(s) ______________________________

   Note: IPCONFIG is not available in Windows 95/98/ME. Instead use the WINIPCFG command from the RUN option. It provides the same information as IPCONFIG.

Hands-on Lab 7-2

Configuring TCP/IP for a Static IP Address on Windows XP

Lab note: The instructions for this lab are written for Windows XP. However, with small modifications, it can also be performed on Vista and Windows Server 2003/2008.

In order to change the IP address of a Windows XP machine, you need to access the properties of the local area network connection.

1. Click on **Start > My Computer.**
2. Select **My Network Places** (under **Other Places** on the left pane of the window). Right-click and select **Properties.**
3. Select **Local Area Connection** and right-click for **Properties.**
4. Select **Internet Protocol (TCP/IP)** protocol and then click the **Properties** button. Windows XP defaults to automatic IP configuration (Figure 7-7).

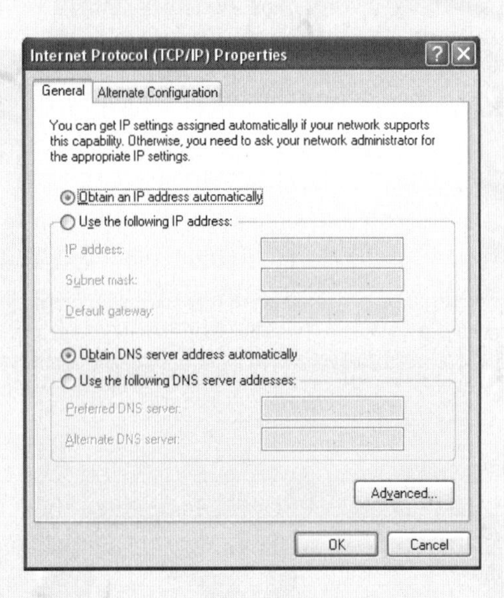

Figure 7-7: IP Configuration screen

Configuring TCP/IP for a Static Address

5. Click the **Use the following IP** address button. You will be asked to provide TCP/IP configuration information. The instructor will provide these numbers to you. Write down the following information:

   IP Address: ____________________

   The numbers for each computer must be unique. Duplicate numbers on the network will lead to communication failure.

   Subnet mask: 255.255.255.0

   Default Gateway: ____________________

   This number represents the router.

6. Type the information into the appropriate fields in the IP address information.
7. Click on **OK** and **OK** again.

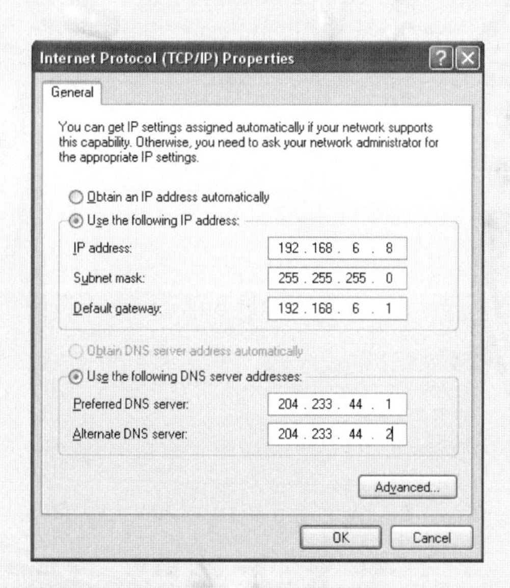

Figure 7-8: IP Configured with a static configuration

Hands-on Lab 7-3

Using PING

PING is the program you use to double check the TCP/IP configuration and confirm that you have a connection to the rest of the network. PING makes a simple request to a remote machine for a response and if the response is received, we know four things: TCP/IP is configured correctly in our computer, it is configured correctly in the remote computer, the remote computer is up and running, and finally, the link between our computers is operating correctly.

1. The PING program provided with Windows runs from the command prompt.

   Open up a command window.

   Use **Start > Run >** type in **CMD > OK**

2. Issue the following command.

   `PING 127.0.0.1`

   IP address 127.x.x.x. is a special address known as a loop-back address. It is used for troubleshooting only and, in this case, indicates that TCP/IP is loaded on your computer.

3. Next PING your own computer to confirm your own configuration.

   `PING 192.168.X.X,` where the IP address is your own address.

4. If successful, PING your neighbor's computer.

   `PING 192.168.Y.Y,` where the IP address is your neighbor's.

   If you don't know the address of any other computer on your network, PING the default gateway's IP address.

5. PING an address that doesn't exist, and familiarize yourself with the different response.

Hands-on Lab 7-4

Duplicate IP Addresses

If two machines have the same IP address, there is an IP address collision. Since this is illegal, one of the machines will have its configuration disabled. How does a machine check for another with the same IP address? The technique is called a "gratuitous ARP." In essence, this means that the host will broadcast to see if another host has its own IP address. The ARP protocol is covered in another section of this book.

Create a Duplicate IP Address

1. For this exercise, two students need to work in pairs. Examine the two IP addresses of the student machines and decide which address should be duplicated.
2. One student leaves his or her computer's configuration as is, but the other student changes his or her IP configuration to match the other computer.
3. Watch for any error messages on both machines. Note the results.

   ____

4. Use IPCONFIG and see what address you have.

Fix the problem.

5. For the machine that was modified, revert its configuration back to the original so that it will function correctly for the remainder of the course.

Review Questions Introduction to TCP/IP

1. TCP/IP is popular because

    a) it is easier to use than the LAN protocols.
    b) it is independent of transmission media and network hardware.
    c) it is one of many transportation protocols used on the Internet.
    d) it requires Ethernet, and Ethernet is the most popular network.

2. Which statement is true?

    a) IP is considered a connectionless protocol.
    b) TCP uses 32-bit addresses.
    c) IP sets up a session with the destination machine before transmitting.
    d) TCP is responsible for routing packets.

3. TCP/IP and the Internet are closely associated with each other because

    a) Bob Metcalfe outlined TCP/IP in his Harvard Ph.D. thesis.
    b) TCP/IP replaced Unix-to-Unix copy program (UUCP) on the Internet.
    c) the DoD model of communications was patterned after TCP/IP.
    d) TCP/IP was designed for ARPANET.

4. TCP/IP is associated with services such as

    a) simple mail transfer protocol (SMTP), an e-mail program.
    b) file-to-print (FTP), a printer utility.
    c) Telnet, a terminal program allowing you to login to a remote host.
    d) PING, a multi-user, network version of the game Pong.

5. TCP/IP does not strictly adhere to the OSI model because

    a) TCP/IP does not have physical and data link layers.
    b) TCP/IP does not use the presentation layer.
    c) The functions of the application, presentation, and session layers are mixed in the TCP/IP services.
    d) TCP/IP was modeled after the DoD model.

6. Which one of the following is not a responsibility of the data link layer?

    a) Node addressing
    b) Guaranteed delivery of the data
    c) Flow control
    d) Error detection

Continues next page

Review Questions: continued

7. Encryption works at the __________ layer of the OSI model.

    a) session
    b) data-link
    c) network
    d) presentation

8. The ________ layer of the OSI model determines the best route for information to take.

    a) data-link
    b) physical
    c) session
    d) network

9. The process by which a header, and optionally, a trailer, is added to data before it is passed onto the next layer is called

    a) encapsulation.
    b) packetizing.
    c) packaging.
    d) fragmenting.

10. Which one of the following statements regarding the OSI model is false?

    a) Layers in the model must be independent of each other.
    b) One goal of the model is to allow the easy substitution of components in a network.
    c) Development of the OSI model was spearheaded by manufacturers who wanted to give consumers a wide choice in their communication options.
    d) A layer must communicate only with the layer above and/or below it.

CHAPTER

8

Data Link Layer

Objectives

Upon completion of this section, the reader will:

- *Understand the data link layer of the OSI model and the specifications for the data link layer as standardized by the IEEE.*
- *Know some of the history of Ethernet, the cable access scheme used by Ethernet, and its packet structure.*
- *Understand how ARP retrieves the MAC address of the destination computer.*
- *Appreciate what a protocol analyzer is, how it can be put to use on a LAN, and why it is useful when learning about TCP/IP.*

INTRODUCTION

VoIP telephones are connected to an Ethernet network and Ethernet exists at the data link layer of the OSI model. If you expect to troubleshoot VoIP telephones, you need an in-depth understanding of Ethernet.

The data link layer of the communications model is responsible for taking the data generated by the upper layers and placing it onto the media (cable or wireless) so that it reaches the destination device. The functions that are required to fulfill this task are implemented as the header that the data link layer adds to the payload that it receives from the layer above it. The goal of this chapter is to see how TCP/IP integrates with the data link layer, how the data link header is structured and how the ARP protocol connects IP to the data link layer.

8.1 DATA LINK LAYER

IEEE Standards

The Institute of Electrical and Electronics Engineers (IEEE) is responsible for developing the data link layer for most LAN technologies. The exception is fiber distributed data interface (FDDI), which is controlled by the American National Standards Institute (ANSI). Here are some well-known standards.

Logical Link Control (LLC)
- 802.2 – Logical Link Control (LLC)

Media Access Control (MAC)
- 802.3 – CSMA/CD – Ethernet (Baseband bus)
- 802.4 – Token-Bus – MAP (Broadband bus)
- 802.5 – Token-Ring (Baseband ring)
- 802.6 – Metropolitan Area Network (MAN, broadband tree)
- 802.8 – Fiber Optic specifications
- 802.11 – Wireless LANs
- 802.16 – Wireless WANs
- ANSI X3T9.5 – Fiber Distributed Data Interface (FDDI)

Data link sub-layers

The IEEE divides the data link layer into two sub-layers, the LLC and the MAC.

The logical link control (LLC) sub-layer provides a standard interface to the network layer above it. This simplifies life for developers of the transport protocols since they don't have to deal with the specifics of each hardware type (Ethernet, token ring, etc.). The standard for the LLC is IEEE 802.2.

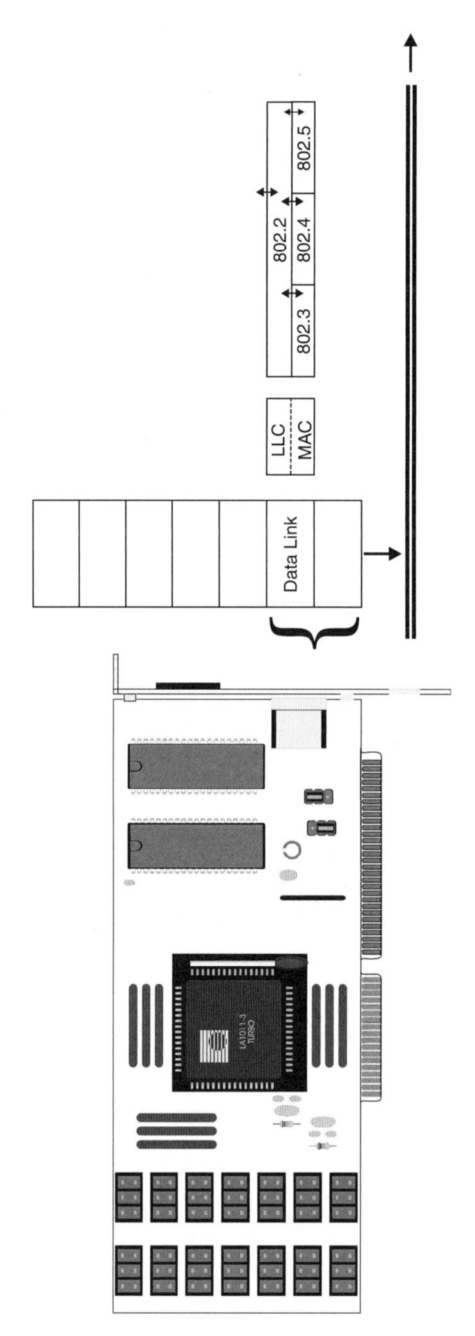

Figure 8-1: The data link layer of the OSI model

The LLC then interfaces to the media access (MAC) sub-layer of each of the hardware types. The details of this interface are the responsibility of the respective protocols and designed into the network interface card. This relationship is illustrated in Figure 8-1. The MAC standard for Ethernet is 802.3, for token ring, 802.5 and so on.

The MAC sub-layer controls the framing, flow control, encoding, and the NIC addresses, which are burnt into the NIC itself. Each card's MAC address is unique.

8.2 ETHERNET

Ethernet was the first widespread LAN technology, is still the most popular and is closely associated with TCP/IP. Some important milestones in its development are listed as follows.

- 1970 – Aloha Net, University of Hawaii, pioneers a radio shared data network, which forms the basis of the carrier sense multiple access (CSMA) scheme used by Ethernet, 1970.
- 1974 – Bob Metcalfe's Harvard Ph.D. thesis outlines improvements to the Aloha Net access scheme
- 1973–75 – Bob Metcalf and David Boggs working for Xerox build Ethernet_I at PARC
- 1982 – DEC, Intel, Xerox (DIX) develop Ethernet_II
- 1982 – 3Com produces first Ethernet NIC for PCs
- 1985 – IEEE 802.3 committee publishes standard

The 802.3 standard made a small change to the Ethernet_II frame. Nevertheless, the Ethernet_II frame type is still in use because it is closely tied in with TCP/IP, despite the fact that the 802.3 standard is newer. Note that Ethernet_II and TCP/IP came into widespread use at the same time, around 1982. The Ethernet_II and 802.3 frame types can mingle on a network. Network adapters have no difficulty telling them apart. Expect 802.3 frames on a network if the IPX/SPX or NetBEUI protocols are used.

Ethernet media access

Ethernet uses a contention-based system called CSMA/CD (carrier sense multiple access/collision detection).

Access is on a first-come, first-served basis. First a workstation "listens" on the cable to make sure there is no traffic. Then it transmits a packet of information down the cable; if a collision is detected, then the station retransmits. Because two stations have to retransmit after a collision, a further collision is likely. Therefore, each generates a random time and the one that has the shortest time will be able to get its packet transmitted first.

The system has little overhead and works best if traffic is light and even among workstations. As traffic increases, performance degrades because of the need to recover from collisions.

Ethernet comes in many varieties depending on frame type, cable, and throughput. Table 8-1 lists some of the common variations of Ethernet.

	Throughput (Mbps)	Cable	Distance
Ethernet_II	10	Thick Coax	500 M
10Base5	10	Thick Coax	500 M
10Base2	10	Thin Coax	185 M
10BaseT	10	UTP cat 3	100 M
100BaseTX	100	UTP cat 5	100 M
10BaseFL	10	Fiber	1.5 KM
1Base5	1	UTP	250M
10Broad36	10	Coax	1,800 M

Table 8-1: Ethernet types

ETHERNET FRAME STRUCTURE NOTES

The structure of the Ethernet frame is simple compared to other protocols; besides the payload, the fields in the frame include the destination and source MAC addresses, the EtherType/length field and a frame check sequence. In addition, a timing signal, the preamble, precedes the frame as it travels along the wire. The structure of the Ethernet frame is important when you need to know the bandwidth and the number of VoIP conversations that can co-exist on the wire. These calculations are made in a later chapter.

In case you were wondering, the correct name for the Ethernet protocol data unit is the *frame*. The word *packet* is too general and doesn't accurately reflect which layer of the OSI model we are discussing. Refer to Figure 8-2 while you follow the discussion of the fields in the Ethernet frame.

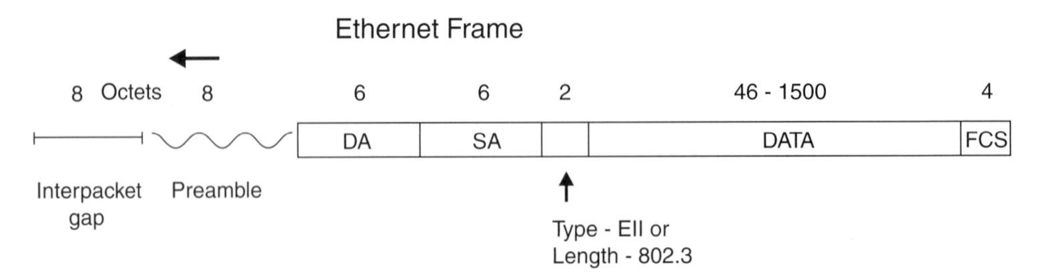

Figure 8-2: Ethernet packet structure

Interpacket gap (IPG)

There must be a quiet time between the successive transmissions of packets on the cable. Because of the Ethernet CSMA/CD access method, a station has to listen before it can transmit. The IPG provides the absence of transmission the network adapter is looking for before it will transmit. This IPG is a time value and the minimum is the time required to transmit 96 bits of raw data. Because the time it takes to transmit 96 bits varies with the throughput of the network, the IPG also varies as follows — 10Mbps Ethernet: 9.6 microseconds, 100Mbps Ethernet .96 microseconds, 1Gbps Ethernet: .096 microseconds.

Preamble/Start frame delimiter (SFD)

Preceding the frame is a seven-octet preamble and one-octet SFD. All 64 bits are composed of alternating 1's and 0's (101010101010...), except the last two bits, which are two 1's (11).

The purpose of the preamble/SFD is to allow the network adapter to synchronize with the incoming data stream. The special pattern of bits allows the adapter to identify the beginning of the frame. The preamble/SFD carries no data and is never included in packet size.

Destination and source address

The first two fields in the Ethernet header are the addresses of the destination and source machines. The address can be referred to as the MAC, physical, or Ethernet address. The MAC address is burned into the chipset found on the network adapter; it is six octets in length, which is enough for a six-digit hexadecimal number. The number must be unique, even between manufacturers. This is a requirement because otherwise, network adapters from different manufacturers could not be mixed on a network. There are two sections to the number—manufacturer's code and unique NIC number. The manufacturer's code is three octets, is called the organizational unique identifier (OUI), and is assigned by the IEEE. The unique NIC number is also three octets in length, is called the serial number, and is assigned by the manufacturer as it makes the chipset. An example of a MAC address is 00 20 AF 23 B2 43. The OUI is 00 20 AF and is assigned to 3Com, whereas 23 B2 43 was assigned by 3Com to this card.

Broadcast and multicast

Ethernet supports both broadcast and multicast transmission. In a broadcast, all machines accept and process the frame. In a multicast, a group (not all and not one, but some) receives the frame. A broadcast, multicast, or unicast frame can be identified by its address. In addition, it should be noted that a MAC address can be overridden by an administrator when there are special circumstances. Some notes on these issues follow.

- All OUIs have an even number as the first octet. The most common is 00 although there are a few 08s and AAs. If the first octet is odd, it indicates a multicast address. A multicast goes to a group of machines, not every machine as in the case of a broadcast.
- In order to turn a manufacturer's unique OUI into a multicast OUI, add 01 00 00. For example, using 3Com's preceding OUI, add 01 00 00 to 00 20 AF to get 01 20 AF, the multicast OUI.
- The first octet of the OUI has extra significance.
  If the least significant bit is a 0, the OUI is a unique address. If it is 1, it is multicast.
  00000000 (0×00) — unique address
  00000001 (0×01) — multicast
  If the second least significant bit is 0, the OUI is universally administered, i.e., purchased from the IEEE. If it is 1, the MAC address is locally administered by the network administrator.
  00000000 (0×00) — universally administered and unicast
  00000001 (0×01) — multicast
  00000010 (0×02) — locally administered and unicast
  00000011 (0×03) — locally administered and multicast
- Destination address FF FF FF FF FF FF indicates a broadcast to all stations.

Length or EtherType

When the IEEE created the 802.3 standard, they revised the meaning of only one field of the original Ethernet_II frame structure. The third field in the frame header is an EtherType field if the frame is of type Ethernet_II, or a length field if it is of type 802.3. It must be one or the other; it can't be both.

Length field - 802.3

If the frame type is 802.3, the third field indicates the number of bytes in the data portion of the frame. The values must fall between a minimum of 46 bytes to a maximum of 1,500 bytes (00 2E to 05 DC in hexadecimal).

Type field - Ethernet_II

If the frame type is Ethernet_II, the third field indicates the protocol that the frame is carrying. This allows Ethernet to pass along its payload to the correct software at the network layer because it is clearly identified. The value in this field is a code expressed as a hexadecimal number. All EtherType codes must be larger than 05 DC hex (1500 decimal). Xerox published 06 00 hex as the cutoff point for Ethernet_II. Table 8-2 lists common EtherType codes.

How can you distinguish between the two frame types? It is a simple matter of looking at the third field, octets 13 and 14. If the number is between 00 2E and 05 DC hex, it is an 802.3 frame. If the value is larger than 05 DC, then it is an Ethernet_II frame.

2-byte Value (hex)	Frame Type	Indicates
00 2 E	802.3	Min. data length (46 bytes)
05 DC	802.3	Max. data length (1,500 bytes)
08 00	E_II	IP
08 06	E_II	ARP
0B AD	E_II	Banyan
80 9B	E_II	AppleTalk
81 37	E_II	NetWare IPX
86 DD	E_II	IP version 6
81 00	E_II	VLAN (802.1Q)
88 47	E_II	MPLS

Table 8-2: Reading the EtherType/length field

Data

The data field represents the number of octets of payload that the frame is carrying and holds a value in the range of 46 to 1,500 decimal (00 2E to 05 DC hex). If the data is less than 46 octets, the data field is padded to bring it up to 46. The complete packet is in the range 64 to 1,518 bytes and this includes the 18 bytes of framing data (DA, SA, EtherType/length, FCS).

Frame check sequence (FCS)

The FCS is used by Ethernet to determine if a frame has been corrupted in transit. The value is a checksum that the source adapter calculates from the data. It is a four-octet value and is appended as a trailer to the frame. The receiving adapter also calculates a checksum from the data and compares its value with the value found in the FCS field. If they match, the frame was not corrupted. If they do not match, the frame is discarded. The checksum used by Ethernet is a cyclic redundancy check (CRC).

Look it up

A comprehensive list of both vendor IDs and Type field codes can be found at http://standards.ieee.org/regauth/ethertype/eth.txt.

Hands-on Lab 8-1

Find the MAC Address

[This exercise works on all modern Windows, including Windows XP, Windows Vista, Windows Server 2003 and 2008]

In order to find the MAC address of the network adapter under Windows, you must use the IPCONFIG command.

IPCONFIG

1. Start a command window

   Use **Start** > **Run** > type in **CMD** > **OK**

2. At the prompt, run the IPCONFIG /ALL command.

   IPCONFIG runs in the Command window.
   Microsoft calls the MAC address the physical address. What is the MAC address of the network adapter in your computer? Which company manufactured it?

Exercise 8-2

Quiz: Ethernet Decode

The following examples show the first 16 bytes of the hex dump of some sample Ethernet packets. The dash in the middle has no meaning and is placed there only to divide the line in two for ease of counting bytes.

1. For each packet, indicate whether it is E_II (Ethernet_II) or 802.3.
2. For E_II, indicate the EtherType and for 802.3, the length in bytes (hex).

   These examples are taken from packet captures. For an example of a packet capture, look at Figure 8-3 later in this chapter.

   a) Packet#1

   `00 aa 00 5f b6 c4 00 00 - d4 62 04 05 08 00 45 00`

   E_II or 802.3 / EtherType or length ________

   b) Packet#2

   `00 00 f3 10 4e 37 02 60 - 8c da e4 9c 86 dd 45 00`

   E_II or 802.3 / EtherType or length ________

   c) Packet#3

   `ff ff ff ff ff ff 08 00 - 09 99 96 cb 08 06 00 01`

   E_II or 802.3 / EtherType or length ________

   d) Packet#4

   `00 00 5a 40 71 60 00 00 - 1b 40 09 52 00 3f e0 e0`

   E_II or 802.3 / EtherType or length ________

   e) Packet#5

   `00 cc 5a 40 71 60 0f 00 - 2c 40 9c 52 81 00 71 c0`

   E_II or 802.3 / EtherType or length ________

   f) Packet#6

   `00 90 27 8a e7 b4 86 00 - a8 06 72 52 88 47 f0 7e`

   E_II or 802.3 / EtherType or length ________

8.3 ADDRESS RESOLUTION PROTOCOL

The address resolution protocol (ARP) is part of the TCP/IP suite of protocols and works hand-in-hand with IP to solve a standard communication problem. IP needs to discover and then give the MAC address of the destination host to the Ethernet driver in order for the Ethernet card to put together an Ethernet frame to deliver the IP datagram. Although Ethernet is used in this discussion, the same principle applies to any data link technology used, including token ring or wireless.

Initially, IP doesn't know the MAC address of the target machine and therefore ARP is pressed into service to discover it. The target machine is an IP host if the target is on the local network or the gateway/router if the ultimate target is on a remote network. Figure 8-3 illustrates the ARP handshake.

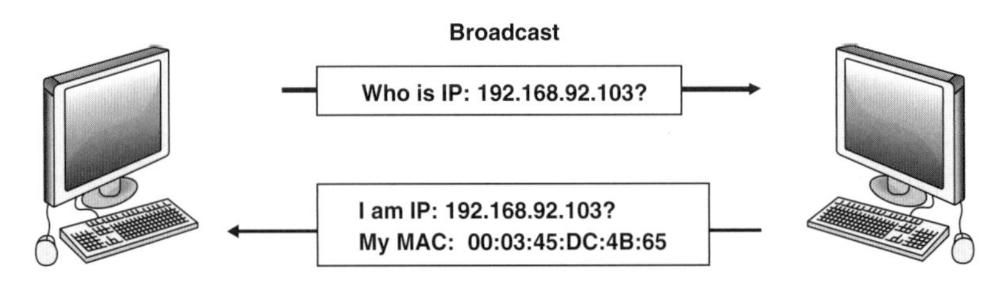

Figure 8-3: The ARP handshake

ARP

Address resolution protocol is distinct from IP but operates at the same layer as IP.

The question invariably comes up: If IP doesn't know the MAC address of the target machine, how does ARP know it? Initially, ARP doesn't know it. Therefore ARP can not use a directed packet to communicate with the target; it must use an Ethernet broadcast.

The ARP broadcast contains the Ethernet broadcast address and the target's IP address. In effect, the message says "If this is your IP address reply with your MAC address." Once known, the MAC address can be passed over to the Ethernet driver for creating frames sent to that target host.

ARP cache

If an ARP broadcast is used first for each IP packet transmitted, the network would quickly become congested. Instead, the MAC address and IP address are stored as a pair in a table in the memory of the host; this table is known as the ARP cache. For subsequent packets, IP merely needs to look in the ARP cache to discover the MAC address of the target host.

In early computers, memory was a scarce resource and in order to conserve it, the ARP cache was dynamic. If there was no activity going to a target machine, the

ARP cache was purged of the address entry. Despite the fact that modern computers have more memory, the practice of purging stale entries from the ARP cache is still used.

The Microsoft algorithm for the ARP cache is as follows.

- If there is no activity, the ARP table entry for a host is purged after 2 minutes.
- If there is activity, the ARP table entry for a host is purged 10 minutes after the last request.

You may view the ARP cache with the ARP -a command.

Gratuitous ARP

Gratuitous ARP is used by a host to discover if its IP address is unique on the network. When a host boots, it sends out an ARP request for its own IP address. It naturally doesn't expect a response. If another host responds, the host can't use the IP address and disables the interface. An error message is generated at both the local and remote host. The user must change the IP address if he expects to communicate.

Hands-on Lab 8-3

Examining the ARP Cache

ARP

The ARP protocol works by broadcasting a request to the network at large requesting that the host identified by a particular IP address respond. When that host responds, it provides its MAC address. The requesting host then adds that IP address-MAC address combination to its ARP table. If this host needs to communicate with the remote host again, it just needs to look up the ARP table instead of issuing a new ARP request.

The ARP table is very dynamic and small. As new entries are created for every new host contacted, old entries are purged. An entry is kept for about 2 minutes unless it is extended because that host is contacted again.

Microsoft's ARP program

In Windows, Microsoft provides an ARP command, which allows you to examine the current ARP table.

1. Use **Start** > **Programs** > **Command Prompt.**
2. Type in this command.

   ```
   ARP -a
   ```

3. In all likelihood, you received a message that there are NO ARP Entries Found.

   This is correct since you have not been in touch with any other host recently.

4. PING as many of the machines on your network as you desire. If you are having trouble finding IP addresses, try the default gateway.
5. Repeat the ARP -a command.

   The ARP table should now be populated with IP address-MAC address pairs.

6. Wait 1 minute and repeat the ARP -a command. Keep repeating every 30 seconds. At what point are the ARP entries purged?

8.4 PROTOCOL ANALYZERS

Protocol analyzers

A protocol analyzer is a software program that allows you to examine the contents of packets as they are transmitted on a LAN or WAN. This allows you to analyze the characteristics of the system or do troubleshooting.

Here are some examples of the type of information available and how they may be put to use.

- Amount of network activity: Measured in packets per second or megabits per second (Mbps), this indicates how much of the network's bandwidth is being used. If the network feels sluggish or performance is unacceptable, it is important to know the bandwidth used. Because activity is uneven throughout the working day, a protocol analyzer can show the pattern of usage. Based on this information, a network administrator can upgrade the system or shift activity to less congested times to increase performance.
- Primary users (top talkers): A protocol analyzer can tell you who generates the most traffic on the network. It could be individuals, particular computers such as servers, or types of usage, such as database or graphics. Based on this information, you can take action to increase performance. You might upgrade the equipment of top talkers or divide the network and segregate them on their own cable system so that they don't impact the rest of the users.
- Troubleshooting: This is the most widely used feature of protocol analyzers. By examining network traffic, you can filter out improperly formed packets. This may indicate hardware problems, such as a failing network interface card or bad cable connections, or software, such as problem drivers or applications.
- Security: Using the protocol analyzer, you may detect abnormal patterns of usage that are the tell-tale signs of hackers. Intrusion detection systems (IDS) are a class of software that provide automatic detection of suspect activity. A protocol analyzer is the basis of IDS software.
- Educational: From the viewpoint of this book, protocol analyzers allow us to capture the workings of the network in action. We will use output from analyzers, sometimes called trace files, to see how the protocols work.

Filtering

Because the network can generate thousands and millions of packets in a very short time, this sheer volume can overwhelm the analyzer and relevant information goes undetected. It is crucial that the analyzer can filter the traffic and present a subset of the packets to us. Some important elements that you may filter on include station addresses, network addresses, and types of protocol or error conditions.

Hardware issues

Not all workstations can act as protocol analyzers. The problem is with the network interface card. Remember that a NIC is designed to "see" packets but it only copies a packet to the workstation's memory if it has its own number as the destination address. A protocol analyzer has to be able to copy all packets. A NIC can accept all packets as long as it has, and is using, "promiscuous mode." Some makes and models have a promiscuous mode, others don't.

Decodes

Protocol analyzers are designed to understand certain protocols. The programmer includes the codes for the values in the different fields of the packet. This allows the program to report back in English the meaning of the fields instead of forcing the operator to refer to tables to figure out the content of the packet. This makes the program much easier to use. It is important that the protocol analyzer have the decodes that you need, particularly for VoIP protocols.

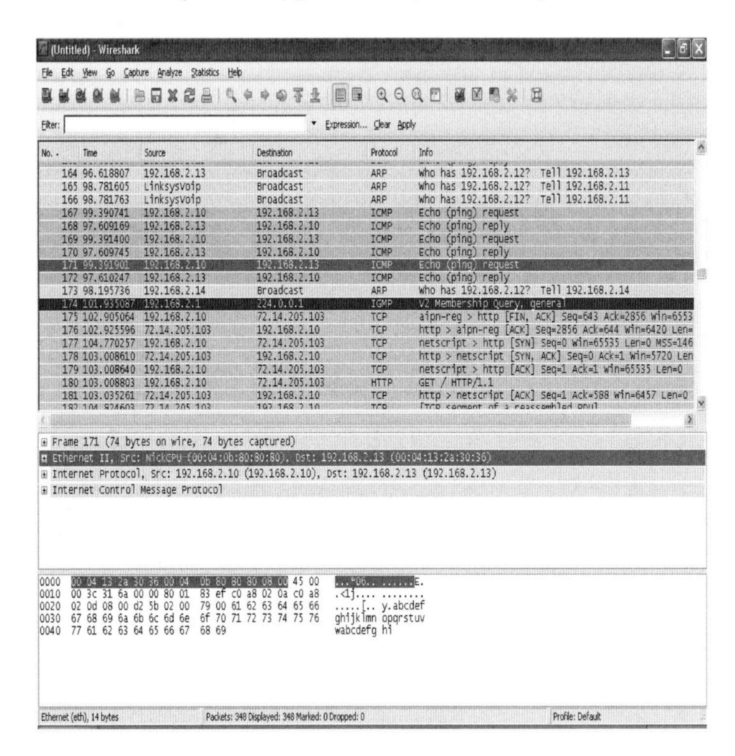

Figure 8-4: Typical layout of a protocol analyzer

Figure 8-4 shows a typical example of a protocol analyzer. The screen is divided into three sections: the capture buffer, the decode, and the hex dump.

Capture buffer

The top section shows the capture buffer, a list of packets that have been captured. This information is useful for deciding which packet to exam further. The columns include the destination and source addresses as well as the major protocol used.

The decode

The middle section of the screen is the decode. It tries to interpret the raw data in a more meaningful way.

Hex dump

The bottom section is called the hex dump because it displays the raw data inside the packet in hexadecimal format. The right-hand portion is an ASCII readout of any text characters.

The screen shot is of Wireshark, an open source protocol analyzer free for the download. Find it at www.wireshark.org.

Hands-on Lab 8-4

Installing Wireshark

Wireshark is the most popular protocol analyzer available from the Internet because it is excellent, it is available for both Windows and Linux, it is open source, and it is free.

Wireshark is a fork of an earlier program called Ethereal. Wireshark is more up to date and is recommended. Download the program from www.wireshark.org.

1. The instructor will indicate where you can find the installation file for Wireshark. Note that Wireshark also requires a driver file called WinPcap, which is included with the installation files. If you are not in a classroom, download Wireshark from its website.
2. Click on the installation file to start the setup. It is safe to accept all defaults. This includes the installation of WinPcap. Wireshark should now be installed.

Capturing packets

3. Start Wireshark. Look for the shortcut on the Desktop, the Start menu or under All Programs. The Wireshark main screen looks similar to Figure 8-4.
4. Familiarize yourself with the interface. Pull down each menu in turn and examine the options. Place the mouse cursor over each button on the button bar and read the description. Some buttons are not active until you start a capture session.
5. Click the Interface button (the first icon on the button bar). The list of interfaces found on your computer is displayed. If your network has any activity, the number under the Packets column will keep changing.
6. Use the interface you want to capture from, such as the Ethernet card. At this point, you can click the Start button to start the capture. However, you will not see the packets captured in real time. Instead, click the Options button, which takes you to the Capture options. Under the Display options, select Update list of packets in real time and Automatic scrolling in live capture and then click the Start button. Wireshark will start capturing packets.
7. You can stop the capture by clicking on the stop button, which is the fourth icon from the left.
8. Select any packet of interest in the capture buffer (top third of the screen) and the details will be displayed in the decode section (middle section of the screen).
9. If there isn't any interesting activity, PING another computer while you are capturing and then examine the PING packets.

SUMMARY

For your convenience, here is a summary of the sections found in this lesson and the material they covered.

Section 8.1: Data Link Layer

This section covers the data link layer, OSI layer 2. The data link layer controls access to the communications channel, controls the flow of data, organizes data into logical frames, identifies specific computers on the network, and detects errors. The Institute of Electrical and Electronics Engineers (IEEE) is responsible for developing the data link layer for most LAN technologies. Some important specifications include, 802.2 Logical Link Control (LLC), 802.3 CSMA/CD - Ethernet (baseband bus), and 802.5 Token-Ring (Baseband ring).

Section 8.2: Ethernet

Ethernet was the first widespread LAN technology, is still the most popular, and is closely associated with TCP/IP. Ethernet uses a contention-based system called CSMA/CD (carrier sense multiple access/collision detection). The frame structure is described in detail and the difference between the Ethernet_II and 802.3 frames is highlighted.

Section 8.3: Address Resolution Protocol

The address resolution protocol is part of the TCP/IP suite and is responsible for finding the MAC address of the destination machine when the IP address is known.

Section 8.4: Protocol Analyzers

This section introduces protocol analyzers, one of the most important tools used for maintaining and troubleshooting networks.

Review Questions Data Link Layer

1. The address that the data link layer uses is that of
   a) TCP/IP.
   b) the network interface card.
   c) the server.
   d) the Internet.
2. The LLC sublayer
   a) is IEEE standard 802.5.
   b) is not required if there is a MAC sublayer.
   c) is different for each MAC sublayer.
   d) provides a standard interface to the network layer.
3. Which of the following statements concerning Ethernet is true?
   a) E_II recognizes IP as a type of packet.
   b) E_II has a length field, whereas, 802.3 has a type field.
   c) The maximum length of the data field is 1,518 bytes.
   d) 802.3 has a field for the LLC sub-layer.
4. What is the function of ARP?
   a) To find the MAC address when the IP address is known.
   b) To find duplicate MAC addresses on the network.
   c) To find the IP address when the MAC address is known.
   d) To retransmit an Ethernet packet if there is a collision.
5. Which of the following functions is not within the capabilities of a protocol analyzer?
   a) Find the machines that have the greatest activity.
   b) Provide a baseline of activity.
   c) Find machines that are transmitting malformed packets.
   d) Configure the IP protocol on a host.

CHAPTER

9

Internet Protocol

Objectives

Upon completion of this section, the reader will:

- *Understand what a connectionless or unreliable protocol is.*
- *Know about the functions that the Internet protocol (IP) performs.*
- *Appreciate what the various fields in the IP datagram do.*
- *Be familiar with the messaging function that the Internet control message protocol (ICMP) brings to the TCP/IP suite of protocols.*

INTRODUCTION

The Internet protocol is the bedrock of the TCP/IP suite of protocols. Because it operates at the network layer of the OSI reference model, it is in charge of network identification and routing. This lesson explains the functions of IP and looks at the IP datagram header to see how these functions are implemented. In addition, ICMP, the companion protocol to IP is examined. Specifically, the different types of messages that ICMP uses are described.

9.1 INTERNET PROTOCOL

IP is the basic packet delivery mechanism on a TCP/IP network. All protocols above IP use IP to deliver the data. IP exists at the network layer, layer 3, of the OSI reference model.

Connectionless service

IP is an example of a connectionless service. A connectionless service or protocol is a service that does not exchange control information (handshaking) to establish an end-to-end connection. In other words, when IP transmits a packet, it doesn't know for a fact whether the destination machine exists (maybe it is turned off) or whether it can receive the packet (it might be busy). IP also doesn't check to see if a packet was received correctly. For all of these reasons, this type of protocol is sometimes said to be "unreliable."

If IP doesn't set up a session with the destination machine and is unreliable, why do we use it? First, it is efficient and speedy. Second, upper-layer protocols (for example, TCP) usually provide session and error correcting services. IP accepts the data that it receives from the upper layer, TCP or UDP, adds a header to it, and relays the resulting datagram to the layer below for transmission (see Figure 9-1).

Internet protocol functions

The functions of IP include the following:

- Defines the datagram, the basic unit of transmission.
- Defines the Internet addressing scheme.
- Moves data between the data-link layer and the transport layer.
- Routes datagrams to remote hosts.
- Performs fragmentation and reassembly of datagrams.

Note: The word "frame" is used to describe the unit of information that is transmitted on the cable, i.e., the unit created at the data link layer. For IP, the unit is called a *datagram*. This avoids some confusion over the unit under discussion.

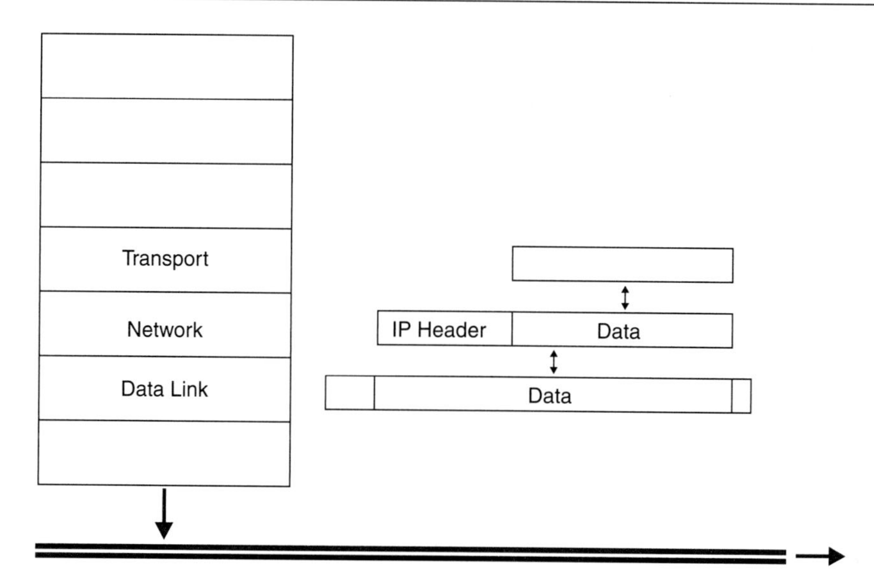

Figure 9-1: IP exists at the network layer

9.2 PACKET STRUCTURE

IP uses various fields in its header to transmit information that is required by routers along the way and the destination host to process the IP datagram. Figures 9-2 and 9-3 give two views of the header structure. Figure 9-2 shows the IP datagram encapsulated in the data link frame. It also lists each field in the header along with its length. Figure 9-3 shows the header divided at 32-bit or 4-octet boundaries. This common depiction is found in the request for comment, which describes the Internet protocol. The standard for the Internet protocol was published as RFC791 in October 1981. A description of each field follows. Keep in mind that each field has a specific purpose and that they all work together to achieve one goal, to deliver the datagram to its final destination.

IP STRUCTURE—THE FIELDS

Version

The version field is the first field in the IP header and is 4 bits in length. It identifies the version of IP under which this datagram was generated. Two versions of IP are currently used. IPv4 has been in use since 1982. IPv6 is now being implemented. The packet structure described in this lesson is IPv4. The receiving station checks the version so that it knows how to handle the packet. If it doesn't understand the version, it will reject the packet.

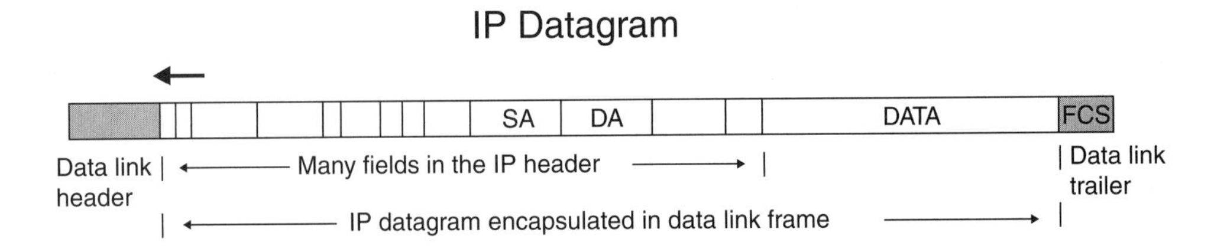

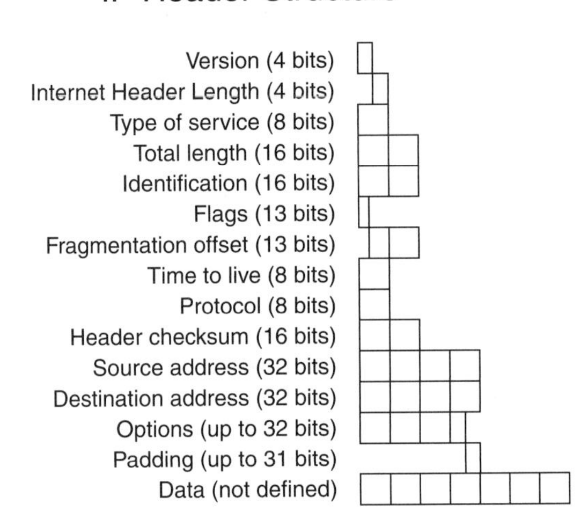

Figure 9-2: The IP header structure

IP Header Length (IHL)

The IHL field indicates the length of the IP header. IP needs to know how long the IP header is in order to figure out where the data starts. There is no start-of-data marker.

The IHL field is 4 bits; however, this presents a problem because the largest number that 4 bits can represent is the number 15. Remember that 4 bits give 2^4, or 16, combinations and these are the numbers from 0 to 15. Therefore, the IHL is actually the header length in 32-bit words. Figure 9-3 illustrates this concept because it shows the IP header divided at the 32-bit boundaries. Notice the strange math here. We normally think in 8-bit words (an octet). Therefore a 32-bit word is actually 4 octets. The number 5 in this field indicates a 20-octet IP header (5×4), a 6 indicates a 24-octet IP header, and so on. The smallest IP header is 20 octets (IHL = 5) while the largest that the header can be is 60 octets (IHL = 15). If the IHL

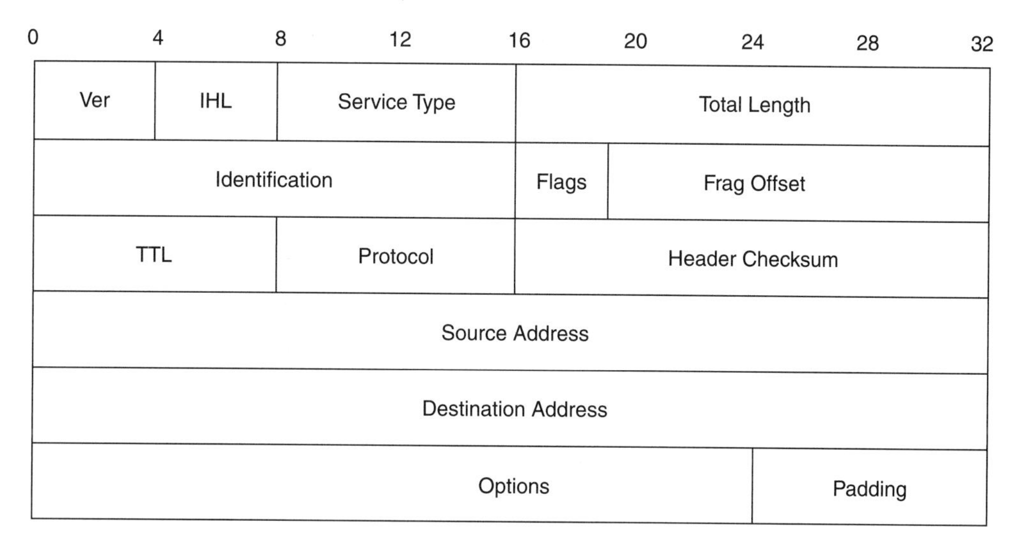

Figure 9-3: The IP header structure divided at 32-bit boundaries

is 6 or larger then the options field is being used. One possible use for the options field is to carry source routing information.

Type of Service (ToS)

The ToS field is used by IP for processing the datagram properly. It marks the datagram for special handling and can indicate precedence, delay, throughput, and reliability. The ToS field is 8 bits in length. Historically, these features were almost never used. If not used, the value will (almost) always be 0, which is defined as "normal." This field has now become relevant with the increased use of multimedia and particularly VoIP. Chapter 16, "Implementing QoS," discusses how DiffServ uses this field.

Total length

This 16-bit field is the total length of the IP datagram in octets. It is the length of the IP header plus the data. Because this field is 16 bits, the largest number it can hold, and hence the largest IP datagram is 2^{16}, or 65,535 (64K), octets. Ethernet doesn't come close to holding this amount of data. Therefore if an IP datagram is larger than the payload that Ethernet can carry, it must be fragmented and carried in multiple Ethernet frames. In order to calculate the length of the data field, subtract the length of the IP header from the total length value.

Identification

This 16-bit field contains a unique identification number identifying the datagram. It is created by the sending node and used to reassemble fragments. All fragments of the same message will have the same identification number.

Flags

This field is 3 bits and the flags are used to indicate whether fragmentation is permitted and/or used. Bit 1 is unused and therefore always set to 0.

Bit 2 is the Don't Fragment flag. If the flag is set to 1, don't fragment under any circumstances. If it is set to 0, fragmentation is allowed.

Bit 3 is the More Fragments flag. If the flag is set to 1, more fragments will follow. If it is set to 0, there are no more fragments and this is the last fragment of the original datagram.

If a message is split into fragments, sometimes called subpackets, all fragments will have the More Fragment field set to 1, except the last one. Since packets can arrive out of order, this field is not a fail-safe method of determining that the last subpacket has arrived. Instead, IP will have to use the next field, Fragmentation Offset, in conjunction with the More Fragments flag to determine if the original datagram has been received in its entirety.

Fragmentation offset

Fragmentation offset is a 13-bit field and is the offset, i.e., the location or distance of the fragment from the start of the data in the original datagram. This field allows IP to reassemble the data in the proper order.

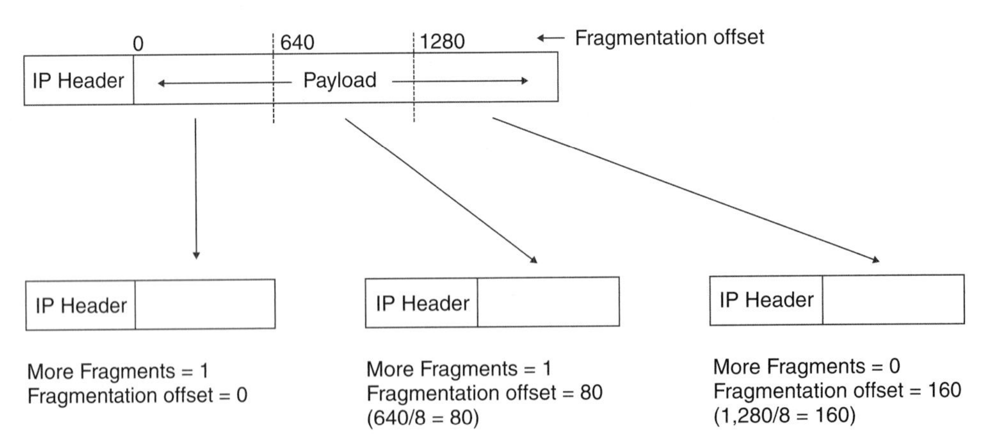

Figure 9-4: Fragmenting the IP datagram

There is one issue with the size of the fragmentation offset field. Because it is only 13 bits, the largest number that it can hold is 2^{13}, or 8,192. However, because the total size field is 16 bits, the largest IP datagram is 65,535 octets. In order to support numbers larger than 8,192, the fragmentation offset field actually holds the offset value divided by 8. This is illustrated in Figure 9-4. In the example, the offset value of 640 is represented in the fragmentation offset field by 80, which is 640 divided by 8. The offset value must always be a multiple of 8.

Time to Live (TTL)

This field gives the time, in seconds, that a datagram may remain on the network before being discarded. It is set by the sending node and because the field is 8 bits, the range is from 1 to 255. Each router that processes the packet must decrement this field by 1 even if it took less than 1 second to process. It must be a whole number. Realistically, the number of seconds is equal to at least the number of hops the packet can make. It may also take longer than 1 second to process the packet. When a packet arrives at a router, the time is tagged. If a router must wait to process the packet, the time is counted against its TTL. Therefore, if a router is particularly overloaded, the packet may expire while it is waiting for processing and be discarded. When the field has decreased to 0, the packet must be discarded. A message is sent back to the sending node alerting it to what happened. The intent of this field is to prevent any packet from circulating endlessly on any network.

The TTL field has a second use. When the first fragment of a fragmented datagram arrives, the receiving node starts a counter with the value set to the TTL. The timer counts down and if the final fragment is not received before the counter reaches 0 the partial datagram is discarded.

Protocol

The protocol field is 8 bits long and identifies the protocol of the next header. These numbers are standard numbers identified by ICANN. For example, the code for TCP is 6 and ICMP is 1. Windows has a file called "Protocol" that lists these codes.

Header checksum

The header checksum field is 16 bits long and it contains the checksum of the IP header only, not the entire datagram. In other words, it does not cover the data portion because IP is not responsible for the correct delivery of data. Other layers may provide error checking or correction. For example, Ethernet calculates a checksum for the complete frame (the FCS) and discards the frame if it finds it corrupt; in contrast TCP will retransmit a packet that was corrupted during transmission.

Because the TTL field in the IP header decrements at each router it passes through, the checksum in the header will also change each time it is processed by a router. For those curious about technical details, the algorithm used for the checksum is the ones-complement of the 16-bit sum of all 16-bit words.

Source and destination addresses

The source and destination address fields are 32 bits in length and hold the familiar 4-octet IP addresses. Their function is to identify the source and destination hosts. These addresses are written into the header at the time of the creation of the datagram and are not altered during the routing of the packet.

Options

The options field can be from nil to a maximum of 40 octets. This is an optional field. It is rarely used for general IP traffic. Instead it is used for network testing because it allows the sender to control routing and time stamping to record response times. One final use for the options field is to provide security information.

Padding

The padding is a variable number of bits if it exists at all. The length of this field depends on the options field. The padding field insures that the options + padding fields equal 32 bits. If there is no options field, there will be no padding field.

Hands-on Lab 9-1

Looking Up the Protocol Field

When TCP/IP is installed on a computer, a file listing the protocol numbers used by IP is created.

Look up this file and list the numbers used by some of the important protocols.

1. Use the find/search feature of windows and search for a file named "protocol."

   Note: This file has no extension in its name, just "protocol."

   The most common location on a Windows system is in \Windows directory\system32\drivers\etc.

   Microsoft Vista Note: Because of Vista security, the search function may not find this file. Instead, use Windows Explorer and navigate directly to the \Windows diretory\system32\drivers\etc folder to find the file.

2. Because this file is a text file, you can open it up with Notepad and read it.
3. What is the protocol number used by the following protocols?

   _______ ICMP

   _______ UDP

   _______ TCP

THE LIFE OF AN IP DATAGRAM

When data from an application needs to be sent out over the network as a datagram, it is passed from the higher levels to the IP software where the IP header is constructed. The IP address of the destination host is used for the destination address. If the destination machine is on the local network, the MAC address of that machine is passed to the data link layer after ARP has retrieved it. If a gateway is the first "hop" on the route, its MAC address is used instead. If routing is controlled by the sender, that information is added to the IP header in the options field. Otherwise, the gateway will decide which is the best route. The IP header checksum is calculated and inserted in the checksum field. The datagram is then passed on to the data link layer for framing and placing on the network.

As a datagram passes along the network, each gateway performs a series of tests. At the data link layer, the gateway checks the FCS and makes sure the checksum matches its own calculations. Stripping off the data link framing information, the datagram is passed up to the network layer. The IP header checksum is calculated and verified. If the checksum fails, the packet is discarded and an error message is sent back to the source machine. Next the time-to-live field is checked. If it has expired it is discarded and an error message returned. The next hop of the route is determined either by examining the destination address or from specific information in the options field. The datagram is rebuilt with a new time-to-live value and a new checksum.

Fragmentation may be required because of software limitations. For example, a datagram is too large to be forwarded onto the next network because of limitations at the data link layer. The datagram is divided and new datagrams are created with the correct header information and a portion of the original payload. If fragmentation is not allowed, the packet is discarded and an error message returned.

When the datagram reaches the destination machine, its checksum is calculated to ensure it matches the checksum in the IP header. The header is checked to see if more fragments are required to assemble the complete datagram. If more are required, the system waits, meanwhile running a timer to ensure that the remainder arrives in a reasonable time. If the balance of the datagrams do not arrive or the fragments can't be reassembled before the time-to-live timer reaches 0, the packet is discarded and an error message is returned.

If everything arrives correctly, the IP header is stripped off of the datagram and the payload is passed up to the transport layer for further processing.

Figure 9-5 illustrates the IP header along with the fields in the header when captured by a protocol analyzer.

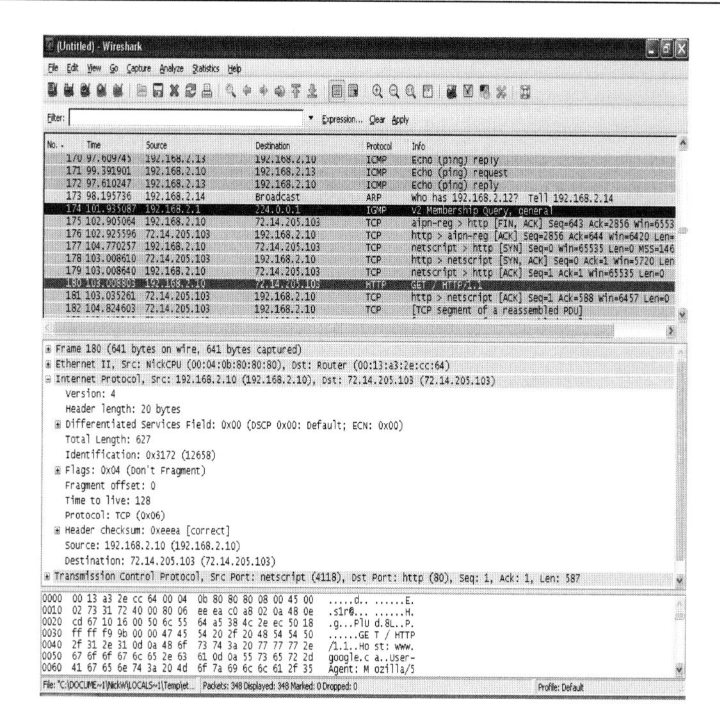

Figure 9-5: IP through the eyes of a protocol analyzer

Review Questions Internet Protocol

1. IP is considered unreliable because

   a) data transmission is not preceded by handshaking to establish a session.
   b) some IP packets can arrive corrupt.
   c) some packets end up at the wrong destination.
   d) it doesn't wait for an acknowledgment before sending the next packet.

2. Which of the following functions does IP perform?

   a) Check the payload in the datagram for possible errors.
   b) Use the type field to identify the next protocol.
   c) Defines the MAC address.
   d) Fragment and reassemble datagrams.

3. How do you calculate the length of the data field in the IP datagram?

   a) Subtract 24 from total length.
   b) Subtract IHL x 4 from total length.
   c) Subtract IHL from total length.
   d) Subtract fragment offset from total length.

4. Fragmentation of an IP datagram may be required if

   a) the must-fragment bit is set to 1.
   b) a datagram is larger than allowed on the next hop of the route.
   c) the more-fragments bit is set to 1.
   d) a router can't accept an oversize datagram.

5. The time-to-live field

   a) keeps track of how many routers a datagram has passed through.
   b) is increased if a router keeps a datagram waiting because of congestion.
   c) keeps decreasing while a machine waits for all sub-packets to arrive.
   d) is ignored once it becomes 0.

Hands-on Lab 9-2

PING

PING is a utility used for troubleshooting. With it, you can determine if a remote computer is active and if the route to it is intact.

In a previous lab, you learned the basic usage of ping. In this lab, you will try out some of the advanced features. Table 9-1 lists some of the options available for the ping command. Type in *ping* with no parameters to get a complete list of the options. Of particular interest are options that reference the fields in the IP header.

Microsoft's PING

Microsoft's implementation of ping is run from the command prompt. Microsoft does not provide a graphical version of ping, although many free utilities are available from the Internet.

1. Use **Start** > **Programs** > **Command Prompt**
2. Find out the options with the ping command.

Type *ping* without an address.

You should recognize many options that were discussed with IP and are listed in Table 9-1.

3. Try out some of the options with the ping command and see the results.

PING *ip_address* -T	Will ping continuously. Use CTRL-C to stop.
PING *ip_address* -N *count*	Set count to a number.
	Will send the *count* number of ping packets.
PING *ip_address* -L size	Set size to the number of octets in the ping packet.
	The default number of octets is 32.
	Try 100, 1000, etc.
	Experiment until you find the largest number allowed. How does this correlate with the total length field of the IP header?
PING *ip_address* -F	Prevents the packet from being fragmented.
	To see this option in action, use it in conjunction with the –L flag to generate a packet that is too big for Ethernet. Remember the largest amount of data that Ethernet can carry is 1,500 octets.

	| Experiment with sizes until you find the largest number of bytes that will get through.
PING ***ip_address*** **-I seconds**	| Sets the TTL for the packet.
	| Although you set the TTL for the outgoing packet, the response you receive reflects the TTL set by the responding host
	| To better illustrate the use of TTL, you need to send your *ping*s through a router that can discard the packets.

-T	Ping the specified host until interrupted.
-A	Resolve addresses to hostnames.
-N count	Number of PINGs to send.
-L size (octets)	Send buffer size.
-F	Set Don't Fragment flag in packet.
-I TTL (seconds)	Time to Live.
-V TOS	Type of Service.
-R count	Record route for count hops.
-S count	Timestamp for count hops.
-J host-list	Loose source route along host-list.
-K host-list	Strict source route along host-list.
-W timeout	Timeout in milliseconds to wait for each reply.

Table 9-1: Options for the PING command

Note: PING stands for packet Internet groper. The early TCP/IP programmers had a sense of humor.

9.3 INTERNET CONTROL MESSAGE PROTOCOL (ICMP)

Because TCP/IP is implemented in software and is independent of the hardware, it cannot depend on the data link layer to provide information on failures or other status data.

The Internet control message protocol is a special-purpose messaging mechanism to report errors or provide other information about unexpected circumstances.

Both gateways and hosts need to report errors. After examining IP, it is easy to see how these errors can come about. For example, the time-to-live field can decrement to 0 and the packet has to be discarded, a checksum could indicate a corrupted packet, or the wrong version of IP is encountered.

ICMP is an error reporting mechanism, not an error correcting mechanism. It cannot fix any problems. It just reports them and then lets the upper level protocols deal with them.

ICMP exists at the network layer of the OSI model. ICMP is itself carried in the data field of the IP datagram. See Figure 9-6. ICMP is always a part of the IP code.

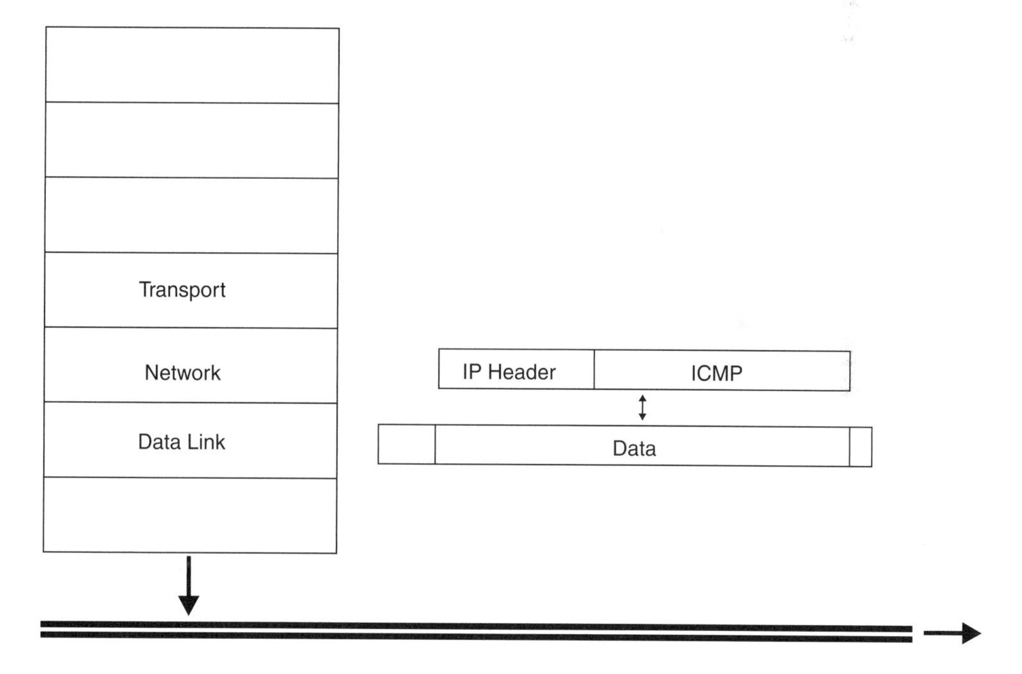

Figure 9-6: ICMP Is carried in the data field of IP

ICMP may be considered the message mechanism of IP itself, i.e., the method by which IP stacks in different machines communicate amongst themselves. Therefore, the unit of ICMP is called a message.

Because ICMP relies on IP and IP is "unreliable," there is no guarantee that an ICMP message will be delivered. Problems with ICMP do not generate further ICMP messages. In addition, for fragmented datagrams, only problems with the first fragment (fragment 0) are reported.

Types of ICMP messages

The protocol data unit of ICMP is called a *message*. These are well defined and each message type has a code assigned to it. The message types can be grouped into status messages, request and reply messages, and router messages.

Destination unreachable

This message is self-explanatory. The code field in the ICMP header may indicate what was unreachable (network, host, protocol, or port). It may indicate that the route specified in the IP header had failed. Remember, the route is usually determined by the routers but in special circumstances may be specified by the sending machine. Finally, if a datagram must be fragmented to get across a router but the Don't Fragment bit was set, then the datagram has to be discarded. This message may be sent by a router. The destination host may also send a Destination Unreachable message if an upper-layer protocol or port number is not available.

Source Quench

The Source Quench message is used to control the rate at which datagrams are transmitted, although it is a very primitive method of flow control. The message is generated by a host or router that has full buffers or needs to slow the processing of incoming packets for other reasons. For each packet discarded, the receiver sends a Source Quench message. When the originating device receives the Source Quench messages, it is supposed to reduce the rate at which it is transmitting until the Source Quench messages cease. It is also possible for this message to be generated when a host or router is approaching its buffer capacity, but before it is exceeded. In this case, datagrams are still being accepted and not discarded.

Time Exceeded

This message is also self-explanatory. The time-to-live (TTL) field has decremented to 0 and the packet must be discarded. The Time Exceeded message can also indicate whether the time was exceeded in transit or whether the time for reassembly of fragmented packets in the host was exceeded.

Redirection

Redirection messages are sent to a gateway in the path when a better route is available. For example, if a gateway has just received a datagram from another gateway but on checking its routing tables finds a better route, then it sends a Redirection message back to that gateway indicating the IP address of the better gateway. A Redirection message may also be sent to a host under the following circumstance: there are multiple routers on a local network and the host directed the datagram to a less efficient router.

Parameter Problem

This type of message is used when a problem has been discovered in the IP header. This can happen when options are used with incorrect arguments. When the Parameter Problem message is sent back to the originating device, a pointer to the problem octet in the IP header is specified. The pointer could range from 1 (type of service) to 20 (options field, if it exists). If you count the version/IHL octet as the first one in the IP header, then the pointer is always one less than the real octet. The Parameter Problem message is returned only if the datagram was discarded because of the problem.

Request and reply

This type of message is always found in pairs, the request and the reply. When a request is sent, a host or gateway down the path will send a reply back.

Timestamp request and reply

These types of messages enable the timing of datagrams passing along the network to be monitored. Coupled with strict routing, this can identify bottlenecks. Three times are tracked. The originate timestamp is when the sender last touched the message, the receive timestamp is when the destination machine first touched the message, and the transmit timestamp is when the destination machine last touched the message as it sends off the reply. Using these, the hosts can calculate response times.

Address mask request and reply

This is used by a host that needs to know the subnet mask of the network.

Echo request and reply

This type of message requests a specific host to reply. In fact, this is the mechanism used by the ping command to test for the presence of a remote host on the system.

The identifier in the request may be used to match up the reply. The sequence number will be used if the ping command makes multiple requests. Any data in the echo request must be returned in the data field of the echo reply. An example of data is the text string "HELLO". Microsoft ping packets always use the alphabet as their data. Echo request and reply messages can be used for debugging routing problems, failed gateways, or network cabling problems.

Router discovery

Router discovery is a new function added to ICMP since 1991. If a host needs to send a datagram beyond the local network, it must address the frame, at the data link layer, to a router. Traditionally, a configuration file, for example "GATEWAYS", holds this information. Maintaining these files is a labor-intensive job for the administrator plus static files cannot immediately respond if conditions change, such as a router is down and a new one is started. Modern networks use the dynamic host configuration protocol (DHCP) to configure the default gateway for the host. It too doesn't respond if a router fails unless there is administrative intervention. Router Discovery messages are an alternative and more dynamic way for hosts to be alerted to the existence of a local router.

Note that Router Discovery is not a router protocol, meaning that it is not used between routers. It is used only by hosts to find the router on their local network. In addition, router discovery does not guarantee that the best router for a particular destination becomes the default router. There may be multiple routers on a local network, but the one discovered first is used. Nevertheless, if the wrong router is used, an ICMP Redirect message will be issued and the host can switch to the correct router. Router Discovery is implemented through two types of messages.

Router advertisements

A router periodically advertises its existence on each of its interfaces. The default interval is 7 to 10 minutes. Included in the message is a "lifetime" field indicating how long the router is to be considered valid, typically 30 minutes. This allows a host to eventually forget a failed router. A "preference" field is also included. This allows multiple routers on a local network to be "rated" in terms of preferred use by hosts.

Router solicitations

When a host starts up, it can send a Router Solicitation message instead of waiting for the periodic Router Advertisement message. A local router should respond. If there is no response, several Router Solicitation messages can be issued. However, if there is still no response, the host desists and instead waits for a Router Advertisement message.

References: The original standard and source for ICMP is RFC 792, dated 09/81. The extension to ICMP for Router Advertisement and Router Solicitation is RFC 1256, dated 09/91.

SUMMARY

Section 9.1: Internet Protocol

IP is the basic packet delivery mechanism on a TCP/IP network. All protocols above IP use IP to deliver the data. IP exists at the network layer, layer 3, of the OSI model.

IP is an example of a connectionless service.

The functions of IP include the following: defines the datagram, the basic unit of transmission; defines the IP addressing scheme; moves data between the data link and the transport layer; routes datagrams to remote hosts and performs fragmentation and reassembly of datagrams.

Section 9.2: Packet Structure

The IP header is a complex construction of a minimum of twelve or maximum fourteen fields. It will be from 20 to 60 octets long. The fields are the IP version, the IP header length, the service type, total length of the datagram, ID field, flags, fragmentation offset, time-to-live, protocol, header checksum, source and destination addresses, options and padding fields.

Section 9.3: Internet Control Message Protocol (ICMP)

Because TCP/IP is implemented in software and is independent of the hardware, it cannot depend on the data link layer to provide information on failures. The Internet control message protocol is a special-purpose messaging mechanism to report errors or provide other information about unexpected circumstances.

ICMP is an error reporting mechanism, not an error correcting mechanism. It cannot fix any problems. It just reports them and then lets the upper level protocols deal with them.

The types of ICMP messages include the following: Destination Unreachable, Time Exceeded, Source Quench, Redirection Message, Parameter Problem Message, and Router Discovery. Request and Reply may be for a Timestamp, Address Mask, and Echo.

Review Questions ICMP

1. ICMP is used for error conditions but is not ideal because

   a) it reports errors to humans in machine code, not in English.
   b) it can fix errors only in IP but not TCP.
   c) it can report errors at the datalink level but not the network level.
   d) it can report errors but can't fix them.

2. A router may use ICMP

   a) to notify another router that a line is clear.
   b) to notify another router that a different route is better.
   c) to tell a host that a packet is still alive.
   d) to notify another router that it is going down.

3. A device will return a Destination Unreachable ICMP message if

   a) the host refuses a packet.
   b) the destination host hasn't any available buffers.
   c) fragmented packets are received even though the don't-fragment bit is set.
   d) a datagram must be fragmented to pass through a router but the don't-fragment bit is set.

4. If a destination device runs out of buffer space,

   a) it will send Source Quench messages to the router.
   b) it will send Destination Unreachable messages to the source device.
   c) it will send a Redirection message to the router to send the packets by a longer route.
   d) it will send Source Quench messages to the source host.

5. Request and Reply

   a) is used with a timestamp to determine the fastest route to a destination host.
   b) is used by the ping command to verify the existence of a target host.
   c) is used with an address mask to find nonresponding hosts.
   d) is one example of an echo message.

CHAPTER

10

IP Addressing

Objectives

Upon completion of this section, the reader will:

- *Understand the format of IP addressing and the reason for the different classes.*
- *Be able to identify different classes of IP addresses.*
- *Appreciate the situations in which subnetting is appropriate and know how to divide up a network into subnets.*
- *Understand how private IP addresses expand the available supply of IP addresses for an organization.*
- *Be able to use the classless IP addressing system.*

INTRODUCTION

IP telephones are identified by IP addresses just as any host attached to a TCP/IP network is. IP is responsible for the identification of networks and hosts for the TCP/IP protocol suite. This lesson examines the structure and composition of IP addresses, particularly its two forms, class based and classless. The scheme by which a large network is divided into subnets connected by routers is also examined. These topics are required knowledge if a student expects to be able to design the IP numbering scheme for any IP network.

10.1 IP ADDRESSING

Unique addresses required

Each host on a network must have a unique identity. Each network must also have a unique identity. Both are satisfied with the IP addressing scheme.

IP addressing scheme

An IP address is a 32-bit value that contains enough information to uniquely identify the network and the host address of a computer.

The IP address is broken down into two parts, the network portion and the host portion. Unfortunately, the format of the two portions is not the same for each address because it varies with the class of IP address. Figure 10-1 illustrates the two parts to the IP address: the network identification, which all hosts on the network share, and the host identification, which is unique to that machine.

The IP addressing scheme was designed to accommodate various sizes of networks. Some organizations are so large that they may have millions of computers on their networks. Others are still large enough to have thousands. The majority of networks may only have hundreds of computers on them.

Network addresses are registered with regional address registrars. This ensures that all of the networks on the Internet have a unique network address. The only exception is networks that are not connected to the outside world. They may choose any internal network numbers they please. If, however, such a network were ever connected to the Internet, it would cause a major disruption.

The IP address identifies the location of the host, not the host itself. If a machine is moved, it may need to be reassigned an IP number. Changing the address configuration of a host can be done manually or automatically through the DHCP service. Nevertheless, maintenance of IP numbers can be a burden to the administrator.

IP addresses are usually written as four decimal numbers separated by dots. This is known as a dotted quad notation.

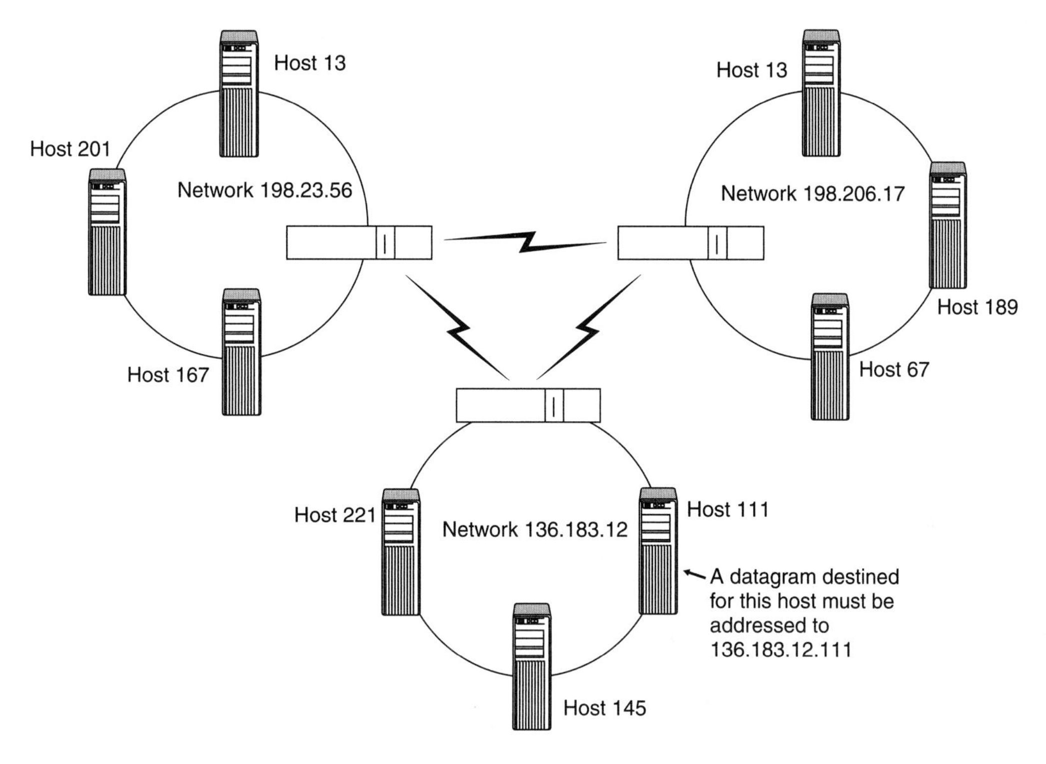

Figure 10-1: Assigning IP addresses to networks and hosts

The four decimal numbers are each in the range of 0–255. This way, the address breaks are on octet boundaries. An IP address is written as follows: 198.54.9.200.

Rules for IP addresses

The rules for creating IP addresses are as follows:

- Each octet must be within the range of 0–255.
- Negative numbers not allowed.
- Four octets are required.

Why is each octet restricted to the range 0–255?

An octet is the basic building block of computer information since it is the unit used to process and store data. An octet is stored in binary as 8 bits. Each bit can have only two values, 0 or 1. Therefore an octet can have a maximum value 2^8, or 255, decimal.

The ranges of values of an octet are as follows:

Decimal (base 10)	0–255
Binary (base 2)	0000 0000–1111 1111
Hexadecimal (base 16)	00–FF

Registering an IP address

In order to obtain an IP address for an organization, contact your ISP, who receives them from regional registrars. The regional registrars are listed in Table 10-1. They, in turn, receive blocks of addresses from the Internet Corporation for Assigned Names and Numbers (ICANN).

ICANN Current top registrar	Internet Corporation for Assigned Names and Numbers
IANA Previous top registrar	Internet Assigned Numbers Authority
Regional registrars	
AFRINIC	African Network Information Center
APNIC	Asia-Pacific Network Information Center
ARIN	American Registry for Internet Numbers
RIPE	Réseaux IP Européens
LACNIC	Latin American and Caribbean Internet Addresses Registry

Table 10-1: Registrars for IP addresses and domain names

Find the geographical location of an IP address

Since IP addresses are assigned on a geographical basis, you should be able to look up the location of a known IP address. Several websites offer this service. Try using the web site www.ip-adress.com to find the location of an IP address. This site uses information provided by the regional registrars to map the location.

Note the spelling: adress, not address!

Look up IP address assignments

You can look up the IP addresses that have been assigned to certain organizations and the blocks of addresses that have been assigned to different registrars at the following URL: www.iana.org > IP address services > IPv4 address space.

Exercise 10-1

IP Addresses

1. An IP address identifies

   a) a host.
   b) a network.
   c) both a host and a network.
   d) neither a host nor a network.

2. If you need to acquire an IP address for your company's network, you would apply to

   a) ICANN.
   b) your ISP.
   c) ARIN.
   d) IANA.

3. Dotted quad notation describes an IP address because

   a) four numbers are separated by dots.
   b) each number is 2 to the power of 4.
   c) the maximum number of IP addresses is 4 million.
   d) there are four periods in the address, although the last one is invisible.

4. Which one of the following is a legal IP address?

   a) 02 60 60
   b) 132.57.89.257
   c) −46.200.56.89
   d) 26.156.200.75
   e) 240.70.177.89.81

10.2 CLASSFUL IP ADDRESSES

In order to accommodate different-sized networks, the IP address space was divided into classes. The intent was to give large networks more addresses than medium-sized or small networks. The composition of an IP address also changes with the class. The IP address is a composite number that identifies the network plus the host on the network. The number of octets used to identify the network and the host changes with the class. The structure of the classful system is too rigid for modern networks and the newer classless system will be discussed in an upcoming section.

IP address classes (classful IP addresses)

The IP address can fall into one of three classes, class A, class B, and class C. There is also a class D address used for multicasting (sending to multiple hosts simultaneously). Each is designed to accommodate a different-size organization. By examining the first three bits of an address, IP can determine the class of address and therefore the rules for decoding it.

Class A address

If the first bit is set to 0, it is a class A address. The next seven bits identify the network and the remainder identify the host. See Figure 10-2.

- Bits in the network address: 7
  Number of possible network addresses: 2^7 (128) minus 2 = 126
  (Addresses 0 and 127 are reserved)
  Octets in the host address: 3
  Possible number of host addresses: $256 \times 256 \times 256 =$
  $16{,}777{,}216 - 2 = 16{,}777{,}214$
- There are only 126 class A networks in existence and each network can have over 16 million hosts. There are no more class A addresses available.
- How to identify a class A address: The first octet must be within the range 1–26.

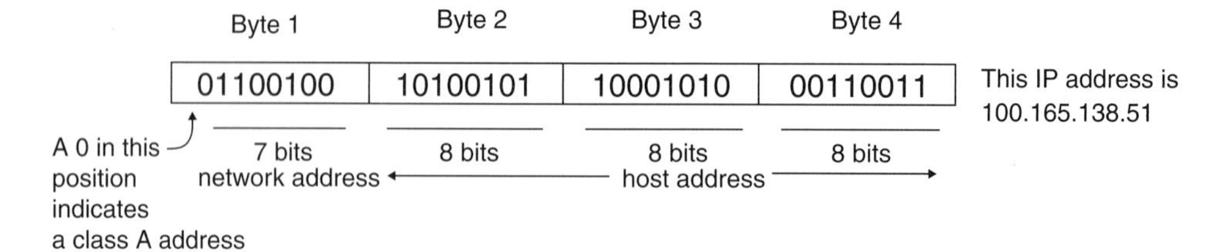

Figure 10-2: The structure of a class A address

Class B address

If the first two bits are set to 10, it is a class B address. The first two octets identify the network and the last two octets identify the host. See Figure 10-3.

- Octets in the network address: 2
  Number of possible network addresses: 2^6 (64) × 2^8 (256) = 16,384
  (The first two bits are fixed, therefore only the remaining 6 bits in the first octet are available for unique combinations.)
  Octets in the host address: 2
  Possible number of host address: 256 × 256 = 65,536 − 2 = 65,534
- There can be more than 16 thousand class B addresses in the world and each can have as many as 65 thousand hosts on them.
- How to identify a class B address: The first octet must be in the range 128–191.

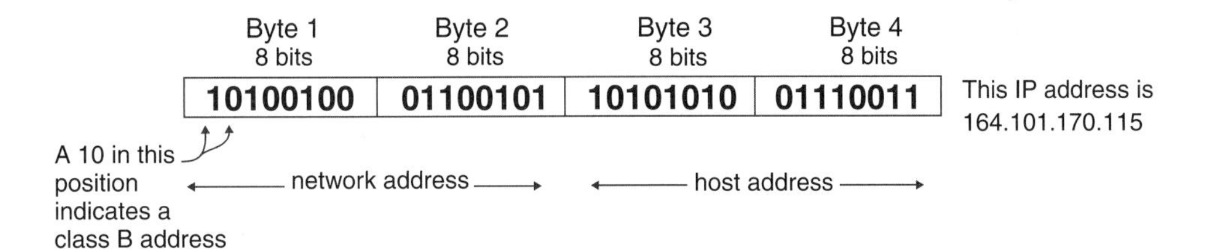

Figure 10-3: The structure of a class B address

Class C address

If the first three bits are set to 110, it is a class C address. The first three octets identify the network and the last octet identifies the host. See Figure 10-4.

- Octets in the network address: 3
  Number of possible network addresses:
  2^5 (32) × 2^8 (256) × 2^8 (256) = 2,097,152
  (The first three bits are fixed, therefore only the remaining five bits in the first octet are available for unique combinations.)
  Octets in the host address: 1
  Possible number of host address: 256 − 2 = 254
- There can be over 2 million class C addresses in the world and each can have as many as 254 hosts on them.
- How to identify a class C address: The first octet must be in the range 192–223.

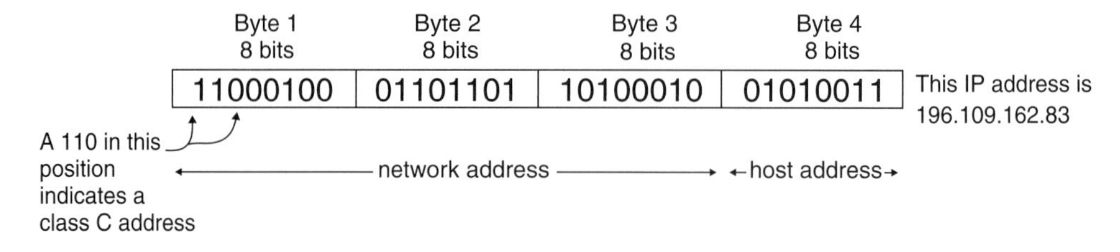

Figure 10-4: The structure of a class C address

Identifying the IP class

Class A: Network address is between 0 and 127.
Class B: The first octet of the network address is between 128 and 191.
Class C: The first octet of the network address is between 192 and 223.
Class D: Is a multicast address and the first octet is between 224 and 239.
Class E: Is considered "experimental" and is not used. The range is 240–254.

Some definitions

You need to be familiar with the following terminology:

Unicast: Sending a datagram or frame to a specific host.

Broadcast: Sending a datagram or frame to all devices. In practice, this means all devices on the local network since routers filter out broadcasts.

Multicast: Sending a frame or datagram to a subset of computing devices, for example all routers or only hosts.

Anycast: Sending a datagram to multiple machines, but only the first one to receive it will take action. This concept is not found in IPv4 but is used in IPv6.

Reserved addresses

Reference has already been made to the fact that certain addresses cannot be used and are reserved for various reasons. Here is a list of these numbers that you need to remember.

Network addresses

127 – Reserved as the "loop-back" addresses, i.e., a packet with this number never leaves the computer, it loops back to itself. This is useful for diagnostics.

0 – Designates the default route and is used to simplify the routing information that IP must handle.

255 – Used for broadcasts.

Host addresses

255 – Reserved as a broadcast address, e.g., 201.10.53.255 is a message destined for all hosts on the network 201.10.53.0.

0 – Reserved to refer to the network itself, e.g., 201.10.53.0 refers to the network 201.10.53.

These reserved addresses have a binary bit pattern that you should note. The general rule is that the network portion of the address cannot be all 1s or 0s. Similarly, the host portion of the address cannot be all 1s or all 0s either. Table 10-2 sums up the information about the classful IP addresses presented here.

	Number of Networks	Number of Host / Network	Range of Network IDs (First Octet)	Network and Host IDs
Class A	126	16,777,214	1–126	N.H.H.H
Class B	16,384	65,534	128–191	N.N.H.H
Class C	2,097,152	254	192–223	N.N.N.H

Table 10-2: Summary of classful network information

Exercise 10-2

IP addresses

1. Which of the following may be assigned to a host? If the address can't be assigned to a host, why not?

Address	Yes or why not
57.200.258.34	
223.223.255.223	
230.57.0.1	
126.0.0.0	
191.80.20.0	
87.167.76.100	
0.40.50.200	
127.100.2.2	
255.255.255.255	
56.0.0.201	
199.21.133.255	

2. What class are the following IP addresses?

223.55.231.5	
76.234.98.123	
134.55.76.105	
225.0.0.56	
191.14.200.98	
240.57.99.1	
126.127.1.254	

ASSIGNING NETWORK NUMBERS

When assigning network numbers, remember the following points:

- All the hosts on a network must have IP addresses with the *same* network ID. They must all have a *different* host ID.
- Routers are also hosts and must have IP addresses. Each interface of a router must have an address that corresponds to the network that the interface is on. With multiple interfaces, a router is on multiple networks.
- Assigning IP addresses is arbitrary but a common numbering convention gives either the largest or smallest IP addresses to the routers and servers.

Figure 10-5 illustrates these rules.

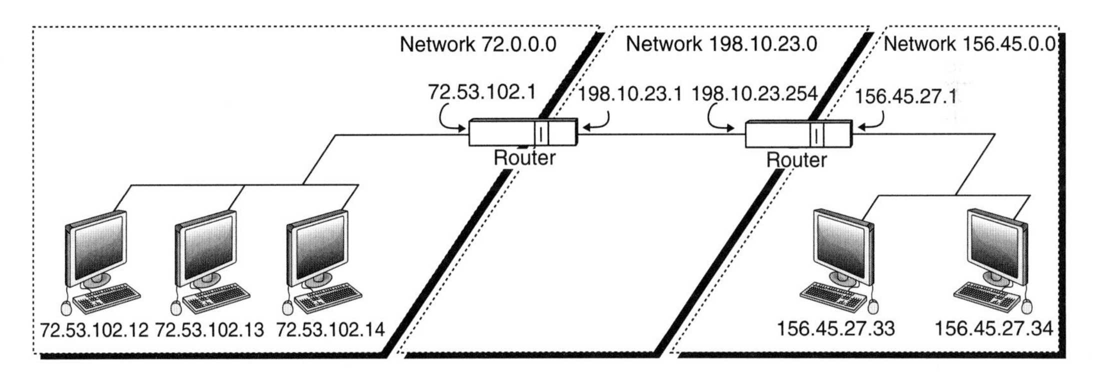

Figure 10-5: Assigning network addresses

Exercise 10-3

Assigning Network Addresses

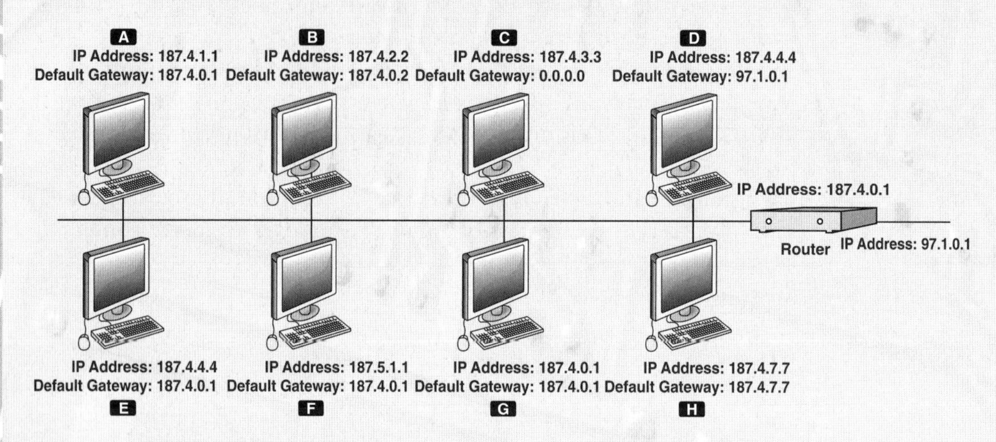

Figure 10-6: Assigning network addresses exercise

In the Figure 10-6 illustration, some hosts may not be able to communicate properly across the router or with their neighbors. For each host, describe the problem if there is one.

A. ______________________________

B. ______________________________

C. ______________________________

D. ______________________________

E. ______________________________

F. ______________________________

G. ______________________________

H. ______________________________

BINARY NUMBERING

Understanding the binary numbering system is important to your mastery of IP addressing, particularly subnets. Therefore, we are taking a time-out in order to understand the mysteries of the binary numbering system.

It should come as no surprise that computers use binary since they are made up of switches and switches have two positions, on and off. The first thing to appreciate is that binary is a positional numbering system.

Positional numbering systems

A positional numbering system uses the position of the numeral to denote its value. For example, the numeral 4 in the number 4 just means 4 but in the number 47 means 40 and in the number 433 means 400.

Decimal, hexadecimal, and binary are all examples of positional numbering systems. Roman numerals are an example of a numbering system that is not positional. In a positional numbering system, each column is a multiple of the column before it. The multiple is called the base.

Let's examine a decimal number since you already know how it works. Decimal is base 10, which means that each column after the first is a multiple of 10. If the first column is the ones column, the second is the 10s, the third is the 100s, and so on.

Figure 10-7 is an illustration of how the number 527 is put together.

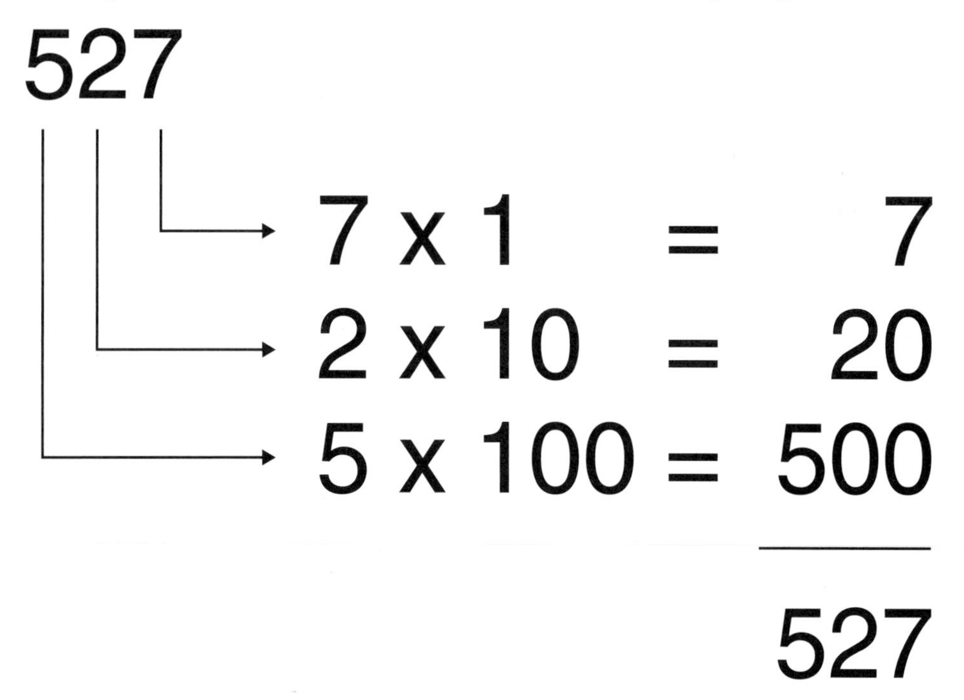

Figure 10-7: How the decimal number 527 is put together

Binary, decimal, and hexadecimal all work the same way. The first column is the ones column and each successive column is multiplied by the base.
Decimal is base 10: Each column is multiplied by 10
Binary is base 2: Each column is multiplied by 2
Hexadecimal is base 16: Each column is multiplied by 16

Binary

Because there are only eight bits in an octet, we only have to worry about eight positions in a binary number. If the number is zero, it has a value of 0. However if the numeral is 1, its value will be as shown in Figure 10-8.

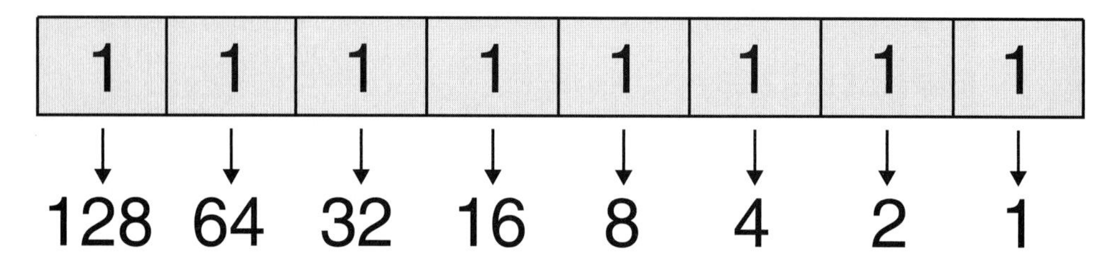

Figure 10-8: The bit values of binary

Figure 10-9 is an illustration of how the binary number 11011011 is converted into decimal.

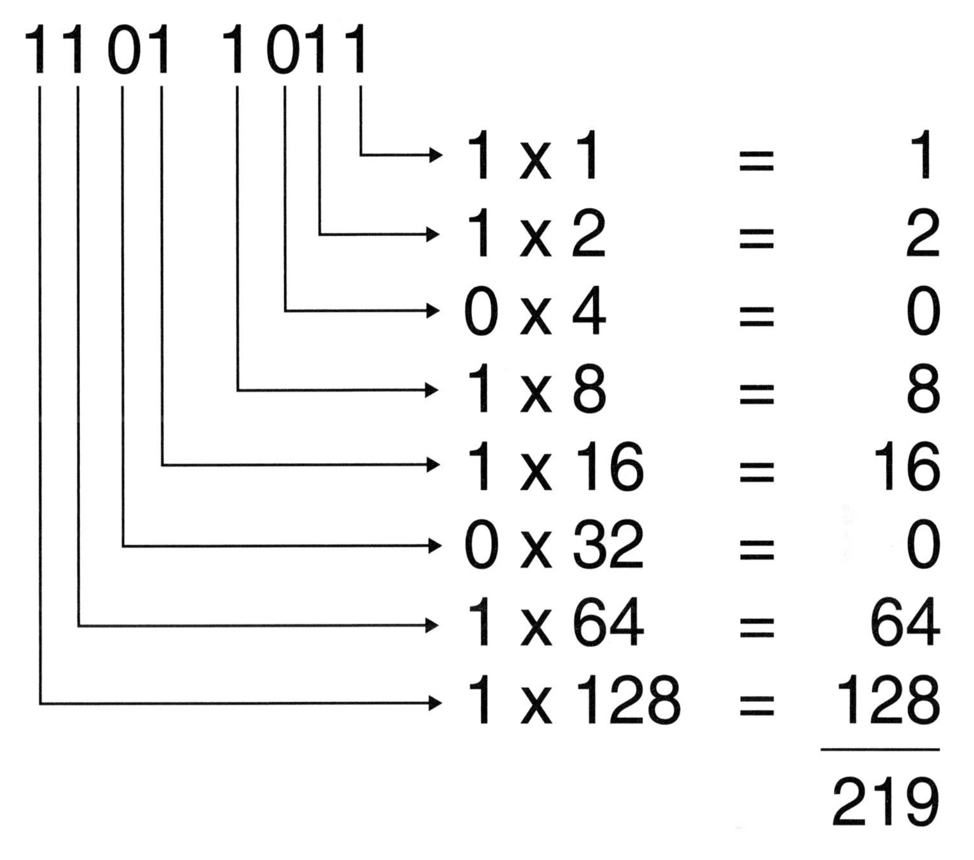

Figure 10-9: Converting binary into decimal

Exercise 10-4

Binary Numbering

1. Convert the following numbers to decimal:

   a) 0011 1001 ___________
   b) 1011 1111 ___________
   c) 0101 0100 ___________

2. Convert the following numbers to binary:

   a) 240 _______________
   b) 101 _______________
   c) 187 _______________

3. What class are the following IP addresses?

   a) 01101001000100100101111010100010 ___________
   b) 10101101001000100100101111010100 ___________
   c) 11010010001001001011110101000100 ___________

10.3 SUBNETTING

What is a subnet?

A subnet is a division of a network that meets the following criteria:

1. All of the IP addresses on it are contiguous.
2. The network ID of all of the IP addresses are the same as the rest of the network.
3. It communicates with the rest of the network through a router. Subnets are distinct cabling systems connected together with routers. Remember, routers work at the network layer and make use of network addresses.

Why subnet?

Subnets require extra administration, decrease the available IP addresses on a network, and are often the cause of failure because of misconfiguration. If you are going to subnet, you need a good reason. Here are a few.

- You need better performance. By dividing the network into separate sections and decreasing the traffic on the cable, you will gain performance.
- You need better security. Sensitive traffic can be localized and not be sent over the general company network. Sniffers will have a hard time capturing it.
- You need more networks but you don't want to apply for an additional IP address from your ISP.
- You have multiple networks that are physically separate because they are at remote sites or on separate cable systems and they are connected together through routers.
- You want to decentralize management, especially of IP numbers, to local administrators on subnets.
- You need a router to connect together two different technologies, such as Ethernet and token ring or between the LAN and the WAN.

IP's subnetting feature

A feature of IP addressing is that the IP address can be locally modified to use host address bits to provide subnet addresses. Essentially what this does is change how the addresses are interpreted locally without changing the global interpretation. This creates additional local networks at the cost of reducing the number of hosts.

The actual process of defining a subnet is accomplished by applying a bit mask, called a subnet mask, to the IP address. If a bit in the mask is "ON," the bit in the address is considered to be a network bit. If it is "OFF" in the mask, the bit in the IP address is considered to be a host bit.

The modified or subnet address is known locally only. To the rest of the Internet, the address appears to be a standard IP address. Figure 10-10 illustrates this concept with a network without subnets and then when it is reconfigured with subnets.

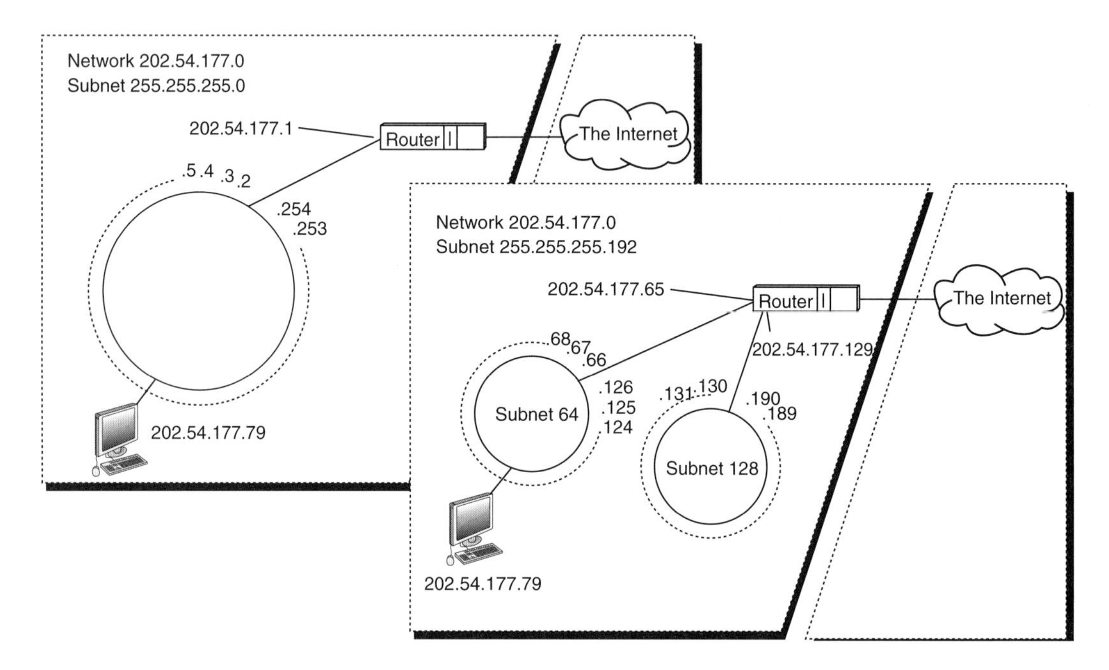

Figure 10-10: Before and after subnetting

Default subnet masks

The IP configuration of a host requires both an IP address and a subnet mask. This applies even if there are no subnets on the network. How can you determine if the network is subnetted? This is easily determined by identifying the network address class and looking at the subnet mask. If the subnet mask matches the mask from the following table, there are no subnets.

The standard subnet masks are:

Class A network mask = 255.0.0.0
Class B network mask = 255.255.0.0
Class C network mask = 255.255.255.0

These masks maintain the standard network/host relationship, meaning that there is no subnet. Note that the number of 255s and 0s matches the number of octets used for the network and host IDs for each address class.

An example of a network without a subnet might be as follows:

IP address: 193.98.14.132
Subnet mask: 255.255.255.0

Nonstandard subnet masks

Change the subnet mask in the previous example and it becomes nonstandard, indicating that the network is subnetted.

An example of a network with subnets might be as follows:

IP address: 193.98.14.132

Subnet mask: 255.255.255.192

The address would be interpreted as network 193.98.14 and subnet 128. How the mask of 192 is used to extract the subnet ID of 128 from the number 132 is not obvious, but one procedure for doing so will be demonstrated in the next section. Host 132 is actually the fourth host on this subnet because hosts 129, 130, and 131 precede it.

Notice that to the outside world, the IP address of the network doesn't change (193.98.14). Only locally does the address change. Now instead of being host 132, the machine is really the fourth host on subnet 128.

Interpreting subnet masks

A basic skill for subnetting is determining the subnet that a host is located on. Since misconfiguration of IP addresses and subnet masks is a common problem, determining the proper values is a troubleshooting requirement.

It is critical that the sending host know if the receiving host is on the same subnet or on a remote subnet or network. The reason is simple enough. If the destination host is on the same subnet, the sending host will use ARP to find its MAC address and forward it directly. On the other hand, if the destination host is on a remote network or subnet, the sending host will use ARP to find the MAC address of the router (default gateway) and send it there to be forwarded to the destination.

Routers also need this skill. For every packet they receive, they must determine which subnet the destination host is on if they are to forward the packet to its proper destination.

What subnet is the host on?

Now that you appreciate why it is important to know the subnet that a host is on, how do you calculate it? The following method is very simple and it works.

Example #1

Using the previous example:

IP address: 193.98.14.132

Subnet mask: 255.255.255.192

Step 1: Subtract the mask from 256.

256 − 192 = 64

Note: This tells you that the subnet IDs are a multiple of 64 except for the first subnet, which is always numbered 0.

Note: The first three 255s belong to the network portion of the address and are ignored.

Step 2: Number the subnets.

The first subnet is 0. All of the other subnets are multiples of 64; therefore they are 64, 128, and 192.

Note: You can easily count up the subnets in this example because there are four.

Step 3: Identify the subnet that the host is on.

Since the host ID of 132 is between 128 and 192, the IP address 193.98.14.132 is on subnet 128.

Example #2

IP Address: 144.12.57.232

Subnet mask 255.255.224.0

Step 1: Subtract the mask from 256.

256 − 224 = 32

Note: Because this is a class B address, the last two octets are the host ID. Because the 224 is the third octet of the mask, it is the third octet of the address that we need to extract the subnet from.

Step 2: Number the subnets.

The first subnet is 0. All of the other subnets are multiples of 32; therefore they are 32, 64, 96, 128, 160, 192, and 224.

Note: The number of subnets in this example is eight.

Step 3: Identify the subnet that the host is on.

Since the host ID of 57 is between 32 and 64, the IP address 144.12.57.232 is on subnet 32.

Exercise 10-5

Subnetting

For the following IP address and subnet mask pairs, determine the subnet that the host is on.

a) IP address: 193.67.108.135 subnet ____________
   Mask: 255.255.255.192

b) IP address: 221.203.86.198 subnet ____________
   Mask: 255.255.255.224

c) IP address: 132.12.203.212 subnet ____________
   Mask: 255.255.255.240

d) IP address: 180.67.175.65 subnet ____________
   Mask: 255.255.240.0

e) IP address: 12.95.212.184 subnet ____________
   Mask: 255.252.0.0

Exercise 10-6

Subnetting

For the following pairs of IP addresses, determine if the hosts are on the same subnet. The subnet mask is the same for both addresses.

Address #1	Address #2	Subnet Mask	Yes or No
192.168.6.45	192.168.6.70	255.255.255.192	
201.56.166.177	201.56.166.190	255.255.255.240	
150.67.24.200	150.67.24.225	255.255.255.224	
185.22.177.88	185.22.175.88	255.255.248.0	
66.12.7.92	66.12.213.79	255.255.0.0	
45.159.254.45	45.129.33.178	255.224.0.0	

IMPLEMENTING SUBNETS

You have been assigned the task of designing the IP addressing scheme for your new VoIP system. How do you go about it?

When designing the subnet architecture for your network, follow these steps:

1. Determine number of required subnets. Normally this is the physical segments of your network plus any WAN links you may have to your remote locations. This does not include the external link to your ISP or the Internet.
2. Determine the number of host IDs you need per subnet. Don't forget to include the router interface in this number.
3. Decide on the subnet mask. A discussion on choosing the mask follows. Problems crop up if there are not enough host IDs for the size of your network. This is particularly acute if the subnets are not evenly sized. Large subnets will waste addresses and small subnets won't have enough host IDs.
4. Assign the subnet ID to the subnets and configure each host and router interface with contiguous IP addresses and the appropriate subnet mask.

Choosing subnet masks

Subnet masks must have a pattern in which the first bits are contiguous 1s and the rest are contiguous 0s. This limits the possibilities to the following:

Decimal	Binary
128	10000000
192	11000000
224	11100000
240	11110000
248	11111000
252	11111100
254	11111110

These numbers plus the numbers 0 and 255 are the only numbers you will use in the mask.

How many subnets, how many hosts?

The subnet mask determines how many subnets can exist. If we use subnet 192 as the example, there can be 4 subnets because the subnet portion of the mask has

2 bits, which translates into 2^2, or 4, combinations. Using these combinations, the subnets would be as follows:

00000000	subnet 0
01000000	subnet 64
10000000	subnet 128
11000000	subnet 192

Because the host portion has 6 bits, the maximum number of hosts is 2^6, or 64. Four subnets with 64 hosts gives us a maximum of 256 hosts, the original number. Subnet masking cannot create additional IP addresses.

Subnet restrictions

The rule is that a subnet cannot be composed of all 0s or all 1s. This has the unfortunate side effect of making a block of addresses unavailable. In the preceding example, subnet 0 represents one quarter, or 64, of the potential hosts that are not now available. In like fashion, subnet 192 is also not available.

This case is one instance of a general rule as follows: When converted to binary, the network portion, subnet portion, or host portion of an IP address can neither be all 1s nor all 0s.

This rule was published in 1985 in RFC950. Because of the severe shrinkage of the available pool of IP addresses this rule causes, many routers make a subnet 0 available (against the rules). However, a more "legal" way to relax or delete these restrictions is by changing numbering systems. IP addresses can be interpreted without regard to their class by using their network prefix. This is a classless IP numbering system, also sometimes called classless inter-domain routing (CIDR). CIDR is described in RF1812 published in 1995. You will explore classless IP addresses in a further section of this chapter.

When deciding on a subnet mask, you will need to convert from decimal to binary to find the number of bits that are 1s and 0s. You will also need to take into account the disallowed subnets. These calculations have been done for you in Tables 10-3 and 10-4 for class B and C addresses. Note that these tables are based on RFC950 and discard two subnets.

Configuring subnets: An example

Another approach to understanding subnets is to try to design a system. Put yourself in the shoes of the network administrator in the following example.

Assume a network with 160 computers. Performance is very sluggish and you decide to break it up into subnets. The original IP address of the network is 202.65.211.0

Step 1: How many subnets?

The number of subnets determines how many hosts are on each subnet. Also, subnet 0 and the subnet with all 1s in its binary address are unavailable. The number of subnets can then be determined. You can use Table 10-4 for this.

Subnet Mask	# of Subnets	Hosts/Subnets	Hosts Available	
128	2 − 2 = 0	126	0	No go
192	4 − 2 = 2	62	124	No go
224	8 − 2 = 6	30	180	OK
240	16 − 2 = 14	14	196	Possible

Step 2: Determine the subnet mask.

Reviewing the information in the preceding table, the mask 224 will provide 6 subnets with a maximum of 180 workstations. This is more than the number required. However, consider the future need for expansion. Will you have to resubnet if you run out of addresses? One other consideration is the router. Each interface to a subnet will require its own IP address. Feeling comfortable with six subnets, the decision is made to use the mask 255.255.255.224. Although using 240 in the mask gives you even more host addresses, it does so by using a hard-to-justify 14 subnets. A simpler network is always preferred.

Step 3: Calculate the subnet numbers.

Using all the combinations of the first three bits, calculate all the subnet numbers. In this example, the subnet IDs will be 32, 64, 96, 128, 160, and 192.

Step 4: Assign host numbers.

Number each machine in turn with a host number. For example, the first machine on the subnet might be 1.

Step 5: Calculate the machine's IP address.

The host portion of the IP address will be the subnet number plus the host number. For example, machine 1 (host 1) on the first subnet (subnet 32) will be 32 + 1, or 33. The complete IP address of this machine is 202.65.211.33, mask 255.255.255.224.

Mask (last 2 octets)	Useable Subnets	Useable Hosts per Subnet	Total Number of Hosts
128.0	2^1=2−2=0	2^{15}=32,768−2=32,766	0
192.0	2^2=4−2=2	2^{14}=16,384−2=16,382	32,764
224.0	2^3=8−2=6	2^{13}=8,192−2=8,190	49,140
240.0	2^4=16−2=14	2^{12}=4,096−2=4,094	57,316
248.0	2^5=32−2=30	2^{11}=2,048−2=2,046	61,380
252.0	2^6=64−2=62	2^{10}=1,024−2=1,022	63,364
254.0	2^7=128−2=126	2^9=512−2=510	64,260
255.0	2^8=256−2=254	2^8=256−2=254	64,516
255.128	2^9=512−2=510	2^7=128−2=126	64,260
255.192	2^{10}=1,024−2=1,022	2^6=64−2=62	63,364
255.224	2^{11}=2,048−2=2,046	2^5=32−2=30	61,380
255.240	2^{12}=4,096−2=4,094	2^4=16−2=14	57,316
255.248	2^{13}=8,192−2=8,190	2^3=8−2=6	49,140
255.252	2^{14}=16,384−2=16,382	2^2=4−2=2	32,764
255.254	2^{15}=32,768−2=32,766	2^1=2−2=0	0

Table 10-3: Useable subnet and hosts per subnet on a class B network (RFC950)

Mask	Useable Subnets	Useable Hosts per Subnet	Total Number of Hosts
128 (1000 0000)	2^1=2−2=0	2^7=128−2=126	0
192 (1100 0000)	2^2=4−2=2	2^6=64−2=62	124
224 (1110 0000)	2^3=8−2=6	2^5=32−2=30	180
240 (1111 0000)	2^4=16−2=14	2^4=16−2=14	196
248 (1111 1000)	2^5=32−2=30	2^3=8−2=6	180
252 (1111 1100)	2^6=64−2=62	2^2=4−2=2	124
254 (1111 1110)	2^7=128−2=126	2^1=2−2=0	0

Table 10-4: Useable subnet and hosts per subnet on a class C network (RFC950)

Exercise 10-7

Subnetting a Network

You are the administrator of network 197.245.12.0. There are 150 hosts on the network. For performance reasons, you need to divide the network into subnets. The company will purchase another 25 computers next year, which will be added to the network. Answer the following questions.

1. How many subnetworks will you divide the network into? _______
2. What subnet mask should be used? _____________
3. Draw a diagram of the network configuration. Calculate all of the subnet numbers and mark them on the diagram.
4. Calculate the IP address of:

   Host thirteen on the first subnet __________

   Host nine on the second subnet __________

10.4 PRIVATE IP ADDRESSES

Now that you understand how the addressing schemes works for the IP protocol, you can appreciate that issues with the system have cropped up since it was devised. Large portions of the addressing space go to waste because organizations cannot use all the addresses assigned to them. A company with a class A address would need 16 million hosts to use all of its numbers.

Organizations are assigned addresses that are inappropriate for their size. A company with 1,000 computers will need at least four class C addresses because a class B address isn't available.

Organizations would like to keep the details of their internal infrastructure a secret. A classful IP address can tip off a hacker about the size of the network.

One type of IP address that offers some relief from these issues is a private IP address. These addresses are neither published nor visible from the Internet. In order to make this scheme work, access to the Internet must be through a device (router, proxy server) that offers network address translation (NAT).

The private IP addresses are:

Class A: 10.0.0.0
Class B: 172.16.0.0 to 172.31.255.255
Class C: 192.168.0.0 to 192.168.255.255

How does it work?

Referring to Figure 10-11, when host 172.16.1.2 sends a request (1), the IP packet will list its address as the sender's address (SA). The proxy server, using NAT, strips out the original address and substitutes its own IP address (198.10.23.1) as the SA (2). When the remote host responds, it uses the sender's address as the destination address (3). Of course, it thinks the original sender is the proxy server. When the proxy server receives the packet, it strips out its own address as the destination and replaces it with the original host's address, 172.16.1.2. It places the packet on the internal network and it makes its way to the original host.

The proxy server represents all of the internal host machines and receives packets for all of them from the Internet. How can it identify the packet stream as belonging to a particular internal host when all of the incoming packets have the same destination IP address? There is further identification inside TCP and UDP packets called port numbers. Port numbers are discussed in detail in the next chapter. NAT also changes the port number of the outgoing packets to a unique number for each connection. NAT recognizes the port number on the incoming packets as belonging to a specific internal host and is able to forward the packet back into the network after it has correctly replaced the IP address and port fields with their original values.

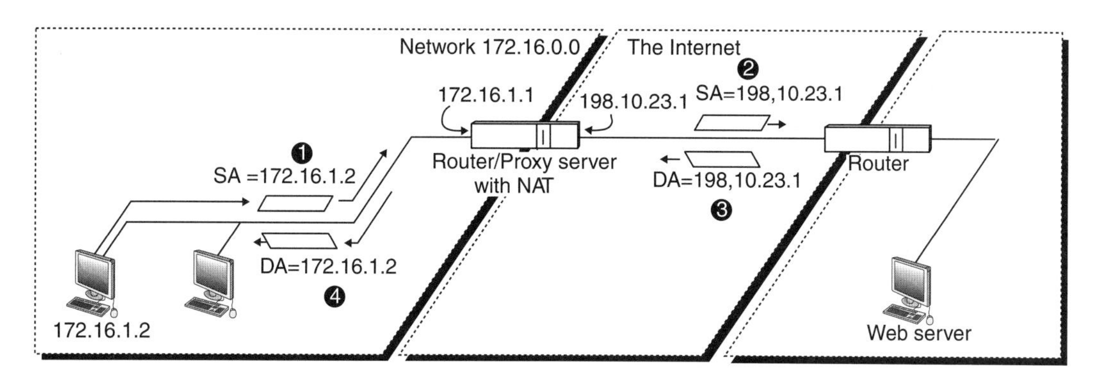

Figure 10-11: Network address translation

The benefits to this scheme are several:

1. The original IP address, the size of the network, or the topology of the internal network are hidden from the Internet.
2. Any address class can be used on an internal network. The address can more closely match the size of the network and precious addresses can be conserved. In the illustration a class B address of 172.16.0.0 was used, possibly because there are more than 254 hosts on the network. In addition, larger subnets can be used with a class B address than a class C address. Since there is only a single host, the proxy server, through which all of the traffic of the internal network passes, only a single class C address is needed for it on the Internet side.

NAT

Network address translation is the service that translates between the private and public addresses. As a service it can run on different devices. NAT is usually found in the following:

- Proxy server
- Broadband router
- Traditional router
- NAT appliance
- Firewall
- ICS—Internet Connection Sharing is a service Microsoft includes with Windows that allows one machine connected to the Internet to share the connection with other computers on a home or small-business network.

10.5 CLASSLESS IP ADDRESSING

In the classful system of IP addresses, the network ID/host IDs fall on octet boundaries. This means that there can only be three classes of addresses for host machines, plus class D and E for other uses. This doesn't give us a method of allocating addresses in small increments. For example a class A address has in excess of 16 million host addresses, even though few organizations could actually use that many.

The answer is to set up network and host IDs on *bit* boundaries instead of *byte* (octet) boundaries. When using classless addresses, the number of bits used for the network ID is part of the address, and using subtraction, the host ID can also be calculated.

Here are some typical IP addresses in their classful and classless forms:

87.45.87.100 = 87.45.87.100/8
145.76.200.23 = 145.76.200.23/16
202.56.31.176 = 202.56.31.176/24

As these illustrate, take the classful address and add the "/" (slash) and the number of bits that are to be used for the network portion of the ID, otherwise known as the network prefix.

The advantage of the classless address is that you are no longer restricted to network ID bits in 8-bit increments. Any number can be used. Take the following examples.

Addresses 64.128.0.0/12 and 64.129.0.0/12 are the same network, but address 64.144.0.0/12 is another network. To understand why this is so, convert the first two octets to binary and compare them.

0100 0000 1000 0000 --> 64.128.0.0/12
0100 0000 1000 0001 --> 64.129.0.0/12
0100 0000 1001 0000 --> 64.144.0.0/12
12 bits-----------

Because the first 12 bits of the first two numbers are identical, they belong to the same network. The first 12 bits of the third number do not match, which is why 64.144.0.0 is not the same network as the others.

The benefit to the classless numbering system is efficiency. The address 64.0.0.0/8 gives us one large network with up to 16,777,214 hosts. The address 64.0.0.0/12 is actually 16 separate networks, each of which can have 1,048,576 hosts.

SUMMARY

Section 10.1: IP Addressing

Each host on a network must have a unique identity. Each network must also have a unique identity. Both are satisfied with the IP addressing scheme. An IP address is a 32-bit value that contains enough information to uniquely identify the network and the host address of a computer. Therefore the IP address is broken down into two parts, the network part and the host part. A network address is normally assigned by the ISP although the overall administration of addresses is provided by regional registrars.

Section 10.2: Classful IP Addresses

The IP address can fall into one of three classes, class A, class B, and class C. Different classes of IP addresses were designed to accommodate different-size networks.

A network can be broken down into smaller subnets. This could be useful if you need more networks but you don't want to apply for an additional IP address from your ISP.

Section 10.3: Subnetting

Subnetting is useful if you have multiple networks that are physically separate because they are at remote sites or on separate cable systems and they are connected together through routers. A feature of IP addressing is that the IP address can be locally modified to use host address bits to provide additional network addresses. Essentially what this does is change how the addresses are interpreted locally without changing the global interpretation. This creates additional local networks at the cost of reducing the number of hosts.

Section 10.4: Private IP Addresses

Private IP addresses can be assigned to hosts on your internal network. This gives you a much larger supply of addresses and great flexibility in assigning them. However, these addresses are not visible on the Internet and internal hosts require the use of NAT to communicate to the Internet.

Section 10.5: Classless IP Addressing

By using a network prefix, you can avoid the use of the classful numbering system altogether. This system is used on the Internet and is known as classless inter-domain routing (CIDR). A big advantage to the system is that the network and host IDs are divided on *bit* boundaries instead of *byte* (octet) boundaries, which provides much more flexibility when designing the network.

Review Questions IP Addressing

1. Your network uses the subnet mask 255.255.255.192. Which computer is on the same subnet as 199.34.82.129?

   a) 199.34.82.189
   b) 199.34.82.128
   c) 199.34.82.193
   d) 199.34.82.126

2. The use of subnets in IP networks is popular because

   a) subnets create extra IP addresses.
   b) subnets allow you to divide a network into smaller, easier-to-manage units.
   c) routers find it easier to work with subnets than large networks.
   d) you can integrate different network classes into one network.

3. Which of the following IP addresses are on the same network?

   a) 157.200.14.25/20 and 157.200.18.200/20
   b) 157.200.25.33/20 and 157.200.32.12/20
   c) 157.200.95.9/20 and 157.200.96.43/20
   d) 157.200.96.25/20 and 157.200.111.100/20

4. Different classes of IP address

   a) are used for priority such that packets from a class A network are routed before packets from other class networks.
   b) were devised to accommodate different-size networks.
   c) are registered with the Network Information Center (NIC) only if they have more than 255 hosts.
   d) have the same length of network and host portion.

5. Which of the following addresses can you use for your network without having to worry that another network may be using it?

   a) 11.0.0.0
   b) 204.12.84.0
   c) 172. 23.0.0
   d) 180.56.0.0

CHAPTER

11

TCP and UDP

Objectives

Upon completion of this section, the reader will:

- *Understand the purpose and uses for transmission control protocol (TCP).*
- *Be able to identify the different fields in the TCP header and what they do.*
- *Know some of the features of TCP including the three-way handshake and sliding window.*
- *Be familiar with ports, their role in multiplexing, and the usage of the NETSTAT command in examining the status of the ports.*
- *Appreciate how UDP fits into the TCP/IP protocol suite.*

INTRODUCTION

The functions of transmission control protocol (TCP) and user datagram protocol (UDP) are crucial to the proper operation of the TCP/IP protocol suite. TCP is entrusted with the guaranteed delivery of data end to end. TCP will resend data that becomes corrupted during transmission. When guaranteed delivery of data is not required or TCP is too slow, UDP is used instead. The decision to use one or the other is up to the programmer, but one or the other must be used.

11.1 TRANSMISSION CONTROL PROTOCOL

TCP operates at the transport layer of the OSI model. See Figure 11-1. In the TCP/IP protocol suite, TCP provides guaranteed end-to-end delivery. It is a reliable, connection-oriented, byte-stream protocol. Delivering data accurately places a heavy burden on TCP, particularly if you consider that some files can be extremely large, in the gigabyte range, and that even if one bit is delivered incorrectly, the complete file will become corrupt. In order to work its magic, TCP uses its header to transmit crucial information to the receiving host. The TCP header is examined in great detail in this chapter.

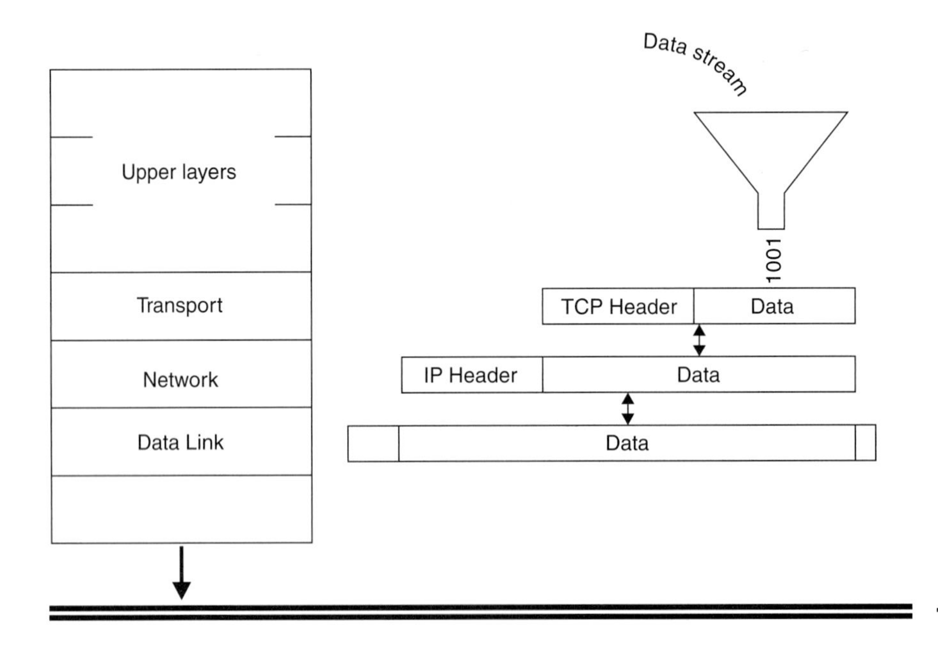

Figure 11-1: TCP exists at the OSI transport layer

The technical name for the protocol data unit (PDU) of TCP is a *segment.* Contrast this with the network layer where IP's PDU name is *datagram* or the data link layer where Ethernet's PDU name is *frame.*

Reliability

TCP uses positive acknowledgment with retransmission. This means that unless TCP gets a positive acknowledgment returned from the destination, it will retransmit the packet. If the destination finds that the packet is corrupt it only has to discard the packet and wait. Without a positive acknowledgment, the sender will continue retransmitting the packet until it receives the acknowledgment or until it times-out. TCP is therefore a "stateful" protocol, meaning that it must track the state or condition of each connection. This topic will be looked at further in the chapter.

Connection-oriented service

TCP is connection oriented because it establishes a logical connection between the hosts that are communicating before transmitting data. By exchanging control information, known as "handshaking," each host synchronizes with the other and the connection has to be established and verified. Error detection and correction is usually handled by connection-oriented services.

The sequence that is used is called a three-way handshake because three segments (packets) are exchanged to set up the connection. The control information is exchanged by setting bits in the Flags field of the segment header.

The three-way handshake is diagrammed in Figure 11-2 and the process can be described this way. Host A wants to open a connection to Host B and starts

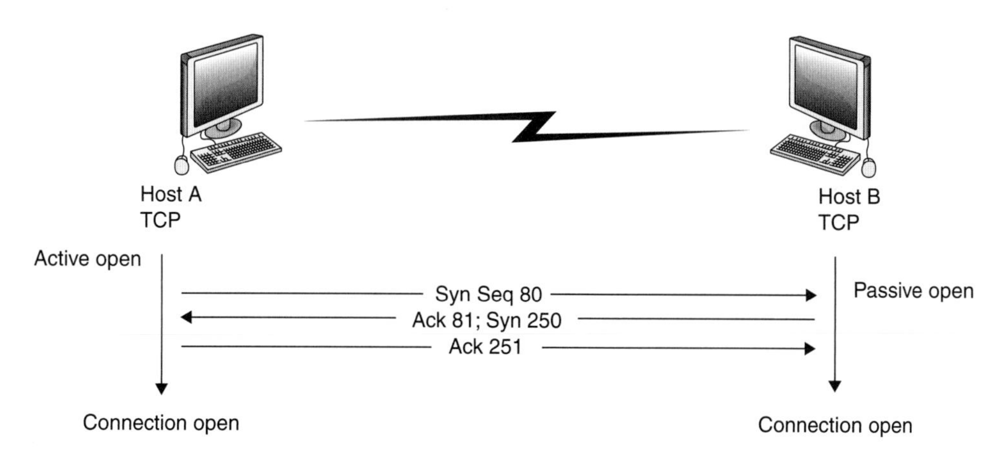

Figure 11-2: The TCP three-way handshake

by allocating buffer space in its memory for the connection, which is called an active open. It sends a packet with the SYNchronization bit set and an arbitrary SYN sequence number of 80. This number is arbitrary except that it can't be in use currently. In order to avoid duplication, the sequence number is chosen from about 4.3 billion possibilities. When Host B receives the packet, it also opens a buffer in memory, which is called a passive open. It replies with a return packet. To answer, it sets the ACKnowledgment bit and sets the ACK sequence number to 81, one more than the SYN sequence number. TCP is a full-duplex protocol, which means that it can communicate in both directions, even in the same packet. Therefore in the return packet, Host B also sets the SYN bit and sets the SYN sequence to some arbitrary number; in our example the number is 250. When Host A receives the return packet it must acknowledge the reverse connection by replying with a final ACK and the SYN sequence number becomes the ACK sequence number incremented by one. The three-way handshake is now complete and the two connections, one in each direction, are now set up.

Byte-stream

TCP is a byte-stream protocol because it views the entire exchange as a continuous stream of data, which it is receiving from the upper layers. TCP is responsible for dividing the data stream into segments that can be delivered by IP; at the receiving end, TCP is responsible for reassembling the segments into the original file.

Sequence numbering becomes very important to maintain the order of the data in the stream. Sequence numbering is used to indicate the position in the total data stream. Sequence numbers are also used for acknowledgment and they indicate how much of the original data stream has been successfully received.

Flow control

TCP can send multiple packets without an acknowledgment. This is crucial to maintaining some performance over a wide area network because each packet does not have to be acknowledged before the next one is sent. Typically, the receiver controls the data flow rate depending on the number and size of its receiving buffers plus the rate at which it can copy the data to the higher layers.

11.2 PORT ASSIGNMENTS

Since data moving over the network, whatever its origin, look the same to the receiver, some method is needed to make sure that they get to the right application in the host. TCP and UDP use port numbers to provide correct delivery of data.

Port numbers are not unique across protocols. Therefore, to completely identify an application, the port number is paired with the protocol. For example, an application using user datagram protocol (UDP) to move data to port 178 would be different than an application using TCP to move data to port 178.

Multiplexing data

Port numbers solve the problem of multiplexing at the transport layer. The computer you are sitting in front of is running many services and applications at the same time. You are transferring files with FTP, browsing a website, controlling a remote host with Telnet, receiving e-mail and using DNS to resolve names, all at the same time. When data is being received by your computer to update a Web page, how can it be distinguished from the e-mail being received by Outlook? The answer is the port numbers that are unique for each connection. Multiplexing is a requirement at many layers of the communication system. Figure 11-3 illustrates how multiplexing is handled by the data link and network layers as well as the transport layer. The Type field in the Ethernet header, the Protocol field in the IP header, and the Port field in the TCP and UDP headers all identify the payload that they are carrying.

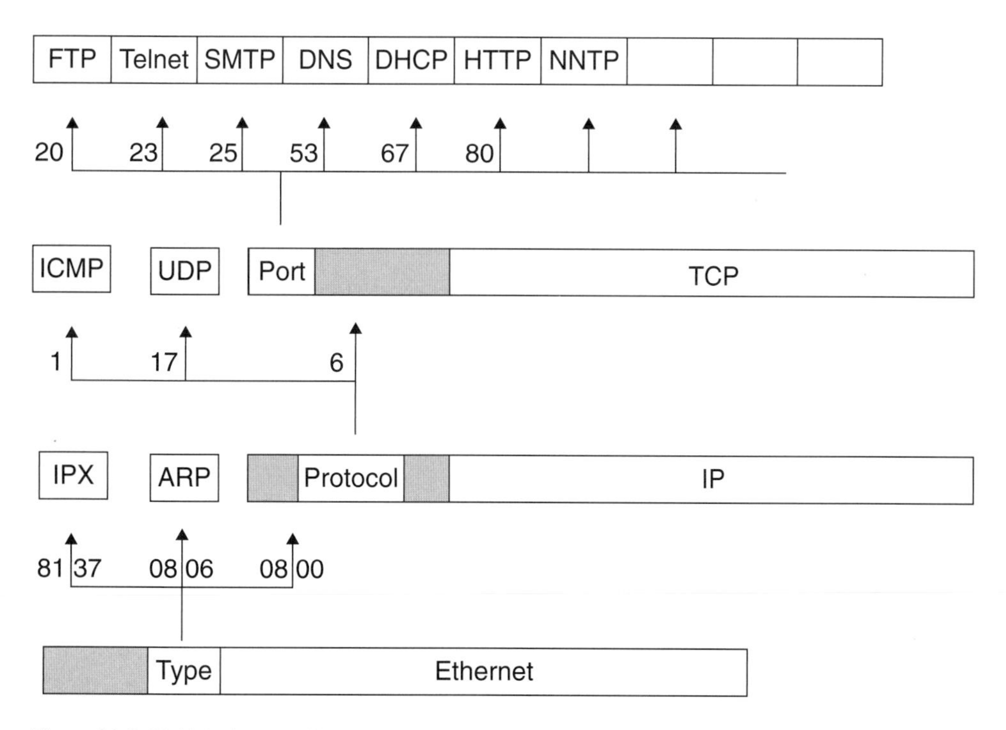

Figure 11-3: Multiplexing at different layers

Well-known ports

Ports can be chosen by an application's developer when programming the software. As long as both the sender and receiver applications understand the use of the common port number, they can communicate. There is, however, a large number of existing applications that use certain port numbers by default.

Well-known services are applications that use a widely known port/protocol pair.

Some examples:

FTP - ports 20 and 21/TCP
Telnet - port 23/TCP
DNS - port 53/UDP and 53/TCP
HTTP - port 80/TCP

Refer to Table 11-1 for a longer list of ports. A still longer list is found on your computer in the file named "services." The definitive list, however, belongs to the organization that controls the numbers, which is IANA (Internet Assigned Numbers Authority). It should be mentioned that the IANA function is controlled by the Internet Corporation for Assigned Names and Number (ICANN). If you need to look up this document, use the URL: http://www.iana.org/assignments/port-numbers.

Because the port fields in the TCP header are 16-bit fields, the number of ports that are available is 65,536. These are broken down as follows:

Ports 0-1023 : Well-known ports
Ports 1024-49151 : Registered ports
Ports 49152-65535 : Dynamic and/or private ports

20	FTP (File Transfer Protocol) — data
21	FTP (File Transfer Protocol) — control
22	SSH (Secure Shell)
23	Telnet
25	SMTP (Simple Mail Transfer Protocol)
53	DNS (Domain Name Service)
69	TFTP (Trivial File Transfer Protocol)
80	HTTP (Hypertext Transfer Protocol)
110	POP3 (Post Office Protocol version 3)
119	NNTP (Network News Transport Protocol)
123	NTP (Network Time Protocol)
137	NETBIOS Name Service
138	NETBIOS Datagram Service
139	NETBIOS Session Service
143	IMAP4 (Internet Message Access Protocol version 4)
443	HTTPS (Hypertext Transfer Protocol Secure) SSL (Secure Sockets Layer)

Table 11-1: Well-known ports

Only servers use well-known ports

Two ports are used in a connection. When a client host initializes a connection, it places two port numbers in the TCP header. The destination port is the well-known port number that the server uses. The source port number identifies the connection on the client end. When a host powers on, the source port number used for the first connection is 1024. As each further connection is created, the number increments by one. When a connection is ended, the port number is no longer used. A port number is never reused until the host is rebooted.

A server is always on the lookout for packets that contain the port number of the service that it is running. This is referred to as "listening" on a port. For example, port number 80 is the well-known port number for the HTTP protocol used by Web servers. If a Web server receives a TCP segment with port 80, it forwards the request to the Web service for further processing.

TCP services

TCP is at the core of the TCP/IP protocol suite and has the following characteristics.

- Full duplex: Enables either end of a connection to transmit at any time, even simultaneously.
- Timeliness: The use of timers ensures that data is transmitted within a reasonable amount of time.
- Ordered: The data flowing to the upper layers is always in the correct order. This despite the fact that datagrams may be received in a different order by IP. TCP always reorders data correctly.
- Stateful: The current state of each connection is tracked.
- Controlled flow: TCP can regulate the flow of data through the use of buffers called a sliding window.
- Error correction: Checksums ensure that the data is free from errors (within the checksum algorithm's limits).

Hands-on Lab 11-1

Look Up Port Numbers

A text file called "Services" is added to the host when the TCP/IP protocol is installed. It has a lengthy, but still incomplete, list of services along with the services'port numbers and the protocol, TCP or UDP, that uses them.

The file is typically located on a Windows machine in the \Windows \system32\drivers\etc folder. However, it is just as easy to use the search function to locate it.

Microsoft Vista Note: Because of Vista security, the search function may not find this file. Instead, use Windows Explorer and navigate directly to the \Windows directory\system32\drivers\etc folder to find the file.

1. Use the **Start** button > **Search.**
2. Search for files and use the name "**services**". Note that this file has no extension.
3. Although many files and folders may be listed, you are interested only in the file located in C:\windows\system32\drivers\etc.
4. Double-click the correct services file and open it with Notepad.
5. Examine the file by scrolling through it. Can you find the ports listed in Table 11-1?

11.3 TCP STRUCTURE — THE FIELDS

The TCP header structure contains the fields that allow TCP to achieve the function assigned to it, namely the reliable end-to-end delivery of data. The fields and their use are described in this section. Figure 11-4 illustrates the TCP segment encapsulated within the IP datagram. It also shows the TCP fields along with their size. Figure 11-5 shows the TCP header divided along 32-bit or four-octet boundaries. This is the depiction of the TCP header found in the document that defines the TCP standard, which is RFC 793, dated September 1981.

TCP Segment

MAC | IP | SP | DP | DATA | FCS

TCP Header Structure

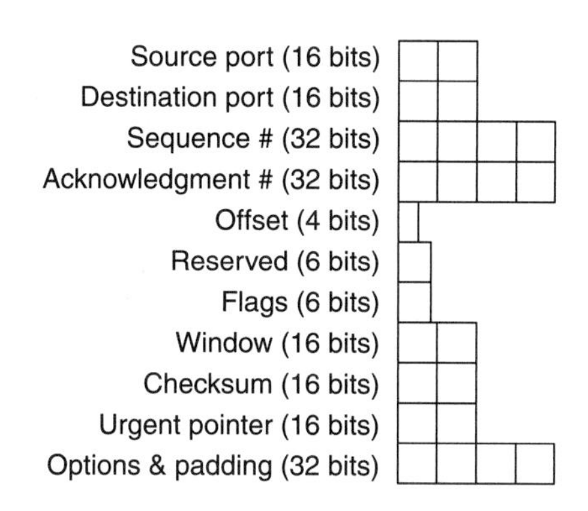

Figure 11-4: TCP header structure (1)

Source port

This 16-bit field identifies the sender's connection point. If this is a server, the port number will be the one that the server is listening on and is associated with the service that the server is running. If this is the client, the port number is dynamic and is assigned when the connection is made. When the host first starts up, the first port number used on the client side is 1024 and increments by one for each connection created thereafter.

The port number is important for multiplexing. As an example, imagine that you have two different Web pages open in two different windows of your Web browser from the same Web server. You have established two different connections but both IP addresses are the same and the Web server's port number of 80 also being the same. The only thing that distinguishes the two connections is the client's port numbers because they are different.

Port numbers were discussed in the preceding section of this chapter. Because the port field is 16 bits in length, the values can be as large as 65,535.

Destination port

Same as the source port but for the receiving machine.

Sequence number

The sequence number is a 32-bit field and the number indicates the position of the octet in the overall data stream. The sequence number is initialized as an arbitrary number, often calculated from an algorithm that references the current time. It is only crucial that all of the sequence numbers used by all of the processes running in this and the destination machine be unique.

This number is also used to provide the initial sequence number (ISN) found in the three-way handshake.

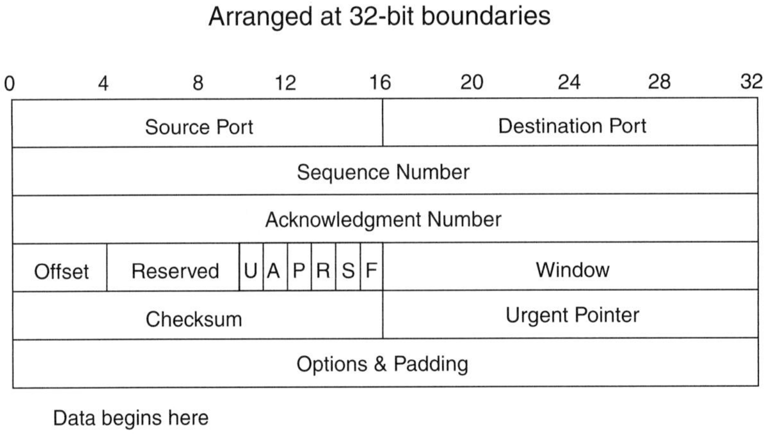

Arranged at 32-bit boundaries

0 4 8 12 16 20 24 28 32

Source Port								Destination Port
Sequence Number								
Acknowledgment Number								
Offset	Reserved	U	A	P	R	S	F	Window
Checksum								Urgent Pointer
Options & Padding								

Data begins here

Figure 11-5: TCP header structure (2)

Acknowledgment number

The acknowledgment number is also 32 bits in length and indicates the next sequence number expected. In a backhanded way, it also shows the sequence number of the last data received. It shows the last sequence number + 1.

In order to demonstrate how the acknowledgment number works, consider that a host has received octets with sequence numbers 150, 151, 152, and 153. The host sends back an acknowledgment with the acknowledgment number set to 154. The other host draws two conclusions from this: all octets up to and including octet 153 have been received correctly and that the next octet that the other host expects to receive will have the sequence number 154.

Data offset

The data offset field is four bits in length and contains the length of the TCP header in 32-bit words. This is exactly the same scheme that IP used for the IP header length field. For example, if the data offset field has the value 5, the TCP header is 5×4, or 20 octets long. It also forces the TCP header to be a multiple of four octets.

Reserved

These six bits are unused and are always set to zeroes.

Flags

This is a six-bit field. Each bit can be set to 0 or 1 and each value has a different significance. Here are the flags.

URG (Urgent)

- Perform urgent processing on data.
- If set (1), indicates that the urgent field is significant.

ACK (Acknowledge)

- If set (1), indicates that the Acknowledgment field contains the next expected sequence number.

PSH (Push)

- If set (1), indicates that the Push function is to be used. If Push is set, the data from this packet is to be copied immediately to the upper layers without waiting for the buffers to fill. Although not efficient from TCP's point of view, it does speed up the movement of urgent data.
- Examples of urgent data include login names and passwords.

RST (Reset)

- If set (1), reset the connection.

SYN (Synchronize)

- If set (1), indicates that the sequence numbers are to be synchronized. This flag is used during the three-way handshake to establish the connection.

FIN (Finish)

- If set, indicates that the sender has no more data to transmit.

The flags may change in future implementations of TCP. RFC 3168, dated September 2001, proposes adding a flag using two bits for explicit congestion notification (ECN) in order that routers can better handle congestion on the network. However, this is currently a proposal and not yet part of the TCP standard.

Window

This 16-bit field is the size of the receive window at the destination machine. It indicates the additional number of octets the buffers can hold before they start dropping data.

The sender also knows that it can send the number of octets indicated in this field before expecting an acknowledgment.

Example: During initial synchronization, the receiver indicated that it could accept 4,096 octets. TCP copies 4,096 octets to IP. IP, however, knows that the datagram is being put out over Ethernet, which can't accommodate this size of packet. Therefore, it fragments the data into four datagrams of 1,024 octets each. This is only an illustration and in the real world, the data probably wouldn't divide so neatly. Even if IP didn't fragment the data, a router along the path may have. In any case, the destination machine receives four segments but doesn't acknowledge until the fourth one is received. It can afford to wait because it told the sender that its window was 4,096 octets. When it does acknowledge, it sends back the current empty buffer as new window size of 3,072. The 3,072 octets have been processed but the buffer still contains 1,024 octets waiting to be processed. TCP in the sender host can now send out 3,072 octets but then must wait for another acknowledgment. Dynamically changing the window size during transmission is known as a "sliding" window and helps adjust transmission for the ever changing conditions within the machines and on the network.

Checksum

The checksum field is 16 bits in length. The checksum covers both the header and data, which is expected since TCP guarantees the transmission of data without errors. Contrast this with IP, which checks only the header for errors. The algorithm used is the 16-bit one's complement of the one's complement sum of the 16-bit words in the header and data.

Urgent pointer

The urgent pointer is a 16-bit field that identifies the octet where urgent data ends. It is only used if the URG flag is set. No specific action is taken by TCP. It is all left up to the application.

Options and padding

This field will always be a multiple of 32 bits if it exists. If the options do not end on a 32 bit boundary, padding will be added. This field is optional. If it does exist, it is used to negotiate various parameters such as segment size.

TCP STATES

TCP states

TCP connections involve synchronizing the connection (SYN), transmitting the data, and then finishing, or closing the connection (FIN). As requests are made between systems, acknowledgments (ACK) are transmitted back and forth to assure that both sides of the connection agree on the state of the connection. As the conversation between the two connections proceeds, TCP is in ever-changing conditions (state).

Examining the state of a TCP connection can give you insight as to how TCP works. As a practical matter, it may also help you troubleshoot problems.

NETSTAT utility

The NETSTAT utility will show you the TCP states of current connections of your machine.

Use the following commands to see the information of interest.

NETSTAT /?

NETSTAT help, see the command line options.

NETSTAT

See the active connections.

NETSTAT -a

See all connections and listening ports.

NETSTAT –an

See all connections and show ports in numeric form.

TCP states

The TCP states are as follows.

CLOSED

A theoretical state where a TCP connection doesn't exist yet.

LISTEN

A state where TCP is waiting for a connection to be made to it. Only servers listen.

SYN_RCVD

A SYN has been received, a SYN ACK has been sent, and we are waiting for an ACK.

SYN_SENT

A SYN has been sent to start a connection; the SYN ACK hasn't yet been received.

ESTABLISHED

A connection has been established.

CLOSE_WAIT

After a TCP connection has been closed, the specification says you should make sure that the connection is not reopened for a few minutes to prevent packets that were delayed in the Internet causing the connection to be reopened. These will disappear by themselves after a while. This can cause problems on servers that have lots of very short connections (such as a Web server) by filling up the TCP's connection table.

LAST_ACK

A FIN has been received, we've sent our own FIN, but we are waiting on the ACK before we can move to the closed state.

FIN_WAIT_1

A FIN has been sent, waiting for an ACK or FIN.

FIN_WAIT_2

A FIN has been sent, and an ACK has been received. More data can be received. This connection is half open.

CLOSING

A FIN has been sent and one received; the FIN hasn't been ACK'd yet.

Figure 11-6 shows the progression of the TCP states as connections are opened and then closed. Refer to it when trying to interpret the output of the NETSTAT command.

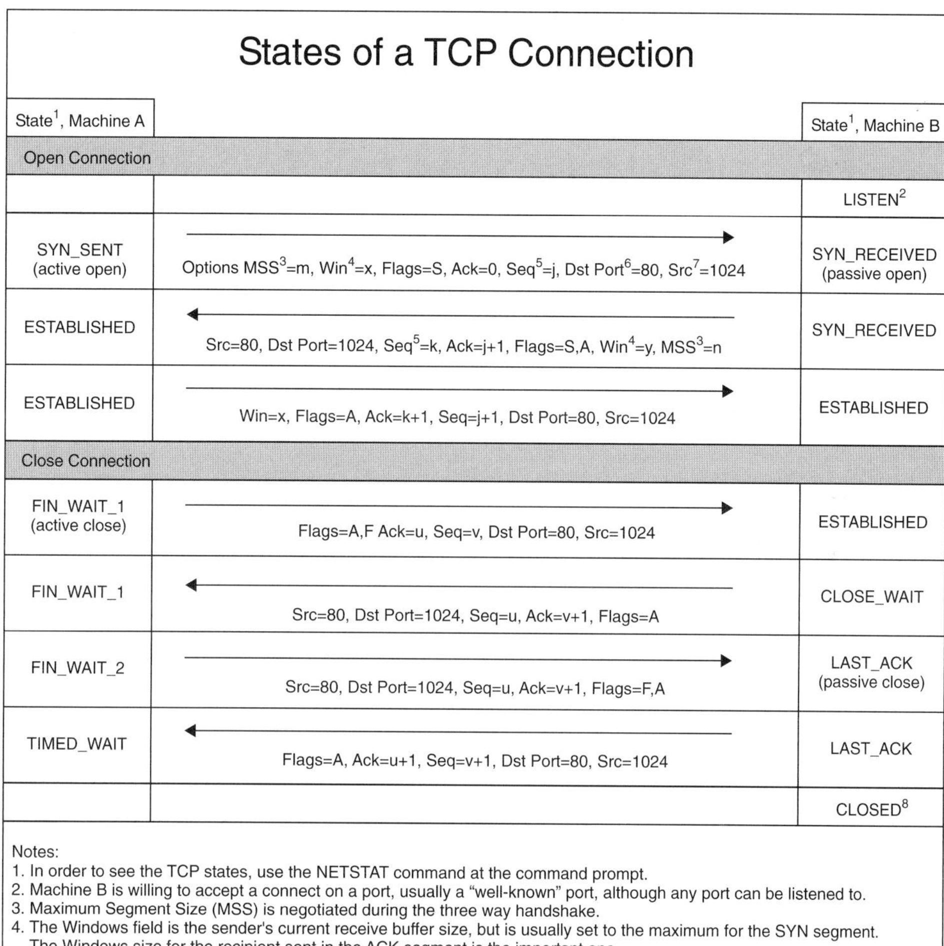

States of a TCP Connection

State[1], Machine A		State[1], Machine B
Open Connection		
		LISTEN[2]
SYN_SENT (active open)	→ Options MSS[3]=m, Win[4]=x, Flags=S, Ack=0, Seq[5]=j, Dst Port[6]=80, Src[7]=1024	SYN_RECEIVED (passive open)
ESTABLISHED	← Src=80, Dst Port=1024, Seq[5]=k, Ack=j+1, Flags=S,A, Win[4]=y, MSS[3]=n	SYN_RECEIVED
ESTABLISHED	→ Win=x, Flags=A, Ack=k+1, Seq=j+1, Dst Port=80, Src=1024	ESTABLISHED
Close Connection		
FIN_WAIT_1 (active close)	→ Flags=A,F Ack=u, Seq=v, Dst Port=80, Src=1024	ESTABLISHED
FIN_WAIT_1	← Src=80, Dst Port=1024, Seq=u, Ack=v+1, Flags=A	CLOSE_WAIT
FIN_WAIT_2	→ Src=80, Dst Port=1024, Seq=u, Ack=v+1, Flags=F,A	LAST_ACK (passive close)
TIMED_WAIT	← Flags=A, Ack=u+1, Seq=v+1, Dst Port=80, Src=1024	LAST_ACK
		CLOSED[8]

Notes:
1. In order to see the TCP states, use the NETSTAT command at the command prompt.
2. Machine B is willing to accept a connect on a port, usually a "well-known" port, although any port can be listened to.
3. Maximum Segment Size (MSS) is negotiated during the three way handshake.
4. The Windows field is the sender's current receive buffer size, but is usually set to the maximum for the SYN segment. The Windows size for the recipient sent in the ACK segment is the important one. This value is specific to each machine and can change during the connection.
5. The Initial Sequence Number (ISN) is calculated to be unique.
6. Port listened to by server. Port 80 is http and is only an example.
7. Source port normally starts at 1024, the first number above the well-known ports (0-1023). This number increments with every connection opened and is not reused until the machine is re-booted.
8. Machine goes into CLOSED state after waiting 2*MSL (Max Segment Lifetime), about 4 minutes.

Figure 11-6: TCP states as connections are opened and closed

Hands-on Lab 11-2

NETSTAT utility

The NETSTAT utility is a standard component of the TCP/IP protocol suite. It should be available with any operating system as long as TCP/IP is installed.

NETSTAT has two main functions: examine the connection states of TCP and UDP and examine statistics of the TCP/IP connections.

1. Open a command line

   START > **Run** > type **CMD** > **<Enter>**

2. Try these commands in turn and note the results.

   Look up NETSTAT help.

   NETSTAT /?

3. The following commands examine the TCP and UDP state conditions.

   Examine active connections.

   NETSTAT

   List active connections and listening ports.

   NETSTAT -a

   Same as above but list all information numerically.

   NETSTAT -an

4. The following commands list TCP/IP statistics.

   Display Ethernet statistics.

   NETSTAT -e

   Display all statistics.

   NETSTAT -s

   Display statistics for only one protocol

   NETSTAT -sp protocol

   where protocol is one of IP, ICMP, TCP, or UDP.

11.4 USER DATAGRAM PROTOCOL

The user datagram protocol (UDP) gives application programs direct access to datagram delivery services. A datagram service is a best effort and unreliable service and this makes UDP similar to IP in this regard.

Nevertheless, UPD still functions at the transport layer as does TCP. Since these are the only two protocols that function at this layer, upper layers must choose one or the other.

Despite being unreliable, UDP may still be a good choice because of its other characteristics. It is simple, small, and fast. This allows applications to exchange data over networks with a minimum of overhead. UDP is a good choice for applications that move only a small amount of data or use a "query/response" model. An example of the former is tiny file transfer protocol (TFTP); an example of the latter is network file system (NFS). In addition, UDP is used by streaming data because lost packets are not retransmitted.

UDP also cuts down on the overhead of establishing a reliable connection when it is not needed. UDP uses a 16-bit source and destination port number to ensure the data is delivered to the proper application when it arrives at the host.

Other than the port addresses and length and checksum, UDP adds very little to an IP datagram. Figure 11-7 diagrams the UDP header structure and illustrates how simple it is.

UDP STRUCTURE – THE FIELDS

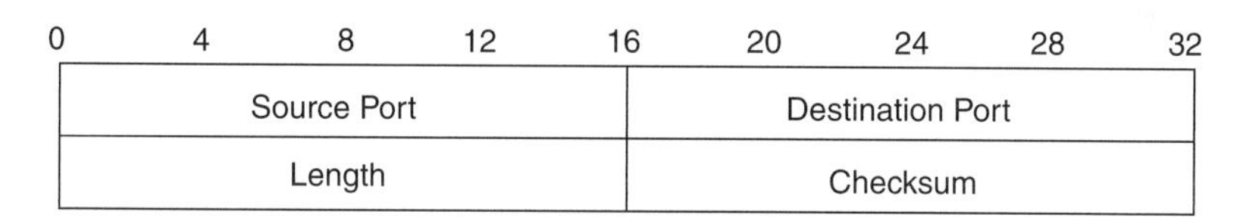

Figure 11-7: UDP header structure

The UDP header is composed of the following fields.

Source port

The source port is a 16-bit field and identifies the sending application process (usually an upper layer application program such as NFS or TFTP).

Destination port

The destination port is the same as source port but for the receiving machine. The port fields in UDP work exactly the same as they do for TCP. Refer to the discussion in the TCP section for more details.

Length

This field is the length of header plus data. Since it is 16 bits, the maximum size UDP datagram is 65,535 octets, the same length as TCP.

Checksum

The UDP checksum is 16 bits in length. The checksum is very strange. It covers both the UDP header and data and it covers some data in the IP header as well.

A pseudo header is created from the IP source and destination addresses, the protocol number (all of which information it can retrieve from IP), and the UDP length. This pseudo header plus the real UDP header plus the data are used to create the checksum.

Note that despite the fact that the UDP data is included in the checksum, there is no data correction capability. However, UDP can detect corrupted data.

Which applications use which protocol?

All upper-layer protocols have to use either TCP or UDP. After considering their characteristics, an application developer can choose the one that fits the application.

Which services use TCP?

- FTP
- Telnet
- SMTP
- Voice over IP control (SIP/H.323)
- And so many more

Which services use UDP?

- TFTP
- DNS queries
- NSF
- IP tunneling
- BOOTP
- SNMP
- Voice over IP media (RTP)

SUMMARY

Section 11.1: Transmission Control Protocol

TCP provides reliable, guaranteed delivery of data. It establishes a connection between the sender and receiver before sending data. By exchanging control information, known as "handshaking," each host synchronizes with the other and the connection can be established and verified.

Section 11.2: Port Assignments

Since data moving over the network, whatever its origin, looks the same to the receiver, some method is needed to make sure that it gets to the right application. TCP, as well as UDP, use port numbers to provide correct delivery of data. To completely identify an application, the port number is paired with the protocol.

Section 11.3: TCP Structure—The Fields

The TCP header contains all of the information that must be conveyed to the remote host in order for the connection to be created and deleted as well as the correct delivery of the data. All of the fields in the TCP header contribute to its function of guaranteed end-to-end delivery of the data.

Section 11.4: User Datagram Protocol

UDP gives application programs direct access to datagram delivery. This allows applications to exchange data over networks with a minimum of overhead. UDP is a good choice for applications that have the following characteristics: they move only a small amount of data or use a "query/response" model, they provide error correction, or they need a small, simple, and fast protocol.

Hands-on Lab 11-3

NMAP Utility

NMAP is a very powerful scanning utility. In this exercise, you will use NMAP to achieve two goals.

1. Find all of the active hosts on a network.
2. Find all of the open, listening TCP ports on the active hosts.

You need to find all of the active hosts on the network. You have already used PING to find active hosts; however, you can't scan a complete network with a single command. Using PING to scan an entire network would be extremely tiresome.

Second, you want to scan the hosts on the network for their open, listening ports. This will tell you what services they are running.

NMAP is open source, free software. Download NMAP for free from http://www.insecure.org. NMAP is traditionally run from the command line, although a graphical version is available.

Install NMAP

If NMAP needs to be installed, the instructor may provide additional instructions.

1. Find the installation file, nmap-4.20-setup.exe (this name may vary with the version). Double-click to start the installation. NMAP requires WinPcap to function and will install it automatically during a default installation. However, if you already have Wireshark installed, you already have WinPcap. There is no need to install it again.

Scan for all hosts on the network

1. Open up a command prompt.
2. Change to the NMAP directory.

```
CD\Program Files\Nmap
```

3. You will use NMAP's PING scan to quickly scan the complete network to find all hosts that are active. Issue the following command (substitute the correct IP address range):

```
Nmap -sP 192.168.1.0/24
```

Hands-on Lab 11-3 cont.

Scan all active hosts for open TCP ports

4. You will use NMAP's SYN scan to find all open TCP ports on all of the active hosts. Issue the following command:

   Nmap -sS 192.168.1.0/24

5. If you want more detailed and reliable information about the service behind the open port, scan for the service version.

   Nmap -sV 192.168.1.0/24

Saving the results to a file

6. You can save the results of an NMAP scan to a file. Several file formats are supported, but saving to an XML file will allow you to open it in a web browser.

   Nmap -sV 192.168.1.0/24 -oX test.xml

7. After the command has finished, type in test.xml. The file will open in your Web browser.

Review Questions TCP and UDP

1. TCP is considered reliable because
   a) it uses a three-way handshake.
   b) it calculates a checksum for the header.
   c) it exists at the transport layer of the OSI model.
   d) it is a connection-oriented protocol.
2. TCP's sliding window implementation of flow control
   a) allows multiple packets to be transmitted before an acknowledgment needs to be returned.
   b) allows a router to decrease transmitted packets if there is congestion.
   c) requires a fixed window size during the file transmission.
   d) is independent of the receive buffers at the destination machine.
3. Port assignments
   a) vary each time a TCP connection is made.
   b) must be paired with a protocol, such as TCP or UDP, to be meaningful.
   c) are well known if they are listed in a file called HOSTS.
   d) include LPT1: and COM1:.
4. If data should be processed immediately by TCP
   a) the Urgent flag is set and a pointer is placed in the Urgent field.
   b) the Urgent flag is set and an Acknowledge is returned by the other host.
   c) the Push flag is set.
   d) the sender creates a TCP packet without waiting for its buffers to fill.
5. User datagram protocol (UDP) is used
   a) when it is important to have fast communications with little overhead.
   b) when guaranteed delivery is not required.
   c) when upper layer protocols or services can do error correction.
   d) All of the above.

CHAPTER

12

DHCP

Objectives

Upon completion of this section, the reader will:

- *Understand the purposes of DHCP. DHCP was designed to ease the burden of managing TCP/IP configuration on client machines.*
- *Appreciate the requirements for DHCP servers and how DHCP works.*
- *Know how to configure DHCP clients.*

INTRODUCTION

One of the weaknesses of TCP/IP is the need to manually configure the IP host. The configuration includes the IP address, subnet mask, and default gateway at a minimum. In addition, domain names, DNS servers, and other information may need to be configured as well. Configuring a single IP host is not a burden to the administrator, configuring a thousand hosts may be intolerable. The dynamic host configuration protocol (DHCP) is a solution to this problem.

12.1 DYNAMIC HOST CONFIGURATION PROTOCOL

The dynamic host configuration protocol (DHCP) is a method of configuring an IP host with its configuration parameters through the use of a DHCP server. See Figure 12-1. When the DHCP client boots up, it broadcasts a request on to the net work to try to contact any available DHCP server. A responding DHCP server will return the IP parameters to the client, which can then proceed to initialize.

Why DHCP?

The benefits to using a DHCP server instead of manually assigning IP addresses are as follows:

- Lower administration overhead. The administrator does not have to physically go to each workstation in order to configure it.
- Centralized management of IP addresses. By using a DHCP server to assign addresses, the administrator can manage his pool of addresses and change parameters as required.
- Less chance for error. Since the administrator doesn't have to type in the IP parameters at each workstation, there is less chance of typing errors or missing parameters.
- Faster parameter changes. An IP parameter may have to be changed quickly. For example, the default gateway (router) fails. The DHCP parameter in the DHCP server can be changed and then the administrator can ask all the users to reboot their workstations for the change to take effect.
- Easy to move computers to other subnets. When a computer is moved to another subnet, DHCP will provide it with a new configuration. No administration involvement is required.

How does DHCP work?

Each time a DHCP client starts, it requests IP address information from the DHCP server. Because the client doesn't know the address of the DHCP server, it has to use a broadcast to try to reach the server.

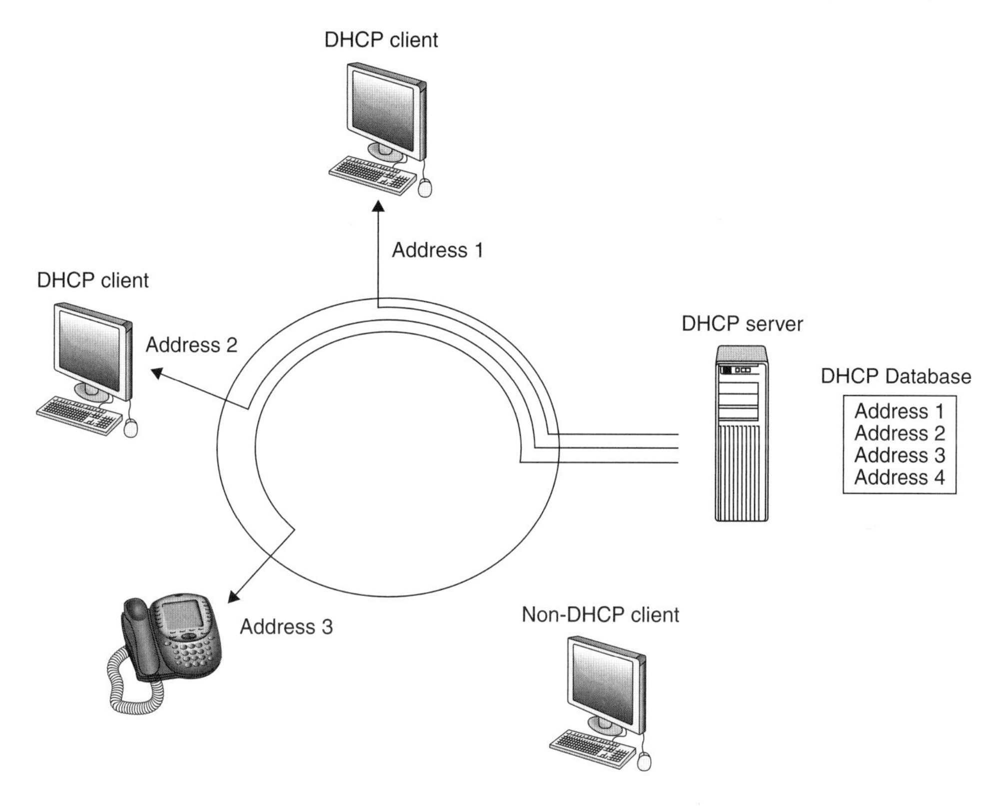

Figure 12-1: DHCP at work

The type of information that the client is looking for includes an IP address, subnet mask, default gateway, domain name, the IP address of DNS servers, and the IP address of WINS servers.

When the DHCP server receives a request, it chooses an IP address from a pool of addresses that the administrator has defined. The server offers this address to the client, and if the client accepts, it receives a lease for the address for a specified period of time. Once the client receives the IP information, it initializes TCP/IP.

APIPA — What happens if DHCP fails?

If there is no available IP address in the pool or if DHCP fails for another reason, TCP/IP initialization by the client cannot proceed normally.

For Windows NT and Windows 95/98, TCP/IP cannot initialize and the results of the IPCONFIG command will show an IP address of 0.0.0.0.

For Windows 2000/2003/XP and Windows ME, TCP/IP will initialize with the reserved private address 169.254.x.y. This private address has been reserved with

ICANN for this specialized use and Microsoft clients take advantage of it. Because the client has an IP address, it can commence normal functions. Note, however, that only other clients that have also initialized with this range of addresses can communicate with this client. In addition, since this client has not been configured for a default gateway, it cannot communicate beyond the local network. Self-assigning an IP address is called automatic private Internet protocol addressing (APIPA).

Boot protocol (BOOTP)

DHCP is actually an enhancement to an older protocol called the boot protocol, or BOOTP. BOOTP was designed for old networks in an age when IP addresses were in short supply. BOOTP would lend an IP address to a host for a very short interval, such as 30 seconds. When the host no longer needed it, BOOTP could assign it to another host. With the introduction of private IP addresses, the supply of addresses was no longer a concern. BOOTP was enhanced to provide options to the host and renamed to DHCP. Now, DHCP leases an IP address to a host for a lengthy period; it does not lend it as BOOTP does.

BOOTP and DHCP are fully interoperable. DHCP is actually a BOOTP packet with a DHCP options section tacked on. They even use the same UDP ports, 67 and 68.

12.2 DHCP COMMUNICATIONS

DHCP uses a four-phase process to configure the DHCP client. These constitute four verbs in DHCP's vocabulary, namely request a lease, offer a lease, select a lease, and acknowledge. DHCP running on a Windows 2003 system with Active Directory adds a fifth verb that is used by the system, *validate* registered DHCP servers, which adds another layer of security to it. DHCP also has a negative acknowledgment, which is used by a server to deny a renewal of an IP address.

If a computer has multiple adapters, DHCP operates independently on each. Unique addresses will be assigned through each interface. Note that it is also possible for one network adapter to use DHCP while the other adapter has a static address assigned. All DHCP communications are done using UDP over ports 67 and 68.

DHCP MESSAGES

IP lease request (DHCPDISCOVER)

When the TCP/IP client initializes, it loads a limited version of TCP/IP. At this point its IP Address is 0.0.0.0. It requests an IP address by broadcasting a request to all DHCP servers. Because its own address is undefined and it doesn't know the address of the DHCP server, the DHCPDISCOVER packet uses a source address of 0.0.0.0 and a destination address of 255.255.255.255. The packet does include the client's hardware (MAC) address so that the server does know who sent the request.

The IP lease request is used when one of the following occurs:

- TCP/IP is initialized for the first time at a client, for example, when the client powers on.
- When the client requests a specific IP address and is denied, possibly because the DHCP server dropped the lease. This might happen when a client is trying to renew its lease.
- The client previously leased an IP address, but released the lease and now requires a new lease. This can be accomplished with the IPCONFIG /RELEASE and /RENEW commands.

IP lease offer (DHCPOFFER)

If a DHCP server receives a lease request and it has a valid configuration for the client, it offers an IP address to the client. It returns the offer by way of a broadcast because the client machine doesn't yet have an IP address. The client will recognize the offer packet because it will contain the client's hardware address.

The offer packet will contain the following information: client's hardware address, offered IP address, subnet mask, length of the lease, and DHCP server's IP address.

The DHCP server will reserve the IP address it offered to the client so that it doesn't offer it again until the offer has been accepted or rejected.

If an offer is not forthcoming because there is no DHCP server available, the client will broadcast four more times (at 9, 13, and 16-second intervals plus a random time between 0 and 1,000 milliseconds). If no offer is received at this point, the client will continue to broadcast every 5 minutes.

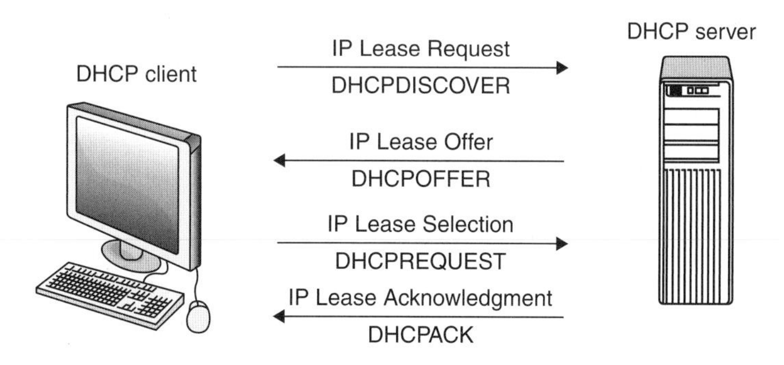

Figure 12-2: DHCP messages

IP lease selection (DHCPREQUEST)

The client will select an IP address by sending a request back to the server. Note that there may be more than one DHCP server on the network and in that case, the client received more than one offer. Normally, the client will select the first offer it receives. The selection is made by broadcasting a DHCPREQUEST to the network. A broadcast is used because all the DHCP servers have to be notified about the selection. If a directed packet was sent only to the selected DHCP server, the other DHCP servers wouldn't know that their offer had been declined. The DHCPREQUEST broadcast does, however, contain the IP address of the DHCP server whose offer was accepted.

The DHCP servers whose offer was not accepted will retract their offer and free up the IP address they had reserved for another client.

IP lease acknowledgment — successful (DHCPACK)

The DHCP server whose lease offer was accepted will finalize the offer by returning a DHCPACK packet. It will include the IP address, the lease period and other information.

When the client receives the DHCPACK packet, it initializes TCP/IP and binds the IP address to the adapter.

The client stores the TCP/IP configuration in the registry. It can be found at the following key in a Windows machine: HKEY_LOCAL_MACHINES\SYSTEM\CurrentControlSet\Services\adapter\Parameters\Tcpip

IP lease acknowledgment — unsuccessful (DHCPNACK)

Occasionally, the DHCP server cannot agree to the lease request. For example, a client is trying to renew the lease of an IP address it previously leased, but it is no longer available. This might occur if a machine is physically moved to another subnet. In this case, the request is declined by sending a negative acknowledgment, a DHCPNACK packet.

IP lease renewal

A DHCP client doesn't wait until the lease on the IP address has lapsed. Instead it tries to renew it when 50% of the lease time has elapsed. It does this by sending a DHCPREQUEST packet directly to the DHCP server that provided the lease originally. The DHCP server will confirm the extension with a DHCPACK packet that includes the new lease time.

It is common for a computer to be restarted before the lease has expired, for example, when an employee shuts down his or her computer on leaving for home after a days work. When the computer is restarted, it attempts to lease the same address that it had before.

It is possible that the DHCP server will not respond to a renewal request because it is powered down. In that case, the DHCP client will make another attempt to renew the lease when 7/8 of the lease time has expired. In this case, the renewal request is sent as a broadcast to any listening DHCP server. If the original DHCP server is available, the lease will probably be renewed. If the original DHCP server is still unavailable, another DHCP server may hear the request. Since the alternate DHCP server did not issue the original lease, it will probably decline to renew the same IP address by sending a negative acknowledgment, DHCPNACK. However, this clears the way for the new DHCP server to issue a lease for a new IP address.

If no DHCP server answers the lease renewal request, the client can still use the address for the balance of the lease period. At the end of that, the client must cease using the address, and without a new one, the use of TCP/IP and hence of communications, must be suspended.

On a Windows machine you can expect the following behavior. If the lease expires, the host starts the DHCP process from the beginning. If a DHCP server cannot be contacted, the IP address reverts to 0.0.0.0. During this period the host continues to try to contact a DHCP server. At the end of this period, approximately 5 minutes, the host self configures using APIPA. The host can now communicate with other similarly configured hosts on the same subnet, but not across the router. Every 5 minutes thereafter, the host will continue to try to contact the DHCP server. If it is successful, it will abandon the APIPA configuration and use the DHCP provided one.

SUMMARY

Section 12.1: Dynamic Host Configuration Protocol

The Dynamic Host Configuration Protocol (DHCP) is a method of configuring an IP host with its configuration parameters through the use of a DHCP server. The benefits to the network administrator are many. They include central management of IP addresses and faster and more accurate configuration changes. A laptop computer will receive a configuration change quickly if it is moved to a new subnet.

Section 12.2: DHCP Communications

DHCP uses a four phase process to configure the DHCP client. These constitute four verbs in DHCP's vocabulary, namely a lease request, lease offer, lease selection and acknowledgment. A DHCP lease is given for a set time; after that, the lease is renewed on a regular basis so that a host is never without a valid IP configuration unless the DHCP server itself fails.

Exercise 12-1

Examining the DHCP packet

DHCP packet exercise

Figure 12-3 is a screen shot of a protocol analyzer that is displaying a DHCP packet. From the information you see, answer the following questions.

1. Which DHCP packet type is being displayed? ______________________
2. Which IP address has been assigned to the client? ______________________
3. What is the client's MAC address? ______________________
4. What is the DHCP server's IP address? ______________________
5. How long is the lease? ______________________
6. What is the default gateway for the subnet? ______________________
7. What are the IP addresses for the 2 DNS servers ? ______________________

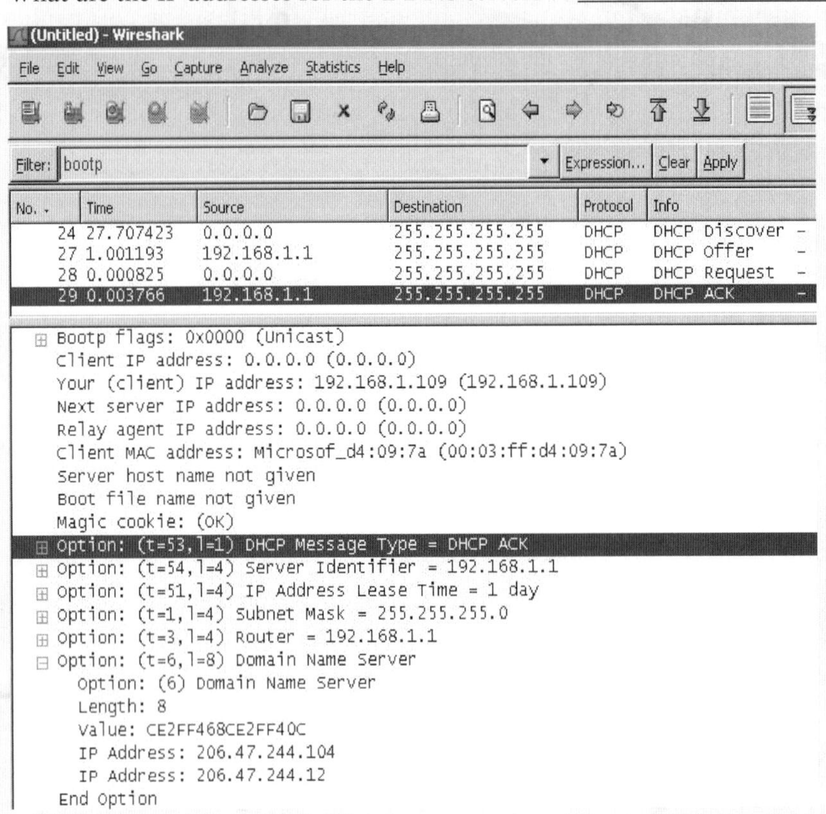

Figure 12-3: Capture of a DHCP message

Hands-on Lab 12-2

DHCP Client Setup

Setting up the client machine for DHCP

Perform this operation at the DHCP client.

1. Open Internet protocol (TCP/IP) Properties.

There are many ways to find this dialog box depending on the configuration of the machine. However, the following is the most predictable way and should work on all Windows XP machines. If there is a problem, see your instructor for guidance.

a) Click the **Start** button > Click on **My Computer** > Under **Other Places**, Right-click on **My Network Places** > Select **Properties** > Right-click **Local Area Connection** > Select **Properties** > Select **Internet Protocol (TCP/IP)** > Click the **Properties** button.
b) Click Obtain an IP Address from a DHCP server.
c) Click Yes when prompted, click OK as required to leave network properties.

Testing DHCP

1. Issue the IPCONFIG /ALL command.

    The results should indicate that your IP address was the one assigned by the DHCP server.

    The DHCP Server's IP address should be listed.

    Your lease expiry should be listed.

Hands-on Lab 12-3

Releasing and Renewing Leases

IP addresses can be released and renewed with the IPCONFIG command. This can be useful in troubleshooting.

These actions will be performed at a DHCP client computer.

Release the IP address

1. Issue the command: IPCONFIG /RELEASE
2. Issue the command: IPCONFIG /ALL
3. The IP address should be 0.0.0.0
4. PING another computer. The command should fail.

Renew the IP address

5. Issue the command: IPCONFIG /RENEW
6. Issue the command: IPCONFIG /ALL
7. You should have an IP address. It will probably be the same address that you released. A DHCP server normally renews the same IP address as long as it is within the lease period.
8. PING another computer. It should succeed.

Review Questions
DHCP

1. Which type of IP client should not use DHCP?

    a) A VoIP telephone
    b) A host that must always have the same IP address
    c) A UNIX host
    d) A host on a subnet

2. DHCP is a variation of which protocol?

    a) BOOTP
    b) ICMP
    c) DNS
    d) ARP

3. Computer B can't communicate with Computer A. Where should you go looking for the problem? See Figure 12-4.

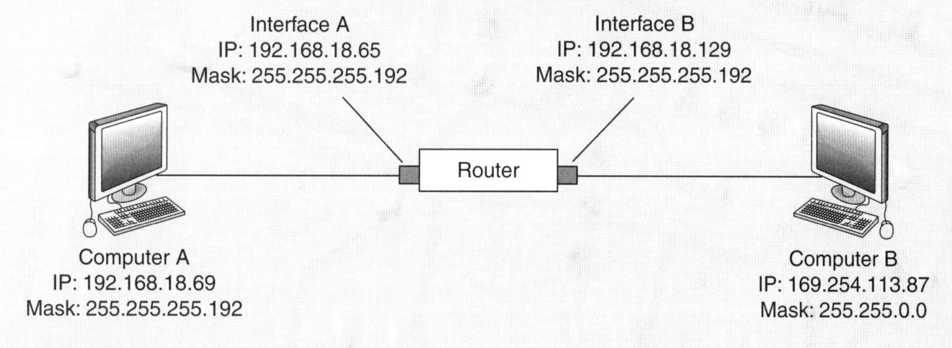

Figure 12-4:

    a) The DHCP server
    b) The router
    c) The DNS server
    d) The hosts file of Computer B

Continues next page

Review Questions: continued

4. Why does a DHCP client use a broadcast to request an IP address and configuration?

   a) The DHCP server's IP address is missing from the client's IP configuration.
   b) The DHCP server is on a different subnet than the client.
   c) The client hopes to receive multiple offers from multiple DHCP servers.
   d) The client doesn't have an IP address and can't receive a direct packet from the DHCP server.

5. At what point does the DHCP client officially have its IP configuration?

   a) When the server sends the DHCPACK packet.
   b) When the client sends the DHCPREQUEST packet.
   c) When the client sends a DHCPDISCOVER packet.
   d) When the server sends the DHCPOFFER packet.

CHAPTER

13

Name Resolution

Objectives

Upon completion of this section, the reader will:

- *Know how host names can be assigned to IP addresses in the HOSTS file.*
- *Understand the function of the domain name service (DNS).*
- *Appreciate how the resolver in the host and the DNS servers work together to resolve domain names to IP addresses.*

INTRODUCTION

IP can use IP addresses only to identify hosts and networks. Unfortunately, the human users of the network have great difficulty remembering IP addresses. However, hosts can be given friendly names and these names, along with their IP addresses, can be stored in a database. When the user needs to access a remote host by using its name, the system can look up the IP address and substitute it for the name during communications. This is called name resolution and is the topic of this chapter.

VoIP also has a need for DNS. Although IP telephones are identified by their IP addresses, people want to use a standard telephone number, an extension number, an e-mail address or some other identifier to call another person. Forwarding voice messages to e-mail is another service that needs DNS.

13.1 NAME RESOLUTION

Although IP requires IP addresses to identify hosts on the local network and the Internet, these numbers are hard for human beings to remember and use. As an alternative, IP allows names to be assigned to host computers and then used just like IP addresses within commands.

For example, instead of issuing the command PING 198.201.56.23, an operator may issue the command PING FREDS_CPU instead.

Naming schemes

Using computer names to refer to network resources can be tricky since there are two commonly used schemes. Both IP and Microsoft networks use names for machines. IP can use either a simple host name or a complex DNS name. Windows can use the simple NetBIOS name of a computer or the complex distinguished name used by Active Directory.

In any case, always keep in mind that a machine's name must be "resolved" to an IP address simply because the IP protocol can only use an IP address, never a name. The act of "resolving" means to look up an address for a name.

Two parallel universes

Both the IP and the Microsoft worlds need to use names and have evolved similar mechanisms to manage the names. However, it can be difficult to keep these systems separate. A computer can have both a simple name and a complex name in both systems. For example, in the IP world, a computer would have a host name and a fully qualified domain name (FQDN) if it is listed in DNS. For Microsoft, a computer would have a simple NetBIOS name and a distinguished name if it was listed in Active Directory.

Both systems have evolved similar mechanisms to deal with name resolution. For a simple text file used to do a lookup, IP uses the HOSTS file, while Microsoft uses the LMHOSTS file. For a database used to hold names, IP uses DNS, while Microsoft uses WINS. Table 13-1 summarizes name resolution in both systems.

To make matters even more complicated, Windows now uses IP as its standard transportation protocol; therefore expect to see and use both systems at the same time. For the balance of this book, name resolution will be restricted to the IP space.

	IP Space	Microsoft
Host Name	COMPUTER01	
NetBIOS Name		SERVER01
FQDN	www.acme.com	
Active Directory		www.acme.com
Text file	HOSTS	LMHOSTS
Name Service	DNS	WINS

Table 13-1: Comparing IP and Microsoft name spaces

13.2 THE HOSTS FILE

To illustrate the concept of name resolution, the HOSTS file is described in more detail. A simple text file is used to resolve the host name to the IP address. The file name is HOSTS. This is a reserved name and cannot be changed. Note that it is a pure text (ASCII) file and has no extension. In other words, don't add ".txt" to the name.

The format of the file is simple. Each line refers to a single host. The line starts with the IP address, at least one space and then the name of the host. The line cannot be indented and no space is allowed in front of the IP address. Although one space is required between the IP address and the name, multiple spaces are usually used for readability. A tab is acceptable. The name can be a simple computer name or a DNS-style fully qualified domain name. Additional names or aliases may also be specified on the same line if they are separated by a space from the previous name. A line that starts with a "#" is considered a remark and ignored by IP. If IP encounters a "#" in the middle of a line, it ignores the rest of the line. This is useful for adding comments.

Example lines follow.

# IP Address	name and optional aliases
127.0.0.1	localhost
144.52.6.1	sales sa #sa is an alias for sales
144.52.6.2	marketing ma
144.52.6.3	engineering en
144.52.6.4	www.support.acme.com su

Only the entry for localhost (IP address 127.0.0.1) is a default entry.

IP reads the HOSTS file sequentially and one line at a time until it finds a match for the host name. If several entries refer to the same host name, only the first one found will be used. The pound sign (#) is a remark and the line is ignored. Use it for comments. Multiple aliases are allowed for any name.

How does it work?

To illustrate how IP uses the HOSTS file, consider the following example. You issue the command "PING sales" because you can't remember the IP address for the host called "sales." IP reads the HOSTS file line by line until it encounters a name called "sales" and notes that the IP address 144.52.6.1 (from the preceding example) is associated with it. IP then continues with the command in the form PING 144.52.6.1.

The location of the HOSTS file varies with the system, but for Windows, it is found in the %system%\system32\drivers\etc folder. Use your search or find command to locate it in your version of Windows. But be careful. You may find multiple copies of the HOSTS file. Only the copy found in the etc folder is used by Windows.

Rarely used

The HOSTS file is rarely used for the following reasons:

- It is different for each computer. Because the HOSTS file is edited on each computer, it is not the same across the network.
- It is impractical to maintain since it is located on each computer on the network. There is no central management.

Hands-on Lab 13-1

Using the Hosts File

The HOSTS file allows you to set up a reference to a remote computer with a friendly name and an IP address. This exercise illustrates the concept of name resolution.

Find the HOSTS file

1. Use the Search or Find option on the Start menu to find the HOSTS file. If more than one file is found, choose the one in the windows\system32\drivers\etc folder.

Edit the HOSTS file

2. Double-click the HOSTS file. It should open up in Notepad. If you are asked which program to open it up with, choose Notepad.
3. Place the cursor at the end of the last line and press Enter. You need to start a new line.
4. Type in the IP address of a machine on your network. Possibilities include your partner's machine, a server or the default gateway. In any case, this machine has to be able to respond to a PING.
5. After the IP address, put in one or more spaces or press the TAB button and then a name for the remote computer. Do not use the actual name of the computer since this will confuse the experiment.
6. Save the file. Do not change the file name or location. You can keep Notepad open for possible troubleshooting.

Using the HOSTS file

7. PING the remote machine with its IP address. This proves that it is reachable.
8. PING it again, but this time use the host name you typed into the HOSTS file. If it works, it proves that IP resolved the name using the HOSTS file.

   If the PING fails, check the HOSTS file.
   - Is there a space before the IP address? The line must be flush left.
   - Was the HOSTS file renamed, maybe with an extension, or is it in the wrong location?
9. Close all files and exit the command prompt.

13.3 DOMAIN NAME SYSTEM

Assigning names to computers

Human beings have a tough time dealing with four-part IP addresses. Therefore a system has evolved that allows a common name to be assigned to a host. A common name may be substituted for an IP address in many TCP/IP or Internet commands.

The previous section described the HOSTS file. The common name of the host plus its IP address are listed there. Use of a local host file works only on small systems when all the names are known. But what do you do on the Internet?

Domain name system

There are millions of hosts on the Internet and they are constantly changing. Management of these names is left up to special machines known as name servers. Even so, there are too many names for individual servers to deal with. Therefore, all of the names have been grouped into domains. The domains are simply hosts from similar organizations that have been grouped together for easier management. The servers that look after the name databases are the DNS servers.

The DNS system has the following characteristics:

- It is hierarchical. The only way to efficiently organize huge amounts of data for easy lookup is a system in which you start at the top and work your way down a path that leads you ever closer to the piece of information you are looking for. Because of the efficient nature of a hierarchical system it is used in databases, directories (e.g., Active Directory, LDAP), and the file system that manages the hard drive of your computer. It is therefore no surprise that DNS also uses a hierarchical system to organize the millions of host names on the Internet.
- It is distributed. Different portions of the DNS name space, called domains, are managed by different name servers. This provides the benefits of better performance, delegation of management, and fault tolerance.
- It is centrally controlled, but not managed. Central control of the DNS database is crucial in order to prevent duplicate domain names from being registered. The nonprofit organization in charge of DNS is ICANN (the Internet Corporation for Assigned Names and Numbers). It controls the basic rules of the system but delegates the management of the domains to other parties.

DNS components

DNS uses the client/server architecture and is made up of three components: name servers, resolvers, and the DNS name space.

Name servers store and manage portions of the distributed DNS database. The servers, called DNS servers, are responsible for the domain, a section of the name space. The portion they are responsible for is called a zone and they are said to be "authoritative," meaning they control the zone.

A resolver is the client software for DNS. It resides in a machine that has TCP/IP installed. Its job is to make a request for name-to-IP-address resolution whenever a name is used to contact a server. The resolver uses UDP and TCP port 53 to contact the DNS server.

The DNS namespace is a hierarchical, distributed database that stores the IP address–name pairs. It is described next.

DNS namespace

The DNS namespace organizes all of the domain names in DNS. It is hierarchical with a minimum of three levels although it can be very deep if need be. Figure 13-1 illustrates the DNS namespace.

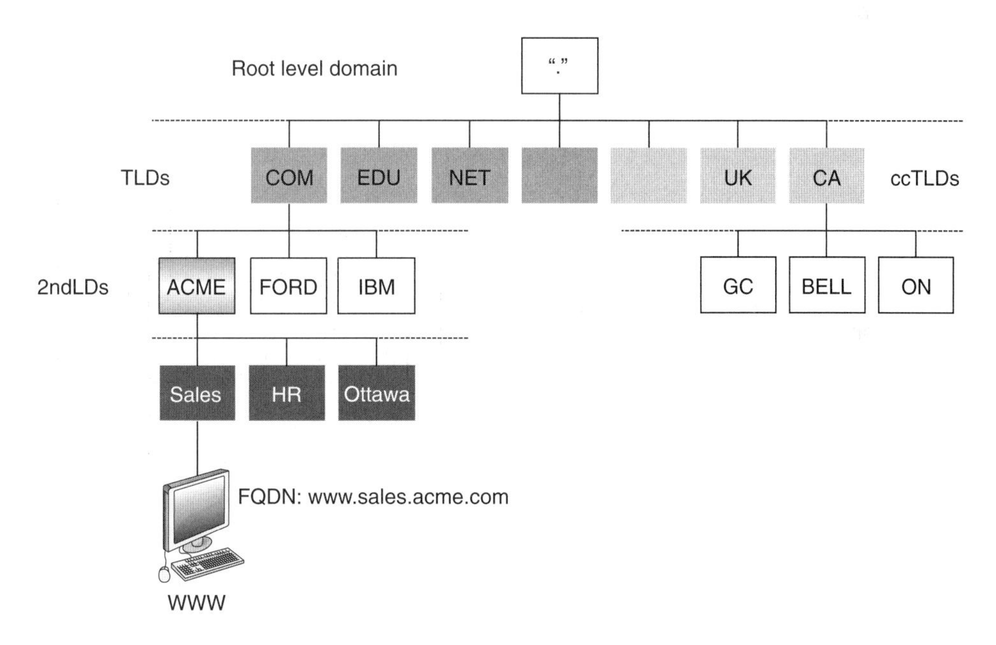

Figure 13-1: The DNS namespace

Root

The root of the namespace is the top and all other domains branch down from it. It is represented as a period, ".". When you resolve a name on the Internet, the search starts at the root.

Top-level domains

Directly below the root are the top-level domains (TLDs). These are either general or country code specific. The domain designation is the last part of a common name and can be easily recognized. Table 13-2 lists the general top-level domains.

ARPA	Special domain used for reverse lookups
COM	Commercial — anyone can register
EDU	U.S. Educational
GOV	U.S. Government
ORG	Noncommercial organization — now no restrictions
MIL	U.S. Military
NET	Network providers — now anyone can register
AERO	Airline industry
BIZ	Commercial
COOP	Cooperatives
INFO	Information providers
INT	International organization
MUSEUM	Museums
NAME	Individuals
PRO	Professionals

Table 13-2: Top-level general domains

Country code top-level domains (ccTLDs)

Countries manage their own domains and a registrar within the country is responsible for registering the domain names. The country domains, for example, would be CA for Canada, FR for France, UK for the United Kingdom, and so on. Even the United States has a country code (US).

Because names have to be unique in a domain, they have to be registered. Many companies provide a registry service. ICANN maintains the InterNIC Web page that provides a list of registrars. Each country also has a national registrar that takes care of the names within its own domain.

Second-level domain

If you want to register a DNS name, you will be registering a second-level domain. Think of a great marketing name or a name that just identifies your organization. Now contact the registrar for the top-level domain. He will do a search to make sure that it is unique. No duplicate names are allowed in a domain. If the name is unique, a small fee will register your second-level domain name. In Figure 13-1, examples of second-level domain names are Acme, Ford, and IBM.

Additional levels

You are not restricted to just two levels in a DNS infrastructure. If your business needs dictate, you can create additional levels in your DNS hierarchy, for example for a department or a geographical location. In Figure 13-1, ACME has third-level domains of Sales, HR, and Ottawa.

Fully qualified domain name (FQDN)

In order to resolve a DNS name, you need to find it in the DNS hierarchy. The FQDN is the location in DNS where the host is located. Once you locate the host, you can find its IP address. In Figure 13-1, the host is named "www," but its FQDN is www.sales.acme.com. It is crucial that you use the FQDN when issuing TCP/IP commands. If you PING www, you will get nowhere; however, if you PING www .sales.acme.com, you will be successful.

Ahem! Don't PING www.sales.acme.com. It is fictitious.

13.4 NAME RESOLUTION WITH DNS

Imagine that you want to go to the website of the United Nations and you type www.un.org into the location/address bar of your Web browser. Since this name can't be used by IP, it will need to be resolved to an IP address before the browser can proceed. How is the name resolution accomplished?

Name resolution proceeds through a query to DNS servers. Two types of queries can be issued: recursive and iterative.

Recursive queries

A recursive query asks for a complete answer. If a complete answer can't be provided, an error is returned. The most common error is that the name does not exist. A host typically sends out a recursive query, although a DNS server can also use one if it is acting as a forwarder and forwarding requests to another DNS server. If the recursive query fails, the host will present an error message to the user and then stop. It will not try another DNS server. It is common practice to list the IP addresses of multiple DNS servers in the IP configuration of a host for fault tolerance purposes. However, the host will try a second DNS server only if there is no response from the first one, not if the query for a name returns a "name not found" message.

Iterative queries

In response to an iterative query, a DNS server can provide the IP address requested or it can provide a referral to another DNS server that may have the complete or partial answer. The iterative query is normally used by one DNS server when it is querying another. Iterative queries are required because the DNS server doesn't know at first where the record for the domain name is stored and must find it by using a multistep process. Refer to Figure 13-1 as we step through the process to resolve the FQDN www.un.org.

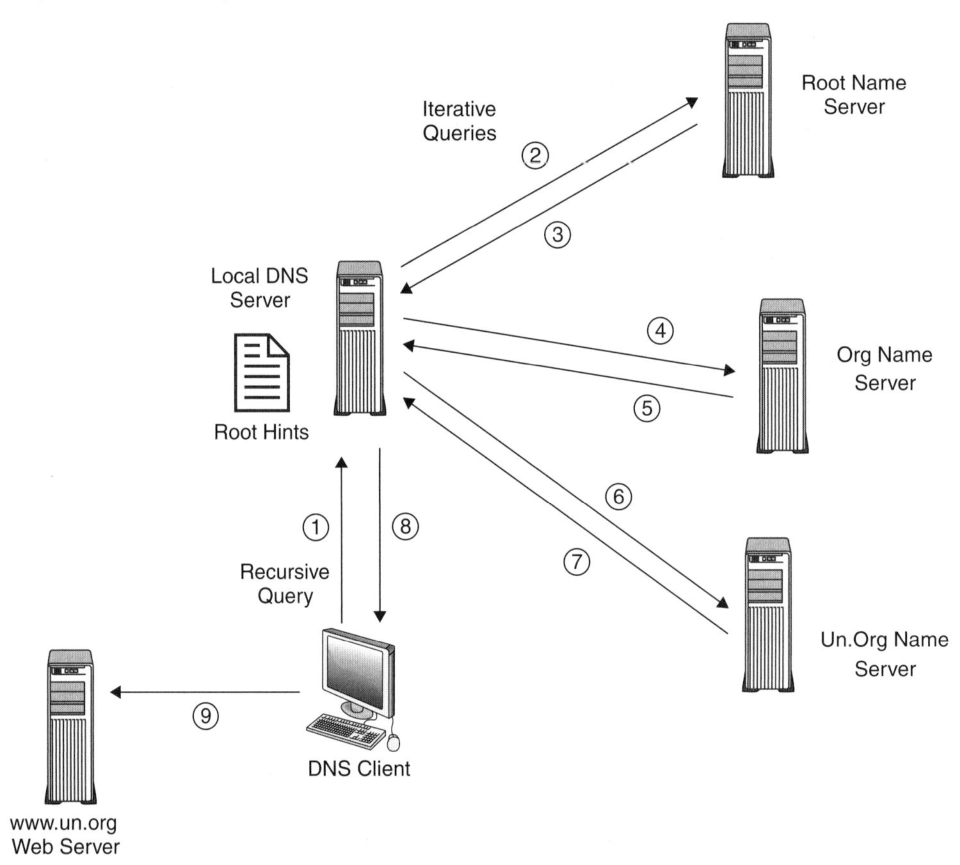

Figure 13-2: Resolving a name using recursive and iterative queries

Stepping through the name resolution process

The numbers in Figure 13-2 correspond to the following steps.

1. The user at the DNS client types www.un.org into the Web browser. The resolver takes over and checks the local cache for the name.

Assuming that it can't find it, it then issues a recursive query for that name to the DNS server whose IP address is listed in the IP configuration of the client. The local DNS server is now responsible for resolving the name and can't ask the client to try another DNS server.
2. The local DNS server first searches its own cache and then any of its own zone files but finds that www.un.org is not listed. The server must now start its search at the root of the DNS namespace. It reads the Root Hints, which list the root servers, and sends an iterative query to the first root server on the list.
3. The root server doesn't have knowledge of www.un.org, but it does send back the IP address of the server that services the org top level domain.
4. The local DNS server sends an iterative query for www.un.org to the name server that holds the org record.
5. The org name server doesn't know the IP address of www.un.org, but it does send back the IP address for the name server that services un.org.
6. The local DNS server sends an iterative query for www.un.org to the name server that holds the un.org record.
7. The name server that holds the un.org record returns the IP address of www.un.org to the local DNS server.
8. The local DNS server now replies to the recursive query from the client with the IP address of www.un.org. It also places this information into cache in case it has to answer a query for the same name.
9. The resolver places the record for www.un.org with its IP address into the local cache and provides the IP address to IP, which now contacts the website using the HTTP protocol.

If this process seems unnecessarily complicated, remember that there are hundreds of millions of names registered in the DNS namespace and yet the name was resolved in under a second. It simply works.

Forward and reverse lookups

Normally you know the name of the server you want to contact and need to look up its IP address. This is called a forward lookup. Occasionally, you have an IP address and want to see if there is a name registered with that address listed in the DNS database. This is called a reverse lookup or an inverse query.

Of what use is doing a reverse lookup? There are some uses but they are specialized. For example, you are receiving malicious packets because someone is trying to break into your network. By capturing these packets, you can see the IP address of the source and as a way of backtracking, can do a reverse lookup in case you can identify the domain they came from. Another potential use is spam filtering. You can instruct your e-mail server to do a reverse lookup when mail is received. If the reverse lookup resolves to a legitimate domain, accept the mail; otherwise, discard the mail.

In-addr.arpa

The problem for a reverse lookup is that in order to get an accurate result, you would need to search through every domain on the Internet looking for the IP address listed for your target host. Clearly, that is impossible. Instead a special top-level domain exists named "arpa" that is used for these searches. In case "arpa" doesn't ring any bells, it is an acronym for Advanced Research Projects Agency and was the network that eventually became the Internet.

If arpa is the top-level domain, then the inverse query is for a record in the form "reversed address.in-addr.arpa." Because IP addresses become more specific as you go from left to right, while FQDNs become less specific as you go from left to right, the IP address in the query must be reversed. An example illustrates this point. You want to do a reverse lookup for the IP address 192.168.6.23. The query will be for the record 23.6.168.19.in-addr.arpa and the domain that holds this record is 6.168.192.in-addr.arpa.

Creating these records, which are called Pointer records, is the job of the DNS administrator. Luckily for you, you can just use the Nslookup utility to do a reverse lookup and you only need to specify a simple IP address.

SUMMARY

Section 13.1: Name Resolution

Although IP requires IP addresses to identify hosts on the network, these numbers are hard for humans to remember and use. As an alternative, IP allows names to be assigned to host computers and then used just like IP addresses within commands.

For example, instead of issuing the command "PING 198.201.56.23", an operator may issue the command "PING FREDS_CPU" instead.

Section 13.2: The HOSTS File

A simple text file named "HOSTS" is used to resolve the name to the IP address. Use of a local host file works only on small systems when all the names are known.

Section 13.3: Domain Name System

There are millions of hosts on the Internet and they are constantly changing. Management of these names is left up to special machines known as name servers. All of the names have been grouped into domains. The domains are simply hosts from similar organizations that have been grouped together for easier management. The servers that look after the name databases are the DNS servers. The domain designation is the last part of a common name and can be easily recognized. Examples of top-level domains include COM, EDU, GOV, and ORG. Examples of country code domains include CA for Canada and FR for France.

Section 13.4: Name Resolution with DNS

Two types of queries are used with DNS. With a recursive query, the DNS server must provide the name resolution or an error message such as "the name doesn't exist." With an iterative query, the name server can also provide a referral to another server.

A forward query resolves a name to an IP address. An inverse or reverse query resolves an IP address to a name.

Hands-on Lab 13-2

Working with the Resolver

The DNS client is called the *resolver*. It is built into the TCP/IP stack. In this lab, you will examine the DNS configuration of your host. There are three possible configurations:

a) The IP address of the DNS server has been statically configured.
b) The IP address of the DNS servers is being provided by a DHCP server.
c) No DNS server is available to your host.

Viewing the DNS client configuration

1. Open up the **Internet Protocol (TCP/IP)** properties.

   The bottom portion of the window is devoted to DNS.

   Has an IP address (or two) been configured for this client? If no address is listed, your host may receive its configuration from DHCP.

2. Open up a command prompt.

   Click the **Start** button > **Run. . .** > type in **CMD**

   Type in **IPCONFIG /ALL**

   Are the IP addresses of any DNS servers listed?

3. Your computer also has a simple host name. Find it as follows:

   Type in **HOSTNAME**

4. PING your computer using its hostname name.

5. PING the DNS server to see if it is available.

6. Write down the following:

   IP address of the DNS server ______________________________

   IP address of the second DNS server (if it exists) ________________

   Your computer's hostname _______________________________

Hands-on Lab 13-3

The DNS Cache

The DNS resolver caches successful lookups in memory. If you need to resolve a name twice in a short period of time, the resolver can use the cache to avoid contacting the DNS server again, thereby speeding up performance. The period during which the name is maintained in the cache is known as the time to live (TTL). In the following exercise, you will view the TTL and the IP address of the DNS names in your host's cache.

1. From the command prompt, issue the following command.

   Type in **IPCONFIG /DISPLAYDNS**

Examine the cache. The contents of the cache will vary with the destinations you have contacted today using TCP/IP. Identify the name (host or DNS), the IP address and the TTL for each entry.

2. If you want to clean out the cache, use the following command:

   Type in **IPCONFIG /FLUSHDNS**

3. Examine the DNS cache again. Notice that the only remaining entries are for your localhost.

The contents of the DNS cache vary greatly with the Internet sites that you visit. If you browse the World Wide Web and then repeat these commands, you will discover how important DNS really is.

Hands-on Lab 13-4

NSLOOKUP

If you need to query the DNS database, use the NSLOOKUP program. It will provide the IP address for a domain name. Some implementations of NSLOOKUP will provide additional information such as a contact name or a mailing address for the domain. NSLOOKUP is the primary troubleshooting tool for DNS and a utility that you need to master.

Microsoft provides an implementation of NSLOOKUP with all versions of Windows. For this lab, you need an available DNS server. In a classroom setting, a DNS server should be available. Your instructor will provide guidance. If you are not in a classroom but are connected to the Internet, your ISP already makes a DNS server available.

1. Open up a command prompt.

   Click the **Start** button > **Run. . .** > type in **CMD**

2. Resolve a domain name with the NSLOOKUP command. For the following command, *domain_name* can be any DNS name. In the classroom, your instructor will provide guidance. If connected to the Internet, use any name that you can contact using the World Wide Web, such as www.whitehouse.gov.

   Type in **NSLOOKUP** *domain_name*

The IP address of this server should be returned.

3. NSLOOKUP will also do a reverse lookup if you provide an IP address. Be aware, however, that you will get a result only if the IP address actually points to a name in DNS. For the following command, try an IP address that was returned by a previous command.

   Type in **NSLOOKUP** *ip_address*

4. For troubleshooting purposes, it can be very useful to look up or confirm the IP address of the DNS server that you are using. Typing NSLOOKUP without any parameters will provide this information.

   Type in **NSLOOKUP**

The IP address of the DNS server will be returned.

If you type NSLOOKUP by itself, it goes into the NSLOOKUP console. You can identify this fact because the prompt is now the greater-than symbol, ">". At this point, you no longer need to type in NSLOOKUP itself. Use HELP to see the commands. To leave the NSLOOKUP console, type in "Exit".

Review Questions DNS

1. Which of the following is a FQDN?

   a) www
   b) www.senecac
   c) www.senecac.on.ca
   d) www.senecac.on.ca/home.htm

2. Your security software warns you that someone is trying to break into your computer by scanning it from the Internet. Your software also shows you the IP address of the computer doing the scanning. You wonder if this computer had a registered domain name. How would you try to find this information?

   a) Do a forward lookup.
   b) Do a DNS search.
   c) Search the domain database.
   d) Do a reverse lookup.

3. You need to create a reverse lookup domain for acme.com. The IP address of the name server for acme.com is 207.81.15.7. The reverse lookup zone will be

   a) 81.207.in-addr.arpa
   b) 7.15.81.207.in-addr.acme
   c) 15.81.207.in-addr.arpa
   d) 15.81.207.acme.com

4. A small company (A) recently merged with a large company (B). Company A used static name resolution. The servers from company A were moved to company B's server farm and readdressed. Company B uses DNS to provide name resolution.

   Users at company A can reach other servers in the server farm but cannot reach the servers that were moved. What should you do to correct the problem?

   a) Delete the HOSTS file.
   b) Correct the LMHOSTS file.
   c) Correct the NETWORKS file.
   d) Create the static NETSTAT entries.

   Continues next page

Review Questions: continued

5. When resolving a DNS name, resolvers and DNS servers typically use different types of queries. Which of the following statements is most correct?

   a) Resolvers use recursive queries, servers use iterative queries.
   b) Resolvers use iterative queries, servers use recursive queries.
   c) Resolvers use recursive queries, servers use iterative queries unless they are a forwarder.
   d) Resolvers use iterative queries, servers use recursive queries unless they are a forwarder.

SECTION

3

Voice Over IP Technical Details

CHAPTER

14

How IP Handles Voice

Objectives

Upon completion of this section, the reader will:

- *Have the big picture of the various protocols involved with transmitting voice over IP networks.*
- *Understand the function of the real-time transport protocol (RTP) in the transmission of voice traffic.*

INTRODUCTION

Traditional voice technology uses dedicated circuits or time division multiplexing (TDM) for transmission of voice. TDM requires that the analog voice signal be converted into a digital signal. Analog or digital, the voice signal has a guaranteed bandwidth for transmission and after a century of fine tuning, the quality of the public system telephone network is excellent. Transmitting voice over an IP network has its own set of challenges. For example, the voice signal has to be divided into discrete sections called packets. In addition, an IP network provides no guarantees for bandwidth. This chapter reviews the protocols of TCP/IP paying special attention to those that VoIP uses. This chapter also looks at the technical details of the real-time transport protocol that carries the voice data.

14.1 TCP/IP AND VOICE PROTOCOLS

In earlier chapters we examined the technology of voice over IP. Then we examined how the TCP/IP protocol worked. This section reviews how the VoIP and the TCP/IP protocols fit together. Refer to Figure 14-1 for this discussion.

VoIP uses the standard TCP/IP protocols including IP, ICMP, TCP, and UDP plus additional protocols as follows:

Signaling protocols: H.323, SIP/SDP, MGCP/Megaco
Quality of service: RSVP, RTCP, IP TOS, DiffServ, MPLS
Media transport: RTP

VoIP makes complete use of the underlying TCP/IP infrastructure. This includes IP at the network layer, which has the job of identifying hosts with IP addresses and routing packets to them. It includes ICMP, which acts as the messenger service for IP and produces the error messages that are needed to troubleshoot the network. TCP guarantees the delivery of packets and UDP provides a best efforts service. Additional protocols are needed as well. These provide functions in signaling, quality of service, and media transport.

VoIP protocols

The protocols needed to make VoIP function are illustrated in Figure 14-1.

- Signaling protocols. The signaling protocols for clients are H.323 and SIP. SIP is usually partnered with SDP. MGCP/Megaco is used by the gateways to communicate with each other.
- Quality of service. The protocols used to provide QoS include RSVP, RTCP, IP TOS, DiffServ, MPLS, and 802.1q.
- Media transport. RTP, the real-time transport protocol, carries the voice conversation.

Note that both TCP and UDP are used alongside each other, but they carry the protocols that best make use of their strengths. Figure 14-1 diagrams all of the protocols involved with VoIP and divides them in two ways. They are divided into Signaling, Quality of Service, and Media Transport categories. In addition, they are placed in the appropriate layers according to the OSI and DoD communication models.

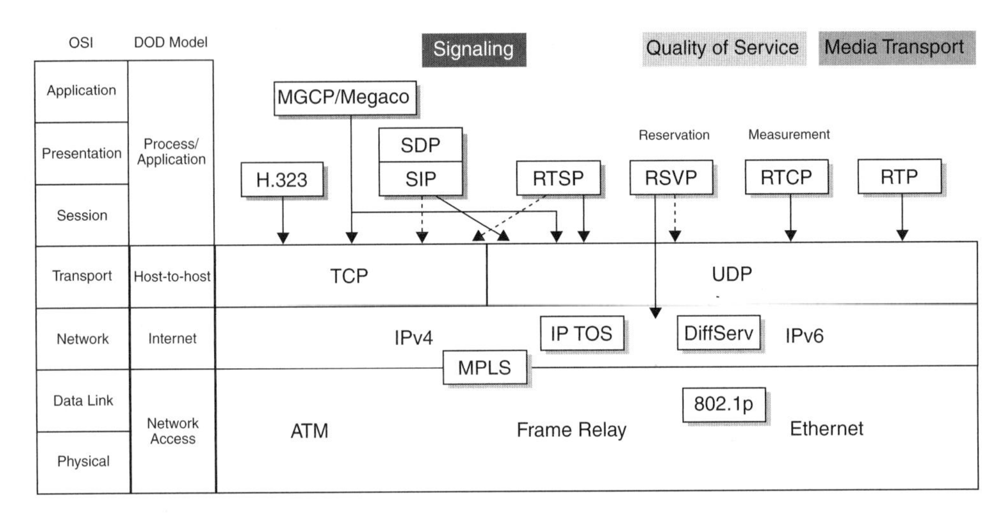

Figure 14-1: The VoIP protocols

Quiz 14-1

VoIP Protocols

VoIP uses many protocols to make a telephone call. The following is a list of many of them. You need to classify them according to their role. Use the following legend to do so. Refer to Figure 14-1 if you need help. Note that some protocols can be classified in two ways.

S - Signaling

Q - Quality of service

M - Media transport

T - Transport layer

N - Network layer

D - Data Link layer

____ DiffServ

____ MGCP/Megaco

____ RSVP

____ Frame Relay

____ IP

____ SIP/SDP

____ UDP

____ 802.1p

____ IP TOS

____ RTP

____ Ethernet

____ H.323

____ ATM

____ RTSP

____ MPLS

____ RTCP

14.2 REAL-TIME TRANSPORT PROTOCOL

The real-time transport protocol (RTP) is used for voice over IP because of the weaknesses of TCP and UDP. RTP is defined in RFC 3550.

Why not TCP?

TCP is used by the control protocols of VoIP, namely, H.323 and SIP. However, it is not used to transmit the voice data itself. TCP is a connection-oriented protocol that promises to deliver the data. If a packet is lost, retransmission of the data is required. Retransmission introduces a lag in the reception of the data. This is not important when transmitting computer data but does produce a noticeable degradation in quality of voice transmission.

Why UDP needs help

Since TCP is unsuitable, UDP is the only alternative. UDP uses a best efforts basis for transmitting data. If a packet is lost, it stays lost. Losing a single packet won't affect voice quality. However, UDP is not ideal either. The major problem with UDP is that it does not track the sequence of packets. If packets arrive out of turn, UDP does not have a mechanism for putting them in the right order. The synchronization and acknowledge fields provide this service in TCP.

What RTP does

RTP corrects the deficiency of UDP by adding two important capabilities as far as VoIP is concerned.

Sequencing

RTP has sequence and timestamp fields that are used to reorder packets if they arrive out of turn.

Payload type identification

RTP has a payload type field that can identify the nature of the payload. As used by VoIP, this field can identify the codec used for the voice data.

The key to understanding RTP is to appreciate that it is not purely a VoIP technology, but a technology that can be used for any real-time application. These include video, broadcasts, and music, in fact, any streaming content. Some applications need multiple streams. For example, a broadcast of a sports game would need video, sound, and maybe a sportcaster's voice-over commentary. Although a mixer would combine these separate sources into one real-time data stream, they would need to be separated out before being presented to the viewer. These separate sources of data are referred to as "contributing sources" and will be mentioned again when the fields of the RTP header are described.

Common requirements for all real-time streams of data include showing up at regularly spaced intervals, being identified correctly, and being in the right order. Following is a description of the RTP header structure and a discussion of the fields in the header.

RTP STRUCTURE – THE FIELDS

The fields in the RTP header are illustrated in Figure 14-2 and are described next. There is a minimum of 12 octets of data in the RTP header up to the Synchronization Source field. Optionally, the header may contain contributing source data, in which case the header may increase by anywhere from 4 to 60 extra octets of data.

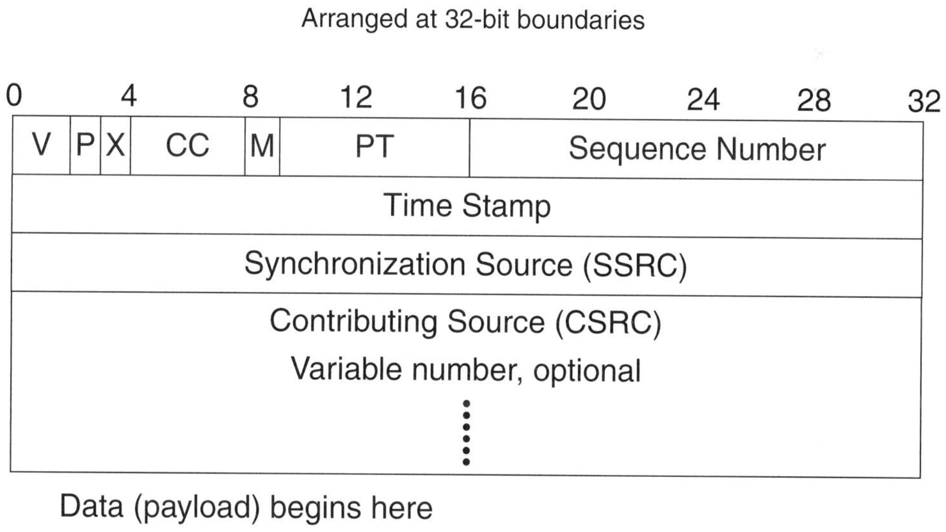

Arranged at 32-bit boundaries

0 4 8 12 16 20 24 28 32

V P X CC M PT Sequence Number

Time Stamp

Synchronization Source (SSRC)

Contributing Source (CSRC)
Variable number, optional

Data (payload) begins here

Figure 14-2: RTP header structure

RTP version number (V)

The version is a two-bit field and holds the version of RTP being used. The version defined by the current specification is 2.

Padding (P)

The padding field is one bit in length. If the padding bit is set, there are one or more octets at the end of the packet that are not part of the payload. The padding is used by some encryption algorithms that require fixed block sizes.

Extension (X)

The extension field is one bit in length. If the extension bit is set, the fixed header is followed by exactly one header extension. This extension mechanism enables implementations to add information to the RTP Header.

CSRC Count (CC)

The CSCR Count field is four bits. CSRC is an acronym for Contributing SouRCe. This is a source of an RTP stream of data. In some circumstances, more than one stream of data is combined. For example, an audio conference between four people would generate four different streams of data. The four streams are combined by a "mixer" to produce a new stream.

The CSRC count (CC) is the number of CSRC identifiers that follow the fixed header. If the CSRC count is zero, the synchronization source is the source of the payload, in other words, there is only one data stream. If the CC field is 0, there is no optional CSRC field.

Marker (M)

The Marker field is one bit in length and is defined by the particular media profile. It can be used to mark significant events in the packet stream. For voice packets, the marker bit can indicate the beginning of a "talk spurt."

Payload Type (PT)

The Payload Type field is seven bits and is an index into a media profile table that describes the payload format. The payload mappings for audio and video are specified in RFC 1890. For VoIP, the payload type identifies the encoding method. Table 14-1 lists some of the PT values for encoding used by VoIP.

Payload Type (PT)	Encoding Name
0	PCMU (G.711)
3	GSM
4	ACELP (G.723)
8	PCMA (G.711)
9	SB-ADPCM (G722)
15	LD-CELP (G.728)
18	CS-CELP (G.729)

Table 14-1: RTP Payload Types for VoIP

Figure 14-3 shows a decode of a sample RTP packet. Find the payload type. Note that the value of binary 0001000, decimal 8, is decoded as ITU-T G.711 PCMA.

Sequence Number

The sequence number is a 16-bit field. It is a unique packet number that identifies this packet's position in the sequence of packets. The sequence number is incremented by one for each packet sent. The sequence number is primarily used to detect lost packets.

Timestamp

The timestamp reflects the sampling instant of the first octet in the payload. Several consecutive packets can have the same timestamp if they are logically generated at the same time, for example, if they are all part of the same video frame. This field is 32 bits in length. The timestamp for subsequent packets of the same stream increases by the time covered by the packet. This field is important in placing the packets in the correct timing order.

Synchronization Source (SSRC)

The SSRC field is 32 bits and it identifies the synchronization source. If the CSRC count is zero, the payload source is the synchronization source. If the CSRC count is nonzero, the SSRC identifies the mixer.

Contributing Source (CSRC)

The CSRC field identifies the contributing sources for the payload. The number of contributing sources is indicated by the CSRC Count field; there can be up to 15 contributing sources. If there are multiple contributing sources, the payload is the mixed data from those sources. Each CSRC field is 32 bits.

UDP Port Number

RTP and its companion protocol, RTCP, use UDP port numbers. However, they are dynamically negotiated during call setup and not fixed. RTP uses an even number and RTCP will use the next higher (odd) number. One reason RTP ports are dynamically assigned is that it is common for an audio visual application to require multiple RTP streams. One fixed UDP port cannot be shared in the same host.

Dynamic port numbers present a problem for the management of firewalls. The administrator can't open a port on the firewall if she doesn't know what it is. A range of ports can be opened, but the more ports opened, the less secure the firewall becomes.

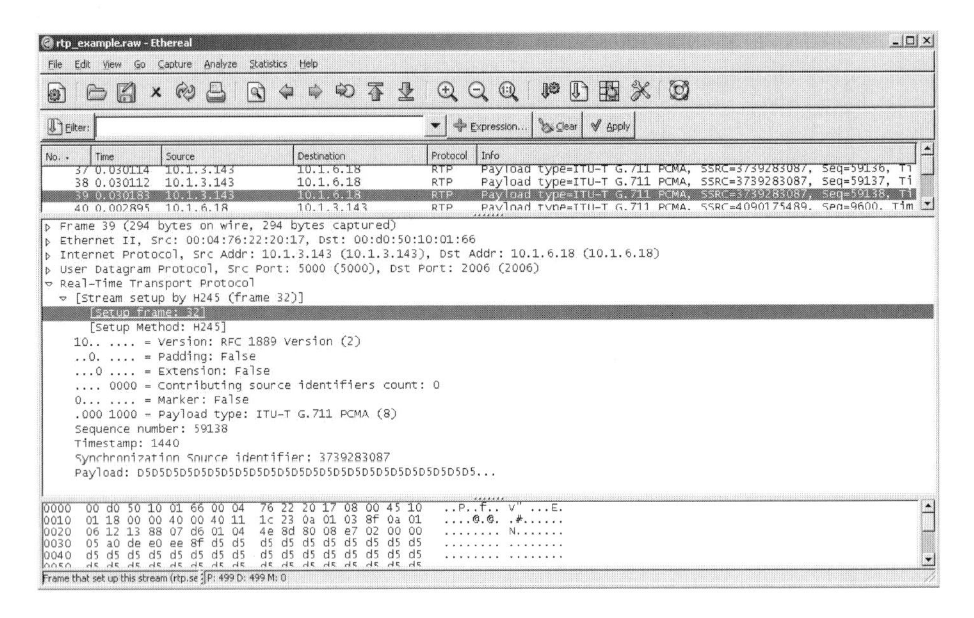

Figure 14-3: Decode of RTP packet

14.3 REAL-TIME CONTROL PROTOCOL

The real-time control protocol (RTCP) is implemented with RTP in the same protocol stack and is described in RFC 3550. RTCP uses periodic transmission of control packets to all participants in the session, using the same distribution mechanism as the data packets. The underlying protocol must provide multiplexing of the data and control packets, for example using separate port numbers with UDP. RTP normally uses an even-numbered UDP port number and RTCP will use the next higher odd-numbered UDP port. RTCP may provide different services depending on the situation.

Feedback on service

RTCP provides feedback on the quality of the RTP service. Appreciate that RTP does not guarantee high-quality, reliable service. That depends on the underlying protocols and physical service. RTCP is used in multicast situations and participants will send and receive RTCP packets to test the response times. These are specific RTCP packets called sender and receiver reports. The timestamp fields in these packets are used to calculate response times.

Source identification

RTCP carries a persistent transport-level identifier for an RTP source called the canonical name, or CNAME, which may be used to associate multiple data streams from a given participant in a set of related RTP sessions, for example to synchronize audio and video. Why is this function needed when the RTP packet identifies the source in the synchronization source (SSRC) field? The reason is that the SSRC identifier may change if a conflict is detected or if a program is restarted. RTCP uses a source description (SDES) packet for this function.

RTCP transmission interval

RTP was designed for conferencing situations where the number of participants could range from a few to thousands. If RTCP sent control packets at a fixed rate, for example as a percentage of RTP packets, it would consume more bandwidth as the number of participants grew. Therefore, RTCP uses an algorithm which takes into account the number of participants as well as the network bandwidth for determining the number of control packets that it transmits. On a VoIP system, you will see RTCP packets only at infrequent intervals if you are monitoring the traffic with a protocol analyzer.

RTCP can also provide information about participants in a conferencing situation. This is useful when participants join or leave the conference dynamically. A site can send the RTCP BYE packet when it leaves a conference.

SUMMARY

Section 14.1: TCP/IP and Voice Protocols

This section looks at the TCP/IP and VoIP protocols and their place in the OSI model.

Section 14.2: Real-Time Transport Protocol

This section looks at the RTP protocol, which carries the voice signal. RTP is used because voice is a real-time data stream. RTP provides identification and the sequencing of packets that UDP lacks.

Section 14.3: Real-Time Control Protocol

RTCP provides the feedback on the quality of the RTP service that VoIP needs for a high-quality telephone conversation.

Hands-on Lab 14-1

Installing a SIP Server

In order to explore VoIP communications further, hands-on experience is beneficial. For this exercise, you will be installing a SIP server and softphone software. Future chapters will take advantage of the VoIP setup you create in this exercise and you will be able to examine or demonstrate important concepts that will be introduced later. The instructions in this chapter are flexible and try to cover both a classroom setup and individual exploration on a home computer. However, a partner is required if you expect to make telephone calls.

In order to make an in-class telephone call, you will need a microphone and headset. However, if your classroom or home computer is not equipped this way, you can still install the software and make the telephone ring through your speakers.

The software you will install is as follows:

The SIP server: 3CX Phone System for Windows (Free edition)
Download: http://www.3cx.com/

Softphone software: 3CX VoIP phone for Windows
Download: http://www.3cx.com/
Download at the same time as you download 3CX Phone System

Softphone software: X-Lite
Download: http://www.counterpath.com/x-lite.html

Softphone software: SJphone
Download: http://www.sjlabs.com/sjp.html

3CX Phone System for Windows is unusual because it is a SIP registration server and SIP proxy server that works on the Windows platform and that is free. There is a larger selection of SIP servers that work on the Linux platform, the best known of which is Asterisk.

X-Lite, SJphone, and 3CX Phone are universal SIP clients that are not tied to any telephony service on the Internet and they are free. All three softphones support the SIP protocol but only SJphone supports the H.323 protocol.

Install 3CX Phone System

1. The instructor has made the 3CX Phone System installation file available or you have downloaded it from the Internet. The following instructions are general and there may be variations depending on the version that you are installing.

2. Start the installation.

3. You will be prompted to provide some information, but most of it will not apply since this is an experimental installation and not connected to any other system. Leave any defaults as the installation progresses and the installation is done.

   Here are the prompts you can expect to see as you progress through the installation.

Installation folder	Leave default
Start Menu Shortcuts Folder	Leave default
SIP domain	Leave default
Digits in extension number	Leave default (3 digits)
Username, Password, Password confirm	You can use your own values but be sure and write them down.
SMTP server	Smtp.mydomain.com (you can choose your own value)
E-mail address	admin@mydomain.com (you can choose your own value)

   It is important that you write down the administrator credentials. You will need this when you log on to the console to manage the system.

4. Start the 3CX console

   a) The installation process will offer to start the management console when the installation is finished.

   or

   b) **Start** > **All programs** > **3CX** > **Phone System** > **3CX PhoneSystem**

   or

   c) Open up your Web browser and use the URL http://localhost:5481.

   Log in with the administrator's credentials.

5. You will now register some users by adding their telephone extensions. Telephone extensions must be added to 3CX before any phone can register with the SIP server.

On the left-hand menu under **Extensions**, click **Add**

Hands-on Lab 14-1 cont.

Add the information for some users. In order to cooperate with other members of the class, add some accounts for them as well. They will need to know their extension and password when they attempt to register with your server.

The only required field is the first name. The system can provide the extension, user ID, and password. Notice that by default these are all the same value. Of course, you can change any value you want.

6. As minimal as this is, it is all you need to start using the system.

Check the Line status

Phone System > Line Status

This screen shows the users (extensions) that you have added and their status. A red circle indicates a status of Not Registered. After a phone is registered with the SIP server, the circle turns green. An orange circle indicates that the user is in a call.

Install X-Lite Softphone

7. The instructor has made the X-Lite softphone software installation file available or you have downloaded it from the Internet.

8. Start the installation. Use all defaults. At the end of the installation, the computer needs to be rebooted. Allow the reboot when asked.

9. After your computer restarts, X-Lite should automatically start. This was one of the options you selected when you installed the program. You can always start the program from the Start menu.

10. A screen opens up to allow you to create a SIP account. You use the account when you start your softphone and register with the SIP server.

    Click the **Add** button and fill in the account properties as follows:

    Display name: Your name

    The following fields must match exactly the information that you used to set up one of the extensions on the SIP server.

        User name: Use the extension number.
        Password: Must be the password for the account.
        Authorization user name: Use the extension number.

Domain: Use the IP address of the SIP server, which is your computer. You can use 127.0.0.1

11. Click the **OK** button, click the **Close** button. Watch the screen on the softphone for messages. It should progress through the initialization, then registering, and finally finish with a Ready and display the username.

Troubleshooting: If you did not receive the Ready message, review the user properties screen and confirm that all of the values are correct. It is crucial that the values match the properties that are set up for the extension in the SIP server.

12. View the status of the extension in the 3CX manager console. If the line status of the extension shows green, than you have additional evidence that the softphone has registered properly. However, because the console is slow to update, the status may still show as red. Making a phone call will also change the status.
13. Using X-Lite, dial another extension that you have configured on the SIP server. You should receive a call failed message plus a voice recording should tell you that the person is unavailable to receive a call. This indicates that you have configured the SIP server and the softphone successfully.

Using the Classroom Call Manager

Since all of your fellow students are installing their own copy of 3CX phone system, you won't be able to call each other. In order to do so, you need to use the same call manager. For the remainder of the exercise, you will use the instructor's copy of 3CX phone system.

14. Use your Web browser and type the following into the Address field: http://192.168.1.200:5481

    Note: 192.168.1.200 is the instructor's computer. If this address is incorrect, substitute the correct IP address.
15. In order to log in to the instructor's copy of 3CX, you will need the administrator's name and password for the instructor's machine. You are now attached to the console of 3CX on the instructor's machine.

    Add an extension for yourself.
16. You must change the SIP server that your copy of X-Lite is using. Change the user properties to reflect the extension you just created on the

instructor's SIP server. The domain must be the IP address of the instructor's computer. When you save this information, your softphone will register with the new SIP server.

17. You should now be able to telephone any other student in the classroom as long as you know the extension of his or her phone.

Review Questions How IP Handles Voice

1. Which port number is used by RTP?

   a) 80
   b) 546
   c) Port numbers are dynamically assigned.
   d) RTP doesn't use port numbers.

2. TCP is not used to transmit voice because

   a) retransmission of lost data degrades voice quality.
   b) it is too slow.
   c) it provides guaranteed delivery.
   d) it is a connection-oriented protocol.

3. Which weakness of UDP is rectified by RTP?

   a) The inability to multiplex more than one data stream
   b) The inability to place packets in the right order
   c) The inability to correct errors in the data
   d) The inability to set up a connection

4. By using IP, UDP and RTP, a minimum of how many bytes of overhead is added to every packet that carries voice data?

   a) 20
   b) 28
   c) 24
   d) 40

5. How does RTP receive feedback on the quality of service of voice transmission?

   a) 802.1p
   b) diffserv
   c) RTCP
   d) MPLS

CHAPTER

15

Voice to Digital

Objectives

Upon completion of this section, the reader will:

- *Understand the process of sampling a voice signal and quantizing it into discrete digital values.*
- *Appreciate how waveform coding and hybrid voice coding perform the encoding function and how they differ from each other.*
- *Know which factors affect the choice of codec, including voice quality, processing requirements, and bandwidth.*
- *Be able to calculate VoIP bandwidth and determine how many voice conversations a system can support.*

INTRODUCTION

The job of converting an analog voice signal into an electronic digital signal is performed by the codec. This is a complicated task and there are many variations. The system designer or administrator needs to decide on which codec to use depending on voice quality desired and available bandwidth. This chapter looks at the technology involved and the characteristics of the various codecs.

VOICE TO DIGITAL

Acoustic signals have been turned into electrical signals since the time of Alexander Graham Bell. But for most of telephone history, these electrical signals were analog and an analog signal can closely model the human voice. With modern digital systems, the infinite variety of voice has to be converted into discrete digital signals. How is this done?

The steps required include sampling, quantizing and companding, encoding, and packetizing. These are the topics covered in this chapter, each with its own section.

There are two approaches to converting voice into digital signals. Sampling the speech is the start of both. The simple approach just converts the samples into digital data and then sends it. However, the data stream can't be compressed below 16 Kbps. Therefore, there is room for a second technique using voice coders, which can compress a voice stream down to 4 Kbps and less. This chapter will explore both approaches.

Encoding the human voice into an electrical signal is done by a codec (code/decode). You may wonder why a whole chapter is devoted to the topic. When you design a VoIP system, you will need to choose the telephones, and a telephone will support only specific codecs. Naturally, the source and destination telephones will need to support at least one codec in common.

Selecting an appropriate codec is complicated and consideration must be given to the factors depicted in Figure 15-1. On the one hand, you want to choose a codec that produces a narrow data stream so that you can fit as many conversations on a line as possible. On the other hand, a codec that compresses the conversation into a very small bandwidth will also probably produce an unacceptably poor sound.

The standards that govern the encoding of speech are mostly controlled by the ITU with its G.xxx series of standards.

The International Telecommunication Union (ITU) is headquartered in Geneva, Switzerland, and is an international organization within the United Nations System where governments and the private sector coordinate global telecom networks and services.

On a rare occasion, an organization besides the ITU designs a codec standard. For example, the IETF has standardized the iLBC codec.

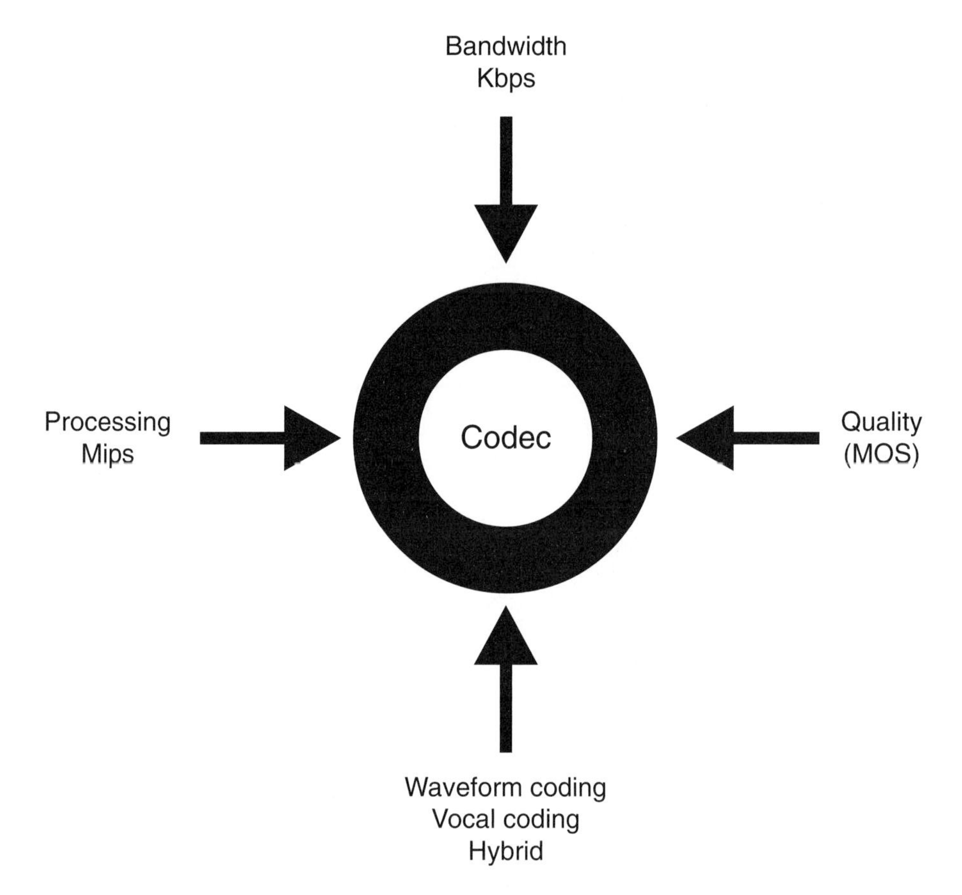

Figure 15-1: Deciding on a codec

15.1 SAMPLING

Voice is a sound wave made by compressing and decompressing air molecules that can be captured by a diaphragm. By attaching wires to the diaphragm and running current through it, the electrical signal can be modulated by the sound wave, that is, change its shape. The electrical wave can be changed by amplitude (height), frequency (number of cycles per time period), or phase.

Sampling means taking a measurement of the state of the electrical wave at a given moment in time.

Why 8,000 per second?

For voice, sampling is done 8,000 times a second. This value derives directly from the Nyquist theorem.

The Nyquist theorem

According to the Nyquist theorem, the discrete time sequence of a sampled continuous function contains enough information to reproduce the function exactly, provided that the sampling rate is at least twice that of the highest frequency contained in the original signal.

In everyday language this means that since the voice range of a person is roughly 4,000 Hz, 8,000 samples per second will accurately capture all of the information in it.

Harry Nyquist was an engineer at Bell Laboratories and published his theory in 1928. It was the basis of further advances by Claude Shannon and the more accurate name for this theorem is the Nyquist-Shannon sampling theorem.

Human hearing and voice

Human voice and human hearing have a narrow range. Typically, human voice can produce sound in the range of 300 Hz to 3,400 Hz. A Hertz (Hz) is one cycle per second. Human hearing is in the range of 20 to 20,000 Hz. However, this range degenerates with age. Older people typically can't hear above 16,000 Hz. In contrast, other species can detect wider ranges. Dogs, for instance, can hear to about 46,000 Hz. This is the basis for dog whistles. Bats use ultrasound above 100,000 Hz for navigation and the Beluga whale uses ultrasound above 136,000 Hz for communications.

Although the human voice has a range of only 4,000 Hz, human hearing has a much wider range, about 20,000 Hz. Music spans this complete sound range. Therefore it follows that if you want to encode music, you will need to sample at a much higher rate. And so it is. Music is sampled at 44,100 Hz in order to achieve CD quality sound.

15.2 QUANTIZATION

An analog signal, by definition, has an infinite number of data points. But a digital signal has a finite number of data values. Modern systems place one sample in one octet, or eight bits.

Quantization is the process by which the sample is placed into the one octet. But more than that, quantization is the process of constraining an infinite number of values into a set of finite values. In simple language, quantization assigns a numerical value to a sample.

Quantization introduces error into the data because no matter how many bits of precision are used, it is impossible to represent an infinite number of amplitude values with a finite number of increments.

The process of encoding sound

When describing the process of converting sound to digital, we need to distinguish between voice and any other sound, particularly music. Voice is special because of its use in telephony. Voice needs to be squeezed into as little bandwidth as possible and yet must still be understandable. Voice can be manipulated in order to meet these goals. Music is different. It must be as pristine and as close to the original as possible.

PAM and PCM

The process of encoding voice and music starts out the same. Pulse amplitude modulation (PAM) assigns a value to the amplitude of the sound sample. Pulse code modulation (PCM) now takes over and quantizes the value into a digital value. In the case of music, quantization is linear, meaning that the range or steps of each value are the same size. This is exactly what we want for music since we don't want to introduce distortion. The standard for CD music is 44,100 samples per second and each sample is encoded in 16 bits using PCM.

Figure 15-2 illustrates how quantizing works if the steps are uniform. High or low amplitudes are given equal value. The graphic also tries to illustrate that quantization error (noise) is introduced because the analog wave form can't exactly match the steps in digital values.

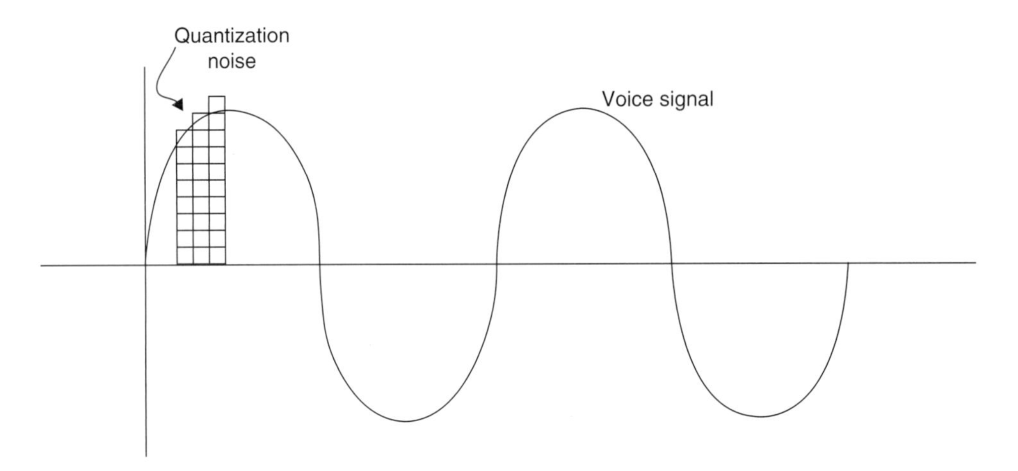

Figure 15-2: Uniform quantizing

Turning away from music and turning to voice instead, it is clear that a linear quantization scheme has a weakness. High amplitudes will have the same weight in the signal as low amplitudes and low amplitudes are where most of the content of human speech resides. Quantizing noise also impacts the Signal to Noise Ratio (SNR) of the voice signal. The SNR is a measure of how strong the primary signal (voice) is compared to the background noise. A higher SNR is better. For all these reasons, voice undergoes one further process that music doesn't.

Companding

As if the word *quantization* wasn't hard enough, how about another one, *companding*. Companding is derived from the words *compression* and *expansion*. It describes how the linear discrete values generated by PCM are compressed during encoding.

The ITU has defined two logarithmic companding schemes that are very similar to each other and are used extensively throughout the world's telephone systems. µ-law (Mu-law) is used in North America and Japan, while A-law is used in the rest of the world.

These companding schemes have the following goals:

- Stress the lower amplitudes in the voice signal.
- Increase the SNR for better voice quality.
- Encode a linear scale into a logarithmic scale.
- Encode and compress 13-bit or 14-bit data into 8-bit data.

A companding operation compresses dynamic range on encode and expands dynamic range on decode. If you are totally scratching your head over this statement, look at it like this. Low-pitch sounds such as "sh" and "th" are more important in speech than high-pitch sounds. Somehow, we have to find a way to stress them while putting less stress on the high-pitch sounds. Companding is the way. The technique is simply explained. Instead of quantizing in equal steps, which gives any data point throughout our 4000 Hz range equal weight, companding quantizes in unequal (logarithmic) steps. Each step is larger as we go higher in amplitude. Therefore, values in the low amplitudes are given more weight than values in the higher amplitudes.

A-law and µ-law

A-law and µ-law are audio compression schemes (codecs) defined by ITU standard G.711 that compress 16-bit linear PCM (pulse code modulation) data down to eight bits of logarithmic data.

Figure 15-3 illustrates the companding of A-law and µ-law and how it differs from equal quantization.

- They both break a dynamic range into a total of 16 segments.
- There are eight positive and eight negative segments.
- Each segment is twice the length of the preceding one.
- Uniform quantization is used *within* each segment.

Both use a similar approach to coding the eight-bit word:

- First bit (most significant bit) identifies polarity.
- Bits two, three, and four identify segment.
- Final four bits quantize the segment.

However, A-law and µ-law are not exactly the same.

A-law provides a greater dynamic range than µ-law. A-law limits the linear sample values to 13 bits and the value scale it uses is −4,096 to 4,096.

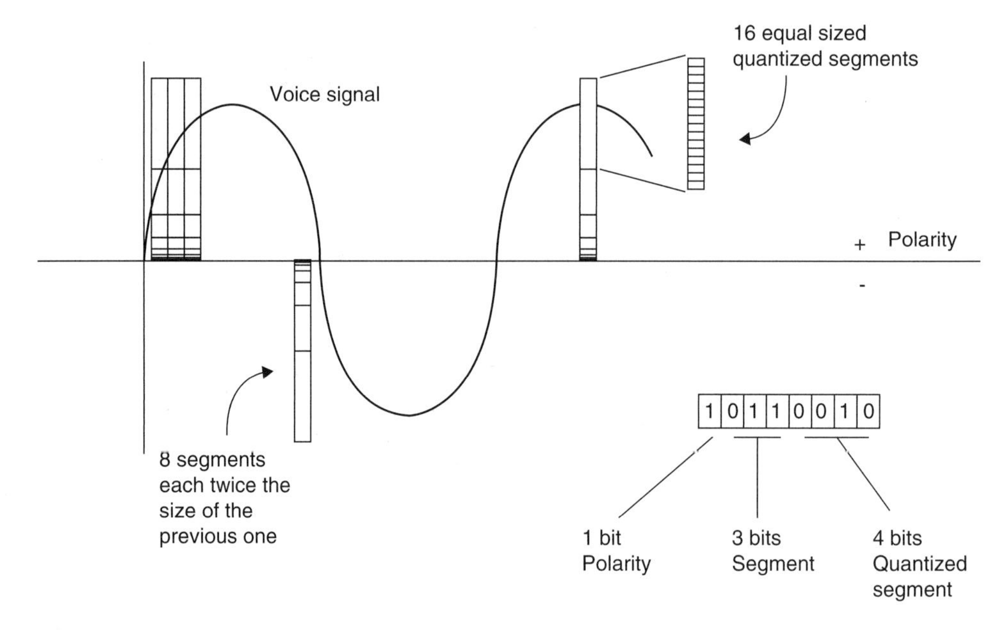

Figure 15-3: Companding with A-law and μ-law

μ-law provides better signal/distortion performance for low level signals than A-law. μ-law limits the linear sample values to 14 bits and the value scale it uses is −8,159 to 8,159.

As already mentioned, both of these companding schemes are defined by and supported in the G.711 specifications. If you are in North America or Japan, you use μ-law. If you are in Europe or the rest of the world, you use A-law.

An international connection needs to use A-law; μ to A conversion is the responsibility of the μ-law country.

It is common to see G.711 written as G.711 a/u which of course refers to A-law and μ-law. By the way, mu is a Greek letter of the alphabet and should be written as μ.

Although we concentrate on voice for this discussion, note that PCM is widely used in music encoding as well. PCM is the basis of music CDs. On a CD, there is no need to compress the sound and therefore PCM is used in its native format and there is no companding. In order to transfer a song from a CD to a computer, it must be re-coded to a different format for storage on the computer's hard disk. If it is to be transmitted across a network, it should also be compressed to a small size format. For your interest, Table 15-1 lists some popular formats and the sampling codec.

Type	Extensions	Codec
AIFF (Mac)	.aif, .aiff	PCM, other
AU (Sun/Next)	.au	µ-law, other
CD audio (CDDA)	N/A	PCM
MP3	.mp3	MPEG Audio Layer-III
Windows Media Audio	.wma	Proprietary (Microsoft)
QuickTime	.qt	Proprietary (Apple)
RealAudio	.ra, ram	Proprietary (Real Networks)
WAV	.wav	PCM, other

Table 15-1: Common digital audio formats

15.3 WAVEFORM CODING

Waveform coding refers to transforming a wave pattern into a bit pattern.

Pulse code modulation

To convert sound into a digital stream using waveform coding, we use pulse code modulation (PCM) and variations of it.

PCM with companding is formalized in ITU specification G.711 using either A-law or µ-law and is an open standard, which anyone can use.

It produces toll quality voice signals, which are the gold standard, and is the quality that every other coding method is judged against. In addition, it requires the least processing overhead of any method, which leads to inexpensive encoding equipment.

The biggest drawback is that it takes the most bandwidth. It requires 64 Kbps on a telephone voice circuit but when used with voice over IP over Ethernet, this increases to 95.2 Kbps. G.711 is the mandatory minimum standard for all ISDN terminal equipment.

Differential PCM

DPCM decreases the bandwidth used by PCM. As you are speaking, your voice doesn't change very fast. With 8,000 samples per second, the difference between succeeding signals is minimal. DPCM calculates the difference and then transmits the difference instead of the value of the sample. This allows for a reduction in the throughput required to transmit voice signals. DPCM can reduce the bit rate of voice transmission down to 48 Kbps.

The reduction in bandwidth is not spectacular. However, a greater problem is that the quality of the voice is not even over the complete range of frequencies. Higher voice signals withstand the compression, but lower ones do not.

Adaptive DPCM

Adaptive differential pulse code modulation (ADPCM) adapts the quantization levels of the difference signal that was generated at the time of the DPCM process. How does ADPCM modify these quantization levels? If the difference signal is low, ADPCM increases the size of the quantization levels. If the difference signal is high, ADPCM decreases the size of the quantization levels. So, ADPCM adapts the quantization level to the size of the input difference signal.

ADPCM reduces the bit rate of voice transmission down to 32 Kbps, half the bit rate of A-law or µ-law PCM. ADPCM produces "toll quality" voice and is comparable to A-law and µ-law PCM.

Here are the steps that ADPCM needs to adapt the quantization levels.

- Turn an A-law or µ-law PCM sample into a linear PCM sample.
- Calculate the predicted value of the next sample.
- Measure the difference between actual sample and predicted value.
- Code difference as four bits, send those bits.
- Feed back four bits to predictor.
- Feed back four bits to quantizer.

ADPCM is formalized as ITU specification G.726. Its nominal bandwidth is 32 Kbps, but when run over Ethernet as VoIP, it uses 55.2 Kbps. In addition to 32 Kbps, ADPCM also supports 40, 24, and 16 Kbps.

15.4 VOICE CODING

The process of encoding voice struggles to shrink the data stream to a manageable bandwidth. Encoding the voice waveform only gets you so far, realistically to 16 Kbps. An alternative approach, called a vocoder, which is a contraction for "voice coder," can do better.

While a waveform encoder works with any sound, music as well as voice, a vocoder works only with voice. It attempts to recognize the different parts of speech and encodes them accordingly. This approach had been recognized as early as 1939 and by now is well established. Although a pure vocoder can encode voice to a very small bandwidth, less than 5 Kbps, the result is also mechanical and robotic. Modern VoIP uses hybrid approaches to this problem, which we will examine as well.

Classes of speech

Much research has been done into how speech is produced and its characteristics. Without going into the mechanics of human anatomy, suffice it to say that three different classes of sound can be produced by the human voice organs. They are as follows.

Voiced sounds

Voiced sounds are created when the vocal cords vibrate open and closed. This way, periodic pulses of air come out of the opening of the vocal cords. The rate at which the opening and closing occurs, determines the pitch of the sound.

Unvoiced sounds

In order to produce unvoiced sounds, the vocal cords do not vibrate—they are held open. Air is then sent at high velocities through a constriction in the vocal tract, creating a noise-like turbulence.

Plosive sounds

Plosive sounds result from building up air pressure behind a closure in the vocal tract and then suddenly releasing this air.

A simple way to think of voice is wind expelled from the lungs, called excitation, followed by a filter, which is the vocal tract but also includes the tongue, nasal cavities, and so on. The filter modifies the air passing through to produce the sound.

Also relevant for a vocoder are these characteristics:

- The type of excitation (the flow of air coming out of the vocal cords) and the shape of the vocal tract change relatively slowly. This means that for short time intervals, for example 20 ms, the speech production system can be considered to be almost stationary.
- Speech signals show a high degree of predictability. This is due to the periodic signal created by the vocal cords and also due to the resonance characteristics of the vocal tract.

Vocoder basics

Vocoding techniques try to determine parameters about how the speech signal was created and use these parameters to encode the signal. To reconstruct the signal, these parameters are fed into a model of the vocal system, which outputs a speech signal.

If the preceding paragraph didn't make any sense, let's try to clarify vocoders with this simplified explanation:

1. The voice signal is sampled just as it is with PCM.
2. A block of samples, typically 5 or 10, are analyzed together and it is determined which kind of sound it is—voiced, unvoiced, or plosive.
3. The block is encoded using the kind of sound and its frequency and then transmitted to the destination.

The most basic vocoder is linear predictive coding (LPC). The LPC algorithm describes each sample as being a linear combination of the previous ones. Such an equation is called a linear predictor, hence the name of the coder.

The LPC method can produce intelligible speech at 2.4 Kbps if you don't mind a synthetic, robotic sound. The person speaking can often not be recognized and the algorithms usually have problems with background noise.

Hybrid coders

Hybrid coders try to produce high-quality speech with low bandwidth requirements. Typically, they extend the LPC algorithm to overcome its simplistic view of the speech classes. The basic problem with vocoders is their simplistic representation of the excitation signal: the signal is considered to be either voiced or unvoiced. It is this representation that causes the synthetic sound of these coders. Hybrid coders try to improve the representation of the excitation signal. Figure 15-4 compares waveform, voice, and hybrid coders in terms of quality and bandwidth.

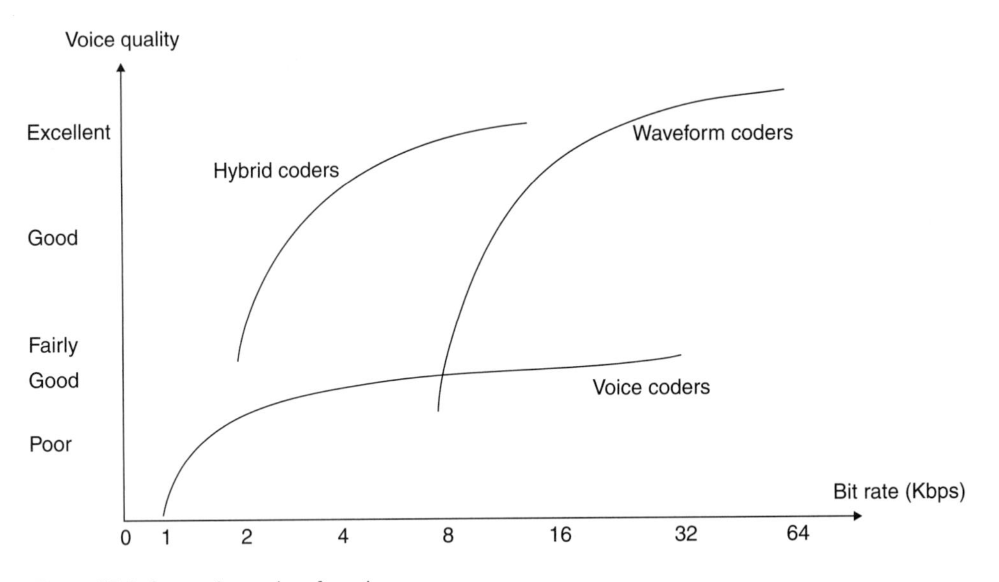

Figure 15-4: Comparing coders for voice

The technical details of hybrid coders are beyond the scope of this book. However, in general, the hybrid coders try to detect the differences or error, between the actual signal and their voice model. The difference is also encoded and sent with the base signal. Following is a list of some of these hybrid coders.

CELP

Code excited linear prediction (CELP) has been adopted as a U.S. federal standard. CELP is pronounced "kelp." CELP's main innovation is the way it handles excitation. CELP coding is based on analysis-by-search procedures. The code in the name refers to a code book that holds indexes to important parameters of the signal and are transmitted to give a better quality. CELP requires intensive computation but it produces very natural speech. It is also very robust to environmental noise.

U.S. Federal Standard 1016 uses CELP and a bandwidth of 4.8 Kbps. Native CELP is not used in VoIP. However, the following variations of it are used extensively.

LD-CELP

Low-delay CELP is ITU G.728 and uses a bandwidth of 16 Kbps.

MP-MLQ/ACELP

This is a dual-rate speech coder that provides either 5.3-Kbps or 6.3-Kbps data streams. This coder was optimized to represent high-quality speech using a limited amount of complexity. It encodes speech in frames using linear predictive analysis-by-synthesis coding. The excitation signal for the high-rate coder is multipulse maximum likelihood quantization (MP-MLQ) and for the low-rate coder is algebraic-code-excited linear prediction (ACELP).

This is ITU standard G.723.1 and a manufacturer must acquire a license in order to incorporate it into equipment.

CS-ACELP

Conjugate-structure algebraic-code-excited linear prediction, the name says it all. Whew! CS-ACELP is formalized under the ITU standard G.729. The encoder functionality includes voice activity detection and comfort noise generation (VAD/CNG) and the decoder is capable of accepting silence frames. G.729 provides near-toll-quality performance under clean channel conditions and is the default codec as prescribed by the Frame Relay Forum; it is also suitable for VoIP applications.

This encoder must be licensed.

GSM

Groupe speciale mobile (GSM) encoding uses a variation of LPC called RPE-LPC (regular pulse excited—linear predictive coder with a long-term predictor loop). GSM began as a European cellular telephone speech encoding standard. For 8 kHz sampling, this means GSM encoded speech requires a bandwidth of 13 Kbps. The original GSM is now called "full rate." Newer versions include EFR (enhanced full rate), which uses ACELP, HR (half rate), which uses CELP-VSELP, and AMR (adaptive multi-rate), which also uses ACELP.

SPEEX

The Speex project is an attempt to create a free software speech codec, unencumbered by patent restrictions. Unlike many other speech codecs, Speex is not targeted at cell phones but rather at VoIP and file-based compression. CELP is the basis for Speex. Speex is well suited to Internet applications. For example, its bandwidth can range from as low as 2.15 Kbps (narrowband) to 44 Kbps (wideband) and with a multi-rate feature it can change its bit rate dynamically as conditions change.

iLBC

iLBC (Internet low bitrate codec) is a FREE speech codec suitable for robust voice communication over IP. The codec is designed for narrowband speech and results in a payload bit rate of 13.33 Kbps. Because iLBC's speech frames are compressed independently, the iLBC codec enables graceful speech quality degradation in the case of lost frames, which occurs in connection with lost or delayed IP packets. iLBC is also unusual in that it is an IETF proposal (RFC 3951). Although the ITU and IETF cross swords occasionally (think H.323 and SIP), this is the only voice coder that the IETF is looking at. iLBC is used by the popular VoIP service, Skype. Although iLBC is free to use, it is not open source.

Every VoIP telephone and softphone offers a choice of codecs. Figure 15-5 shows the codec selection of one softphone. Because VoIP devices negotiate the codec that they use for a conversation, it is import that the devices have a selection to choose from so that they have at least one in common. Table 15-2 summarizes the codecs mentioned in this section.

Name		Specification	Bandwidth
Waveform Encoding			
PCM μ-law/A-law		ITU G.711	64 Kbps
ADPCM		ITU G.726	32 Kbps 40, 24, and 16 Kbps
Hybrid Encoding			
MP-MLQ		ITU G.723.1	6.3 Kbps
ACELP		ITU G.723.1	5.3 Kbps
LD-CELP		ITU G.728	16 Kbps
CS-ACELP		ITU G.729	8 Kbps
GSM	FR	ETSI 06.10	13 Kbps
GSM-HR	HR	ETSI 06.20	5.6 Kpbs
GSM-EFR	EFR	ETSI 06.60	12.2 Kbps
GSM-AMR	AMR	ETSI 06.90	4.75–12.2 Kbps
iLBC		RFC 3952	13.3/15.2 Kbps
Speex (uses CELP)			2.5–44.2 Kbps

Table 15-2: The most common codecs

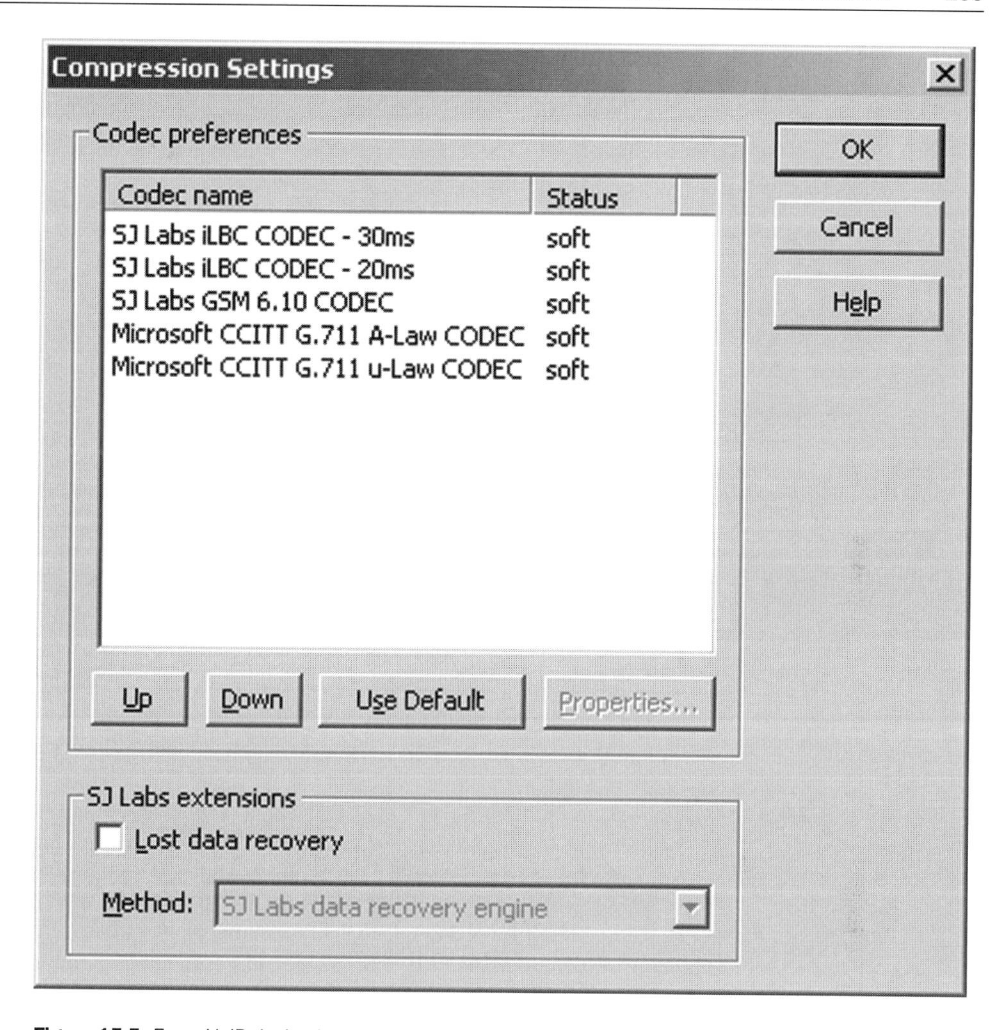

Figure 15-5: Every VoIP device has a selection of codecs

15.5 MORE CODEC CONSIDERATIONS

Mean opinion score

The mean opinion score (MOS) provides a numerical measure of the quality of human speech at the destination end of the circuit. The scheme uses subjective tests (opinionated scores) that are mathematically averaged to obtain a quantitative indicator of the system performance.

To determine MOS, a number of listeners rate the quality of test sentences read aloud over the communications circuit by male and female speakers. A listener gives each sentence a rating as follows: (1) bad; (2) poor; (3) fair; (4) good; (5) excellent.

The MOS is the arithmetic mean of all the individual scores, and can range from 1 (worst) to 5 (best). Figure 15-6 diagrams the MOS scale.

This is a very subjective test. Therefore, the more subjects that take the test, the more "reliable" it becomes. The codec with the highest MOS score should be selected if all other factors are equal.

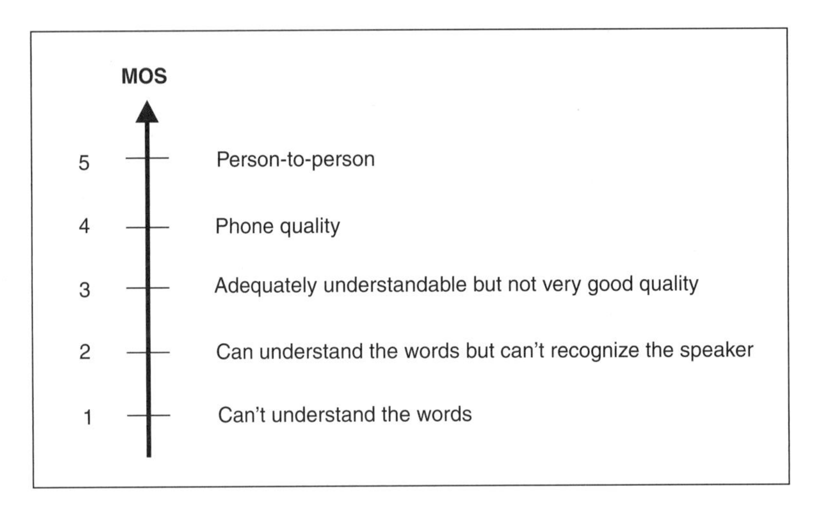

Figure 15-6: The MOS scale

Processing

The encoding of the audio signal into digital and back again requires significant computing resources. The topic is relevant because some of the devices that need to process the voice signal don't have very much processing power. Cell phones are an example. Even computers, until recently, had trouble processing some encoding schemes. In Table 15-3, processing requirements are listed in MIPS (millions of instructions per second), which is a widely used metric for processing power.

PCM has the least burden for processing. After sampling, which all coders need to do anyway, manipulating the signal is trivial. In contrast, the hybrid coders require intensive processing to work. Naturally, the designers of devices such as cell phones, PDAs, and VoIP telephones have to take this into account when developing their devices.

Compression delay

Why do some codecs require more processing power than others? One factor is the compression of the voice data. Although a small, compressed data stream is advantageous when you are trying to cram as many voice calls onto a cable as possible, the quality of the voice may suffer because of the delay introduced into the signal.

Narrowband versus wideband

The codec technology discussed so far in this chapter is narrowband because the process starts by sampling the voice at 8,000 per second. You need to be aware that wideband codecs are also available that have much higher sampling rates, for example 16,000 or even 32,000 times per second. The design of the narrowband codecs derive directly from the narrow frequency range of the human voice and the technology of the public system telephone network that carries it. The limitations of narrowband are only too painfully obvious when listening to music while on hold. Wideband can enhance the quality of sound, both music and, to a lesser extent, voice. Codecs are available for wideband, particularly G.722 and G.722.2 and Speex. G.722 uses a split-band version of ADPCM. G.722.2 is newer and uses the wideband adaptive multi-rate (WB-AMR) compression codec.

The downside to wideband is the larger amount of data generated and the greater bandwidth required. Wideband is suitable only for VoIP systems using LANs or high-speed WANS.

Licensing

Technology is heavily encumbered with intellectual property, ideas that were developed for profit. Contributions to voice technology have been provided by companies, educational institutions, and individuals. The initial ideas for a concept are usually patented. However, the world of patents is also murky. This is particularly true when a refinement to an idea is involved as opposed to the original idea. This can be seen in voice coder technology since so much of it is a variation on just two themes, PCM and LPC. Patents are issued for only 20 years. PCM was developed in 1939 and is no longer under patent protection. In addition, the ability to patent certain technology varies by country. American patents can cover a broader range of technology than European patents.

So why do we care? The target market for a technology determines what can be charged for it. Technology destined for corporate markets can demand the premium that must be paid for licenses. Technology that is targeted to the home or individuals, particularly "free" and open source software, can't recoup license fees. Sometimes the status of a codec has not been tested. Phillips Electronics claims certain patents for GSM-FR, but these have not been tested in court.

PCM, which is universal in all VoIP products, is unencumbered by patents. However, the next two most important encoder technologies, G.723.1 and G.729, are definitely licensed. That is why you won't see them in free or open source products.

To address this issue, two coders have been developed that avoid patent issues. Speex is an open source encoder that is licensed similar to Linux. It is completely open. The second is iLBC. Although originally developed by a private company, it has been submitted to the IETF for ratification as an Internet standard. Since the IETF won't ratify any technology that requires a license, iLBC is also license free.

It is important to distinguish between licensing a patent and licensing source code. Even if a codec is patent free, a programmer has to turn the ideas into computer code. A developer of VoIP products may be more than happy to license the source code for a codec if the result is a quality product that he can get to market faster.

Note that technology standards and patents are not normally connected. The ITU is happy to publish standards such as the G.xxx series, even if the technology is licensed.

Dependable information on the licensing of codecs is hard to come by and is scattered over many sources. Table 15-3 summarizes the codecs along with MOS scores, processing required, and patent information. As the table indicates, choosing a codec is not always a straightforward decision.

Summary of Voice Encoding						
Name	Specification	Bandwidth	MOS	Processing (MIPS)[1]	Compression Delay	Patents
Waveform Encoding						
PCM µ-law/A-law	ITU G.711	64 kbps	4.2	0.34	0.75 ms	Unencumbered
ADPCM	ITU G.726	32 kbps 40, 24 and 16 kbps	3.90	14	1 ms	Original patents, current status unknown
Hybrid Encoding						
MP-MLQ	ITU G.723.1	6.3 kbps	3.9	16	30 ms	Covered by patents, must be licensed
ACELP	ITU G.723.1	5.3 kbps	3.65	16		
LD-CELP	ITU G.728	16 kbps	3.7	33	3 to 5 ms	
CS-ACELP	ITU G.729	8 kbps	3.92	20	10 ms	Covered by patents, must be licensed
GSM FR	ETSI 06.10	13 kbps	2.5			Possible patent dispute
GSM HR	ETSI 06.20	5.6 kpbs				Covered by patents, must be licensed
GSM EFR	ETSI 06.60	12.2 kbps	4.2	14		
GSM AMR	ETSI 06.90	4.75–12.2 kbps	3.8	14		
iLBC	RFC 3952	13.3/15.2 kbps	3.7			Royalty free
CELP	Speex	2.5–24.6 kbps				Royalty free BSD style license

1. MIPS = Millions of Instructions per Second. The MIPS value varies with the DSP chip.

Table 15-3: Codecs and MOS scores, processing power requirements, and licensing

15.6 VoIP BANDWIDTH

Now that our voice signal is digitized, it is time to pour it into a packet. It is important to stress that voice does not have to be packetized in order to be transmitted. The public telephone system does this all the time and that is why T1 lines were invented. Using PCM a/u encoding, one voice stream fits nicely into a single T1 channel of 64 Kbps. In order to traverse an IP network, however, the voice data must be packetized.

There is a downside to packetizing. The data is increased because of the packet headers and this increased overhead can negatively impact the available bandwidth of our network, even to the extent of decreasing the number of telephone calls that can be made simultaneously.

If the voice quality is acceptable, then our most pressing concern is how many telephone conversations we can get on a line. It stands to reason that there is an economic consequence to insufficient bandwidth. We will need to invest to upgrade the equipment in our system.

This section will familiarize you with the bandwidth requirements of various codecs and their configurations.

Packet overhead

The previous chapter laid out how VoIP uses the IP, UDP, and RTP protocols. These protocols add a standard byte count to the voice payload because of their headers. Figure 15-7 diagrams the headers involved using an Ethernet network. The header lengths are as follows:

IP: 20 octets
UDP: 8 octets
RTP: 12 octets
Total: 40 octets

An octet is 8 bits, or a byte.

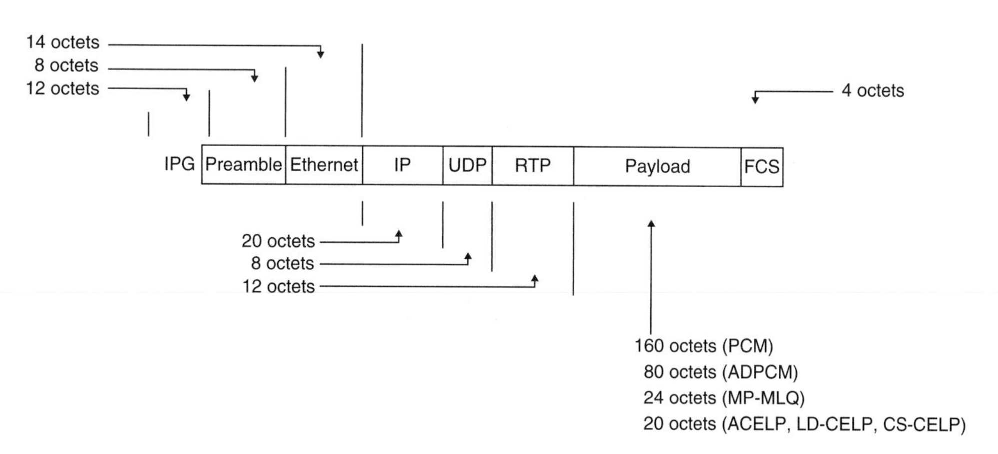

Figure 15-7: VoIP inside Ethernet

Compressed RTP

Because the content of the IP/UDP/RTP headers is very repetitive for a voice conversation, it should come as no surprise that a compression scheme exists. Compressed RTP is described in RFC 2508. This method allows the IP/UDP/RTP headers to be compressed down to 2–4 bytes in most cases. But don't expect to see it in routine VoIP traffic. CRTP was designed for reliable point-to-point links with short delays. It does not perform well over links with a high rate of packet loss, packet reordering, and long delays.

Link overhead

The link that the VoIP packet travels over also adds to the overhead of the communications. The most common links are Ethernet, frame relay, PRI, and ATM. We will look closely only at Ethernet. You can also refer to Figure 15-7 for this discussion.

As described in a previous chapter, Ethernet frames have a 14-octet header and a 4-octet trailer. In addition, an Ethernet frame starts with an 8-octet preamble, which is used by the Ethernet adapter to synchronize with the incoming electrical signal.

Ethernet also has idle time on the cable and this must be taken into account when discussing bandwidth. During idle time, no information is transmitted. Idle time is the time between packets, and for Ethernet this is called the inter-packet gap (IPG). The IPG is specified in time as follows—10 Mbps = 9.6 msec, 100 Mbps = 0.96 msec, GigE = .096 msec. This resolves down to 12 octets in all three cases.

To sum up, Ethernet adds the following fields to the packet overhead:

Ethernet header: 14 octets
Ethernet trailer: 4 octets
Preamble: 8 octets
Inter-packet gap: 12 octets
Total: 38 octets

Voice payload

How much voice data is placed inside each packet? That depends on the encoder and may be under the control of the administrator.

The sample period is the slice of the conversation that is placed inside one packet. Long sample periods can produce high latency, which can affect the perceived quality of the call. There are two reasons for this. A long sample period forces the codec to wait until the entire sample has been received before it can release the sample. Secondly, if a packet is lost, the receiver must artificially fill in a longer period of missing conversation.

Although on the surface it appears that a smaller sample period is best, there is a penalty to pay. With smaller sample periods, the packet overhead becomes a larger portion of the packet, leading to lost efficiency. It is possible for the packet header to be larger than the data it is actually carrying. Therefore a tradeoff is required between voice quality and network efficiency. Here is how some of the codecs deal with sample size. Be aware that the typical sample sizes mentioned are not fixed. They may vary either under the control of the programmer or the administrator.

The meaning of "sample" also varies between waveform coders and hybrid voice coders. Waveform coders take a slice of the signal, quantize it, and digitize it as bits and bytes. Voice coders, however, take a block of PCM samples and analyze it for voice structures. It is this representation that is encoded.

Waveform coders

G.711 - PCM

With G.711, a 20 ms sample period is considered optimum. This is 1/50 of a second, or 160 bytes. If you need to review the math, then remember that 20 ms is 20/1000, or 1/50 of a second. Also that 1/50 of 8000 samples per second is 160.

G.726 - ADPCM

ADPCM is another waveform encoder. Unlike G.711, which encodes each sample, it encodes the difference between samples. Instead of using eight bits per sample, it uses four bits per difference between samples. This accounts for its 32 Kbps rate. Typically, a 20 ms sample period is used. This gives us the same 160 samples as with G.711, but since only four bits are used to encode the sample, the payload is only one half that of G.711, or 80 octets.

Hybrid voice coders

ACELP/MP-MLQ

G.723.1, which covers both these codecs, uses a sample block of 30 ms. The 30 ms sample is encoded into 24 octets for MP-MLQ and 20 octets for ACELP.

CS-ACELP

G.729 uses a sample block of 10 ms. Two resulting frames are transmitted in a packet. The two frames result in a payload of 20 octets.

LD-CELP

G.728 uses a sample block of 5 ms for each frame. However, four frames are sent in each packet that results in a payload of 20 octets.

How much bandwidth?

The bandwidth used by one voice conversation will vary with the link that it is traveling on, Ethernet, frame relay, or ATM. For illustration purposes, let us just consider G.711 running over Ethernet using a 20 ms sample time.

Voice payload: 160 octets
IP/UDP/RTP overhead: 40 octets
Ethernet overhead: 38 octets
Total packet: 238 octets

If one packet holds 20 ms, or 1/50 of a second of voice, then 50 × 238 will hold 1 second of conversation. This works out to 11,900 octets. Since all communication links are rated in bits, not bytes, per second, we multiply the number of octets by 8 to get 95.2 kbps.

This means that 105 VoIP conversations can be placed on a 10 Mbps Ethernet provided that there was no other traffic on it and that there were no collisions. These conditions are unrealistic; however, performance can be enhanced if you segregate voice traffic on virtual LANs (VLANs) using Ethernet switches.

Silence suppression

When there is a pause in the conversation, no signal is generated. The silence suppression feature of modern codecs will prevent null information from being transmitted. This decreases the bandwidth requirements for a conversation from 33% to 50%. It is the responsibility of the receiving end to replace the missing signal in the conversation. Because the missing signal gives pauses in the conversation an eerie, strange effect, the codec will compensate by putting in an artificial background "comfort" noise.

Clearly, determining how many voice conversations that a system will support is a daunting calculation. In a following lab exercise, you will be introduced to a simpler method of making these calculations.

SUMMARY

Section 15.1: Sampling

The tricky transition from an analog voice signal to a series of digital bits starts with sampling and is performed by the codec. The codec can be programmed in software or embedded in a chip called a digital signal processor (DSP).This first section discusses the concept of sampling a voice signal and how many samples are required to be able to reconstruct it accurately.

Section 15.2: Quantization

Being analog, a voice signal has an infinite number of data points. Quantization is the process by which the infinite data points are assigned a value from a finite range.

This is required when you consider that a single sample is encoded in eight bits, and eight bits can have only one of 256 values.

Section 15.3: Waveform Coding

This section looks at the two encoding methods, PCM and ADPCM, which encode the samples as binary data.

Section 15.4: Voice Coding

This section looks at an alternative encoding technique that looks at how the voice signal was made as well as its content. These techniques are based on some variation of CELP.

Section 15.5: More Codec Considerations

When deciding on a codec, other considerations need to be taken into account such as voice quality, as measured on the MOS scale, required processing power, and its status as intellectual property.

Section 15.6: VoIP Bandwidth

This section looks at how the codec and the packetization of the signal impacts the bandwidth required for a voice conversation.

Hands-on Lab 15-1

Measuring Mean Opinion Score

In this exercise, you will have the experience of measuring voice quality using the mean opinion score. This measurement is very subjective and requires a large sample to be meaningful.

The instructor will play five voice samples in the classroom. You will judge their quality using the following scale:

5 – Person-to-person quality

4 – Telephone quality

3 – Adequately understandable, but not very good quality

2 – Can understand the words, but can't recognize the speaker

1 – Can't understand the words

Scores do not have to be whole number; answers in between are allowed. Please use decimals, not fractions.

After each sample is played twice, mark your answer beside the question number in the next column. You should also note the name of the codec.

At the end of the exercise, the ratings will be added up and averaged for the class.

Sample #1 ____________

Sample #2 ____________

Sample #3 ____________

Sample #4 ____________

Sample #5 ____________

Hands-on Lab 15-2

Calculating Bandwidth

The number of voice calls a VoIP system can support is an important parameter in designing any VoIP system.

Although this chapter showed an example of a manual calculation, this is too painful and inefficient for normal use.

VoIP calculators are available on the Internet and you will use one in this exercise.

1. Open up your Web browser to the following URL http://www.newport-networks.com/pages/voip-bandwidth-calculator.html

The workings of this calculator are straightforward. Select the codec type and use the defaults as shown. Read the Ethernet bandwidth from the page. Try without silence suppression and with it.

Record the answers in the following table.

G.711, 64 Kbps

Without silence suppression ____________

With silence suppression ____________

G.723.1A, 6.4 Kbps

Without silence suppression ____________

With silence suppression ____________

G.723.1A, 5.3 Kbps

Without silence suppression ____________

With silence suppression ____________

G.726, 32 Kbps

Without silence suppression ____________

With silence suppression ____________

G.728, 16 Kbps

Without silence suppression ____________

With silence suppression ____________

G.729A, 8 Kbps

Without silence suppression ____________

With silence suppression ____________

Hands-on Lab 15-3

Selecting Codecs

Most VoIP devices and softphones give you control over the selection of codecs that are available for the device. Having a selection of codecs is important because then the negotiating process between the two devices is more likely to find one that they have in common. Alternatively, you may not wish a particular codec to be used, maybe because it takes up too much bandwidth or provides a poor quality of sound.

In this exercise, you change the selection of codecs available for the X-Lite softphone.

1. Open up the X-Lite softphone.
2. Place your mouse cursor anywhere on the X-Lite softphone and right-click. From the menu, select **Options. . .**
3. On the far left of the Options screen, at the bottom of the windows, click on the **Advanced** button. Because of its position on the screen, this button is easy to overlook.
4. Select **Audio Codecs.**
   The codecs are listed as either Enabled or Disabled. In order to change the status of a codec, select it and click on the appropriate arrow.

   Notice that selecting a codec gives you more information about it, specifically, its bit rate and quality index.
5. Disable codecs until G.711 u-law is at the top of the list. Click **OK.**
6. Telephone another student in the class. When the two softphones negotiate the codec they will use, they will then choose G.711 u-law.

Confirming the codec

7. Hang up any telephone calls you have active.

   Start Wireshark.

   Start capturing packets in Wireshark.

   Make a telephone call to another student.

   Talk for a short while and then terminate the call.

   Stop the packet capture in Wireshark.

   In the Filter box, type in RTP.

   You should now have a list of the RTP packets that capture your telephone conversation.

The payload type (PT) will identify the codec. Confirm that it is G.711 (PT = 8).

8. Exit Wireshark and X-Lite.

Review Questions Voice to Digital

1. Companding occurs during which phase of the encoding process?

   a) Packetizing
   b) Quantizing
   c) Sampling
   d) Compressing

2. What is the biggest disadvantage to waveform coding?

   a) It requires more processing power.
   b) It has a low MOS value.
   c) It uses a large bandwidth.
   d) It is protected by patents.

3. If you needed to support the maximum number of voice conversations on an Ethernet cable, which of the following codecs would you choose?

   a) G.723.1
   b) iLBC
   c) G.728
   d) GSM 06.10

4. Code excited linear prediction (CELP) is very important because

   a) it is the most widely used codec in VoIP.
   b) it is a U.S. government standard.
   c) it gives the best voice quality at the lowest bandwidth.
   d) it is the basis for other popular codecs.

5. According to the Nyquist theorem, if you need to capture all of the information in a 8,000 Hz frequency range, you would need to sample __________ times per second.

   a) 8,000
   b) 4,000
   c) 16,000
   d) 44,100

CHAPTER

16

Implementing QoS

Objectives

Upon completion of this section, the reader will:

- *Have been exposed to the many techniques available to provide excellent quality of service (QoS) on a VoIP system.*
- *Know which protocols can provide QoS at the data link layer and specifically, how 802.1Q/p works.*
- *Appreciate the techniques available at the network layer for QoS and how type of service and differentiated services are different from each other.*
- *Understand the role that MPLS can play in the modern carrier network for voice over IP.*

INTRODUCTION

VoIP will compete with traditional telephone systems only if it can deliver the same quality as the public system telephone network. The technology to provide the same quality of service is available and generally used. However, if you do receive poor quality, this chapter will explain why.

16.1 QoS OVERVIEW

What does QoS include?

Quality of service for VoIP means providing voice quality as close to the public switched telephone system as feasible. It should enable as many of the following as possible:

- Response times that are predictable
- Management of delay-sensitive applications
- Management of jitter sensitive applications
- Control of packet loss. This is particularly noticeable when congestion occurs during a burst. Continuous congestion means that a link is oversubscribed.
- Traffic priorities can be set
- Congestion avoidance
- Management of congestion when it occurs

This section on implementing quality of service extends the information provided in Chapter 4, "Quality of Service," which gives an overview of QoS. In this module we expand on QoS and look at the technical details.

Quality of voice over an IP network is dependent on the following five factors: bandwidth, delay, jitter, lost packets, and echo. These factors are examined in the next section.

Bandwidth

Bandwidth refers to the quantity of traffic that can flow over a communications link. The links themselves are rated by bandwidth. For example, Ethernet is rated at 10, 100, or 1000 Mbps, while a T1 line is rated at 1.544 Mbps. Because voice traffic must share links with other data traffic, congestion will result from too much traffic, and voice quality will suffer. You can alleviate bandwidth problems in the following ways.

- Restrict a link to only voice traffic. This is actually feasible and recommended if you set up virtual LANs (VLANs) on your internal network. It will not be feasible once your voice traffic is forwarded onto your service provider's network.
- Over-provision your network. In other words, buy more than you need. Upgrading your LAN is cost effective because Ethernet

components are inexpensive. In practice this means that a 100-Mbps link is established to all end devices on your network, including IP telephones and computers. Only switches are used, not hubs. Finally, all switches and routers are connected using gigabit Ethernet. Over-provisioning on a WAN, however, is difficult to justify since the extra monthly communication charges will hit your bottom line directly.
- Ration your available bandwidth or give certain classes of traffic priority over other classes. This is the subject of the rest of this module.

Delay

There are many points in the transmission of voice at which the voice signal or the packet carrying the voice signal must be processed. Each time this happens, delay is introduced into the transmission. The typical delay for a signal going end to end on the PSTN is 30 to 50 ms. Consider this the best-case scenario. At the other end, delays of 250 ms or longer are intolerable.

Delay can be caused by many factors in a VoIP system, including efficiency of the codec, processing of the packets, router delays, number of routers in the path, and congestion caused by insufficient bandwidth.

While the voice traffic is traversing the internal network, it is under the control of the enterprise and careful analysis of the traffic can point to areas where network improvement can lead to better performance. Once the voice traffic enters the network of the service provider, service level agreements (SLAs) are the primary method that the enterprise can use to ensure acceptable performance.

Jitter

Jitter is defined as variable delay. In other words, the packets are showing up with different intervals between them. Jitter degrades the voice quality, which becomes noticeable to the listener.

Jitter is caused by differences in packet processing, router forwarding, and congestion of the network.

Jitter can be easily dealt with by the use of playout buffers. This is now a standard feature of VoIP equipment. Memory is set aside to receive the voice packets, which arrive at irregular intervals, and then the voice is played out to the listener at regular intervals.

Lost packets

Dropped packets lead to voice degradation on a VoIP network. An IP network guarantees the deliver of packets if the TCP protocol is used. However, the delay caused by TCP retransmission of lost voice packets introduces unacceptable quality degradation to a voice conversation. Therefore, voice transmission uses UDP because

UDP doesn't guarantee delivery of a packet and if a packet is lost, it stays lost. The voice equipment can compensate somewhat by replaying the last packet in the buffer again. Although this is not entirely a satisfactory solution, experience has shown that it is at least more satisfactory than using TCP. This issue highlights the fact that the network has to be configured and maintained properly in order to avoid lost packets in the first place.

Echo

Echo in a telephone circuit refers to the speaker's voice bouncing back from certain disjunctions of the circuit such that the speaker can hear parts of his conversation. Echo occurs even in a traditional switched telephone circuit, but since the round-trip time is less than 50 ms, the effect is masked and not noticeable. When the round trip is longer than 50 ms, such as in a long distance call, echo canceling techniques need to be used. With VoIP, round-trip times are always greater than 50 ms and therefore echo cancellation needs to be employed.

CLASSES OF PACKETS AND THEIR TREATMENT

Voice traffic needs to move across many networks from source caller to destination caller. When many networks exist, many service providers and complex equipment are involved; no one entity has complete control or responsibility for the end-to-end journey of the voice conversation. This is similar to the situation with data networks and the Internet in particular. Therefore voice networks need to employ the same philosophy as the Internet. In its simplest form, the philosophy can be stated as "keep the center simple, move complexity to the edges." The routers in the center of the Internet are responsible only for forwarding packets along the best path to their destination. They need to examine only the IP header of a packet to make a decision. The complex job of processing a packet by TCP, UDP, and higher protocols is left to the host machines at either end of the journey. You can also see this in action with the routers that need to figure out the most efficient paths. Updating router tables is done by the routers when they communicate among themselves using special router protocols. This is independent of the forwarding of the data packets, which is a separate function.

If we apply this philosophy to VoIP, then it is easy to see that two functions are involved. First we need to identify the class that a packet belongs to in terms of special handling. This will allow us to treat all packets of the same class in the same way. Secondly, we have to define what the special treatment is and implement it in the intermediary devices, such as switches and routers. If we can identify and group VoIP packets together, we can give them special treatment as a class, specifically, giving them precedence over other kinds of packets.

Identifying and marking the packets is the first function and must be done at the edge of the network. Marking a voice packet would identify it as belonging to the class of voice packets. This would ideally be done by the VoIP telephone or software phone in the computer. Alternatively, it could be done by the Ethernet switch, or the first router that the packet encounters.

How a packet is handled for the next leg of its journey, is called the per hop behavior, or PHB. An example of a PHB is giving the class of VoIP packets precedence over data packets when they are forwarded by the router. Notice that the service provider is responsible for the PHB of a class of packets on his network. There is no universal definition of PHBs used by service providers. The upshot is that a voice packet will experience different levels of service as it crosses multiple networks from one end of the call to the other. Only if your call stays within the realm of one service provider are you guaranteed a level of service.

16.2 802.1Q AND 802.1P

Virtual LANs (VLANs)

A VLAN is a group of PCs, servers, and other network resources that behave as if they were connected to a single network segment—even though they may not be. Only computers that belong to the same VLAN can talk to each other directly. If a machine needs to communicate with another in a different VLAN, then the packet must be forwarded by a router or a switch. Many switches can make up a VLAN and this brings up a problem when a packet must be forwarded through multiple switches to get to its destination. How does the switch know which VLAN the packet belongs to? A multi-segment network with VLANs and multiple switches is diagrammed in Figure 16-1.

Two methods exist—implicit and explicit—for indicating VLAN membership when a packet travels between switches.

- Implicit—VLAN membership is indicated by the MAC address. In this case, all switches that support a particular VLAN must share a table of member MAC addresses.
- Explicit—A tag is added to the packet to indicate VLAN membership. The IEEE 802.1Q VLAN specification uses this method.

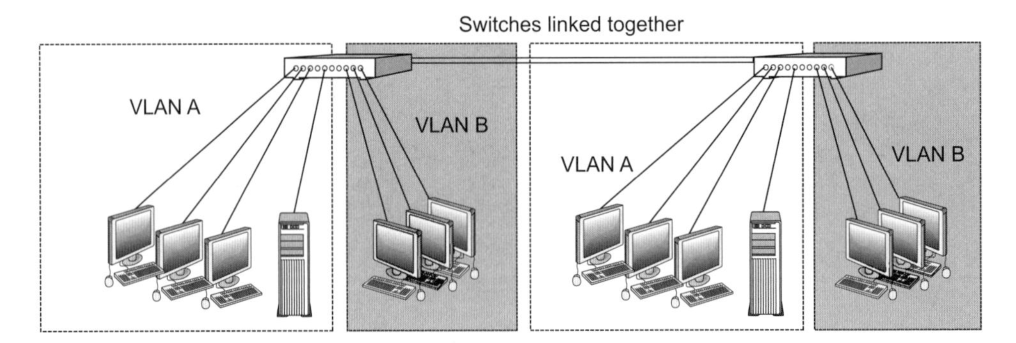

Figure 16-1: Virtual LANs

802.1 specifications

The 802.1 working group of the IEEE develops standards in the area of bridging and management for WANs and LANS. Of the many standards they have developed, three are relevant to this discussion.

IEEE 802.1D standardizes the methods that MAC level bridges can use to interoperate. IEEE 802.1p is an extension to 802.1D and defines how prioritization should be implemented within MAC level bridges.

802.1p provides three bits for prioritization, which yields eight distinct prioritization levels. These levels map directly to the levels built into some MAC types, specifically 802.4 (token bus) and 802.6 (MANs). 802.5 (token ring) and FDDI also support prioritization, but their schemes do not map directly to the eight levels of 802.1p. The one MAC that does not support priority inherently is Ethernet.

IEEE 802.1Q is a general-purpose VLAN implementation that also provides prioritization capabilities to those topologies that do not already support them, such as Ethernet. 802.1p provides the prioritization scheme for 802.1Q.

802.1Q tagging

In order to tag an Ethernet frame for 802.1Q, four octets are inserted into the Ethernet header. 802.1Q tags a frame for VLAN identification, but 802.1p is used to tag the frame for priority.

802.1Q/802.1p fields

Figure 16-2 shows a normal Ethernet frame and its structure after it has been modified with the 802.1Q/p information. The four octets of the 802.1Q/p fields are inserted between the source address field and the Ethertype/length field. The 802.1Q/p fields are described next.

Tag protocol identifier (TPI or TPID)

This two-octet field indicates that this frame contains 802.1Q data. The value will always be 8100h. Don't confuse this field with the type/length field of the Ethernet header. If the device is not 802.1Q compliant, Ethernet will think this is the type/length field and report an unknown type or Wellfleet (Wellfleet used to have the 8100 type number).

Priority (P)

This three-bit field indicates priority. This is the 802.1p specification. With three bits, eight different values are possible, from 0 to 7.

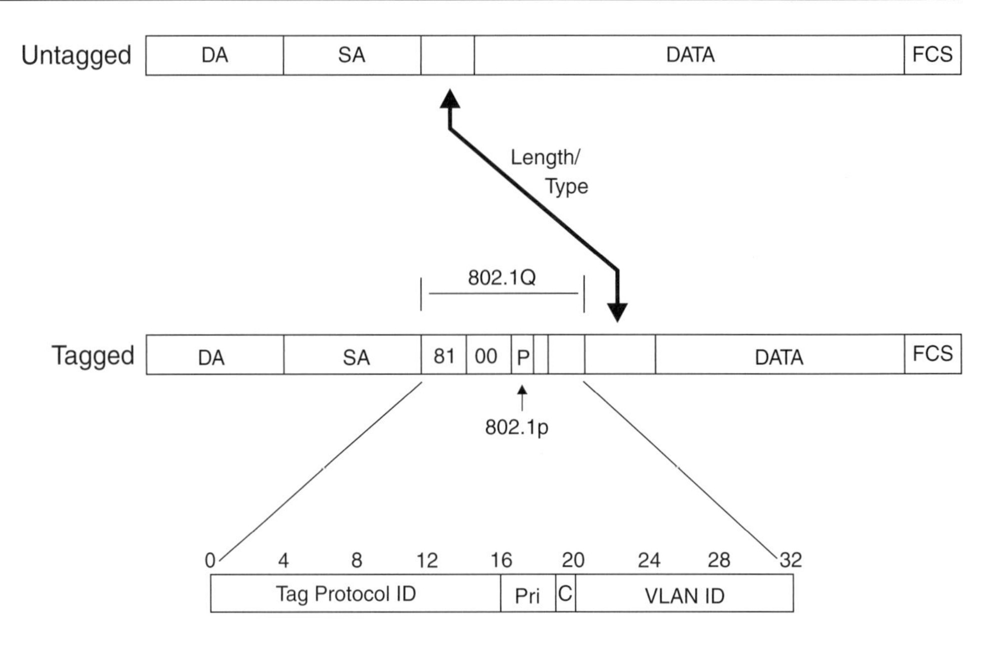

Figure 16-2: 802.1Q, 802.1p Ethernet frames

Canonical format indicator (C)

This one-bit field indicates whether the embedded MAC addresses are in canonical format. Ethernet uses 0. This field is used for compatibility between Ethernet and token ring.

VLAN identifier (VI)

This 12-bit field indicates which specific VLAN this frame belongs to.

Issues

Compatibility can be a problem with 802.1Q/p. The Ethertype and the length are two issues.

Devices that expect to see either an Ethertype field (Ethernet II) or a length field (802.3) in their correct positions will be confused.

With the addition of four extra bytes, what happens to the Ethernet standard length? There are two possibilities: Let an Ethernet frame grow four bytes from 1,518 to 1,522, or decrease the allowed payload length from 1,500 to 1,496 bytes. Remembering that 802.1Q is designed for switches that support VLANS, the simplest solution would be to configure the ports on the switch to either support or not support 802.1Q as required.

Figure 16-3 illustrates a decode of an Ethernet frame with 802.1Q. The protocol identifier is 81 00 hex. The next two octets (40 00) resolve down to a priority of 2 with no VLAN identifier.

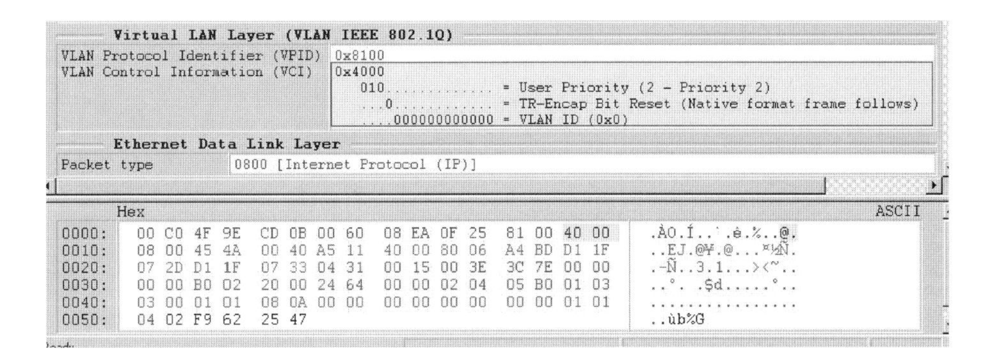

Figure 16-3: VLAN decode

16.3 TYPE OF SERVICE

The original specification for IPv4 required a one-octet field for type of service. The original thinking was that the American military would need priority for some of its data. When the American military created their own network, MILNET, they no longer used the Internet for their data. Nevertheless, the type of service field was defined and redefined several times along the way and Figure 16-4 diagrams its current structure. The fields are:

Three bits = Precedence
Four bits = Type of service
One bit = Must be zero and is not used

Precedence

The three bits of the precedence field define eight different precedence levels and can be used to give a packet priority over others of lesser precedence when they arrive at a router.

The precedence levels are listed as follows, including their bit patterns and a label.

000 (0) – Routine
001 (1) – Priority
010 (2) – Immediate
011 (3) – Flash
100 (4) – Flash Override
101 (5) – Critical

110 (6) – Internetwork Control
111 (7) – Network Control

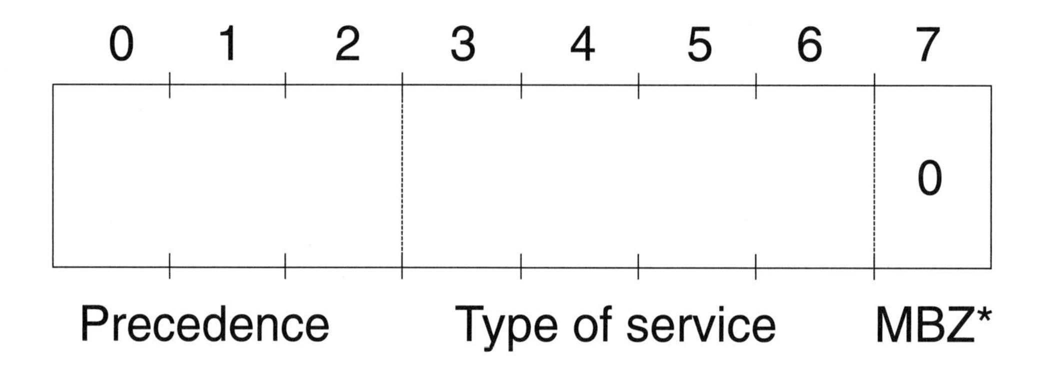

Figure 16-4: The IP type of service field

802.1p and precedence

If the packet has been tagged for 802.1Q, then it has a three-bit precedence field in the 802.1p. The eight precedence levels of 802.1p map exactly to the precedence levels in the IP packet. Therefore the precedence level can be inherited by IP from 802.1p and it will persist when the packet leaves the Ethernet system for the WAN.

Type of Service

Type of Service uses the next four bits of the ToS field. RFC 1349 defines five-bit patterns as follows:

1000 – minimize delay
0100 – maximize throughput
0010 – maximize reliability
0001 – minimize monetary cost
0000 – normal service

Despite these predefined values, any combination of 1s and 0s is acceptable. With four bits, there are 16 possible values.

What's wrong with ToS

The use of the ToS field in the IP header for precedence and type of service has been found inadequate. The implementation has been inconsistent with some routers supporting it while others did not. Even when supporting type of service was deemed to be desirable, router manufacturers often used other mechanisms. Another issue is that eight levels of precedence are not considered enough for modern communication needs. Recognition of these problems has lead to the development of differentiated services, which we examine in the next section.

16.4 DIFFERENTIATED SERVICES

Differentiated services (DiffServ) approaches the problem of QoS by dividing traffic into a small number of classes and allocating resources on a per class basis.

DiffServ extends ToS and addresses its weaknesses. Specifically, there are 64 possible levels of service. In addition, the designers of DiffServ hope that its usage will be universally applied now that a need for it has been demonstrated.

Not wanting to re-architect the IP header, the designers of DiffServ decided to reuse the poorly supported ToS field in the IPv4 header and redefine the IPv6 traffic class field. However, not wanting to abandon backward compatibility with the ToS field, eight levels of DiffServ map directly to the eight levels of precedence of ToS. The type of service in the ToS field has, however, been entirely abandoned. The standard for DiffServ was published as RFC 2474 in 1998.

The DiffServ field definition

The ToS field in the IP header is now renamed the DS (differentiated services) field. The first six bits are used to indicate a DiffServ codepoint (DSCP), which maps to a PHB as defined by the service provider. The last two bits are used to manage congestion. This is the explicit congestion notification (ECN) field defined in RFC 3168 in 2001.

Figure 16-5 illustrates the DiffServ field structure and, for comparison, the IP precedence field structure.

Behavior aggregates

Packets from different sources can have the same DSCP value and so can be grouped together as a behavior aggregate and treated in the same manner. All packets with the same DSCP value can be considered part of the same stream, no matter what their source or payload. Dealing with a small number of behaviors is easier for a router than having to codify each packet individually.

Per hop behaviors

Once we identify distinct streams of packets that have to be handled differently, how do we handle them? This of course is the purpose of the DiffServ exercise. Here we leave the standards track because there is no definition of per hop behaviors (PHBs).

This is left up to the service providers to implement; there is no adherence to a universal definition of how these should be handled.

Two functions that affect PHBs are queuing and dropping packets. Various algorithms are available for giving different data streams precedence. Queuing algorithms were discussed in Chapter 4. In addition, if the link becomes congested, packets will need to be dropped.

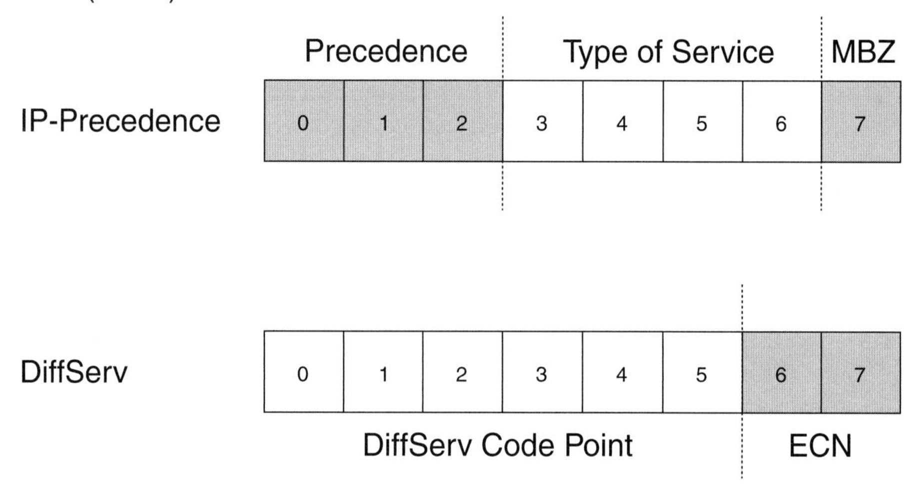

Figure 16-5: Comparing IP ToS to DiffServ

DSCP codepoint

The DSCP codepoint field is six bits and therefore allows 64 combinations, or levels of service. How can there be 64 distinct classes of service? In reality, there aren't. This gets complicated, so follow closely.

There is a major division in service called classes. The class selector identifying them is defined by the first three bits of the DSCP. There is a minor variation within the classes called assured forwarding (AF), which defines the probability that the packet will get dropped if there is bandwidth congestion.

Table 16-1 lists the DSCP values and you will need to refer to it as the discussion progresses.

The levels of the DSCP field are often referred to by names (CS1, AF23, etc.) or by decimal number. The decimal number is tricky, however. The DSCP field is only 6 bits and when converted to decimal goes only as high as 64. The full ToS field in the IP header is actually eight bits. If examining a protocol decode of the IP header, a different number would display. For example, take the level AF23 from Table 16-1. Its binary is 010110, which translates to 22 decimal. However, the actual value in the complete ToS field is 01011000 because the DSCP field is the first six bits and we assume that the last two bits are 00 for this example. This is the number 88 when converted to decimal.

The default value is 000000. This is used if there is no PHB agreement in place or if the system doesn't recognize or use the DSCP field.

There are seven class selectors. These, plus the default, correspond to the eight values in the IP precedence field. These make DiffServ backward compatible with IP precedence for any routers that still recognize the old ToS field.

The last two class selectors, CS6 and CS7, are defined for backward compatibility. They are not subdivided further into assured forwarding levels.

Class Selectors CS1 through CS4 are subdivided further with drop probabilities. In other words, a probability is assigned to the levels if the router needs to drop packets because of congestion. These are named assured forwarding (AF) levels. The numbering is straightforward; AF43 is Class Selector 4 and drop probability 3 (high).

Class Selector 5 only has one division, called Expedited Forwarding. It is the most secure level and is interpreted to mean low-loss, low-latency, low-jitter, and guaranteed bandwidth.

DSCP Values for Per Hop Behavior			
DSCP	Binary	Decimal	
Default	000000	0	
CS1	001000	8	Class Selector 1
AF11	001010	10	Low drop probability
AF12	001100	12	Medium drop probability
AF13	001110	14	High drop probability
CS2	010000	16	Class Selector 2
AF21	010010	18	Low drop probability
AF22	010100	20	Medium drop probability
AF23	010110	22	High drop probability
CS3	011000	24	Class Selector 3
AF31	011010	26	Low drop probability
AF32	011100	28	Medium drop probability
AF33	011110	30	High drop probability
CS4	100000	32	Class Selector 4
AF41	100010	34	Low drop probability
AF42	100100	36	Medium drop probability
AF43	100110	38	High drop probability
CS5	101000	40	Class Selector 5
EF	101110	46	Expedited Forwarding
CS6	110000	48	Class Selector 6
CS7	111000	56	Class Selector 7

Table 16-1: Differentiated code point behaviors

Explicit congestion notification

The explicit congestion notification (ECN) feature was defined later (RFC 3168 in 2001) than DiffServ (RFC 2474 in 1998). Its purpose is to notify intermediate devices, such as routers and end points, about congestion on the link. It is important to point out that if there is actual congestion on the link then some packets will have to be sacrificed. This can be a messy business. If a router has no other information, then packets may be dropped at random. This forces TCP to retransmit packets. UDP packets are lost for good.

ECN has the ability to mark a link as congested. Historically, a router could mark a link as being congested simply by dropping packets. Note that the link might not actually be congested, just close to being congested. If notified of link congestion, TCP in the end points might take some compensating action, such as increasing buffer space.

With the ECN mechanism, a router could notify TCP end-points of the congestion without actually dropping packets.

ECN uses the last two bits (the least significant bits) of the ToS field. There are only four possible combinations that are interpreted as follows:

00 – not ECN
01 – ECN capable
10 – ECN capable
11 – CE (congestion experienced)

The values 01 and 10 are interchangeable and mean that ECN is available. If there is congestion, this value changes to 11. If the packet arrives at a router where there is no buffer space, it is dropped.

16.5 MULTIPROTOCOL LABEL SWITCHING

Why MPLS?

With the QoS techniques already discussed, you may wonder why there is a need for yet another QoS technique. Although QoS techniques are available for IP, they aren't much good if the underlying data link layer doesn't cooperate. We have explored the 802.1p protocol used by Ethernet; however, Ethernet is not used on WAN systems. Indeed, WAN systems use disparate technologies, but the most commonly used by service providers are ATM or frame relay.

Multiprotocol label switching (MPLS) is a technology favored by service providers because it can provide QoS across data link boundaries. The technology allows the service providers to offer service level agreements (SLAs) to their customers.

MPLS

MPLS works very similarly to the other techniques already discussed. At the entrance to the system, a packet is identified, assigned a service class, and marked by a label in the MPLS header. The router/switches in the system can then forward the packet based on its class. MPLS is a blend of traditional IP routing technology and layer 2 switching technology. Indeed, you could say that it belongs to "layer 2.5." To understand how MPLS works, let's look at the details of the system.

Basic components

The basic building blocks of the MPLS system are the label edge router, the label switch router, and the label switch path.

- **Label edge router (LER):** Sometimes called the Edge LSR, this device is the first and last device encountered by the packet on the MPLS system. The ingress LER identifies the class (forward equivalency class or FEC) and attaches a MPLS header with this information plus a label for identification to the packet. The final LER (the egress LER) strips the header from the packet as it exits the system.
- **Label switch router (LSR):** At each hop in the network, a LSR examines the incoming label to figure out the next forwarding hop for the packet. At each hop, the LSR strips off the existing label and applies a new label that tells the next hop LSR how to forward the packet.
- **Label switch path (LSP):** The LSP is the end-to-end MPLS path for the packet.

Refer to Figure 16-6 for a diagram of the MPLS network.

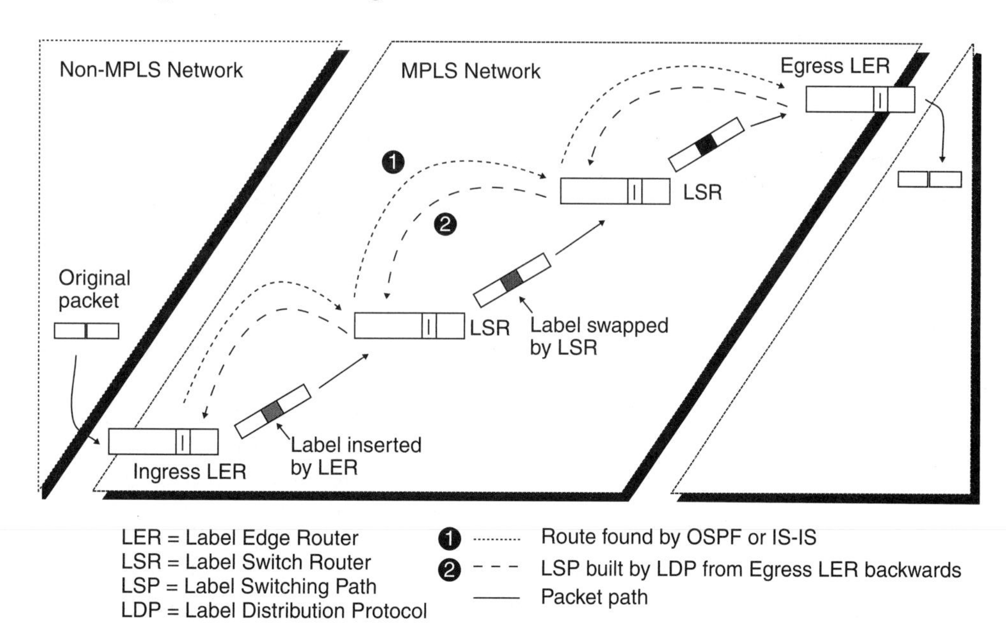

Figure 16-6: Overview of the MPLS network

Switching increases performance

At each hop in the network, the LSR examines the incoming label to figure out the next forwarding hop for the packet. This eliminates resource-intensive address lookups that reduce overall packet throughput and limit scalability. This is not the traditional IP routing in which the router must look up the address tables in order to determine what the next hop is. Like circuit-switched networks, MPLS establishes the end-to-end connection path before transferring information. Another point is that there are few labels because they are dependent on the Forward Equivalency Class and this results in much smaller label tables.

The crux of the performance increase is that each LSR switches the packet based on the MSRP label instead of routing the packet using a complex network layer routing algorithm based on the IP address and the routing table.

Forward Equivalency Class (FEC)

All of the packets are aggregated into a small number of FECs. A FEC is a number of packets that will be forwarded in the same manner, i.e., that follow the same path and are given the same treatment.

One example of a FEC is a set of unicast packets whose network layer destination address matches a particular IP address prefix. A set of multicast packets with the same source and destination network layer addresses is another example of a FEC. Yet another example is a set of unicast packets whose destination addresses match a particular IP address prefix and whose type of service bits are the same.

Besides the destination IP address, other characteristics can be used to assign packets to a particular FEC. Properties from layer 2, layer 3, and layer 4 can be recognized by MPLS. These may include the TCP/UDP port number, VLAN identifier, DiffServ, or 802.1p priority.

How is the label switch path determined?

There are two issues with the label switch path (LSP). First what is the actual path that the packet needs to take through the network and secondly, setting up the LSP with the labels and label tables in the LSRs and LERs.

When a packet enters the network with a new FEC, there are no entries in tables maintained by the LSRs and LERs. A path is first found by the normal discovery process implemented in IP routers by using the router protocols, such as OSPF or IS-IS. The egress LER must now commence building the LSP taking into account any constraints imposed by quality of service or class of service constraints. This requires negotiation between the LSRs and LERs. A special signaling protocol is used for this negotiation and LSP setup. The protocol is called a label distribution protocol (LDP) and in fact, one of these protocols is actually called label distribution protocol. Also available are an extension to LDP called constraint-based routing-LDP (CR-LDP) and the RSVP-TE protocol. TE stands for traffic engineering. These LDPs are used to distribute the label/FEC binding to the LSRs and LERs along the path.

Notice that the LSP is built from the exit router backward to the entrance router. Labels are only locally meaningful in that when the LSR receives the packet, it uses the label to decide how to forward it. Therefore, the upstream LSR must request the label from the downstream LSR. When it receives the label, it attaches it to the packet and then forwards it to the downstream LSR. This procedure is repeated at each LSR until the packet reaches the egress LER, which strips the label from the packet before forwarding it to the non-MPLS system.

MPLS header structure

Figure 16-7 diagrams the MPLS header. It is four octets/32 bits with four fields and has the following structure:

Label

This is a 20 bit field mapping the FEC to an MPLS identifier. This is the label value.

Experimental — EXP

It is generally accepted that this three-bit field differentiates classes of service or per hop behavior for differing classes of traffic traveling within the LSP.

Bottom of stack — S

This single-bit field in position 23 represents the last MPLS label contained in the packet. MPLS labels can be stacked, that is, multiple labels can be nested one after the other. The S bit indicates if there are any additional labels or if this is the final one.

Time-to-live — TTL

The final eight bits of the MPLS shim is analogous to the IP TTL field. When the packet arrives at the ingress LER, it copies the TTL value from the IP TTL field into the TTL field of the label. Each LSR decrements this field by one as it forwards the packet. The egress LER replaces the IP TTL field with the new value it reads from the TTL field in the MPLS label. In this way, the IP TTL value is the same as if it had been forwarded by traditional IP routers.

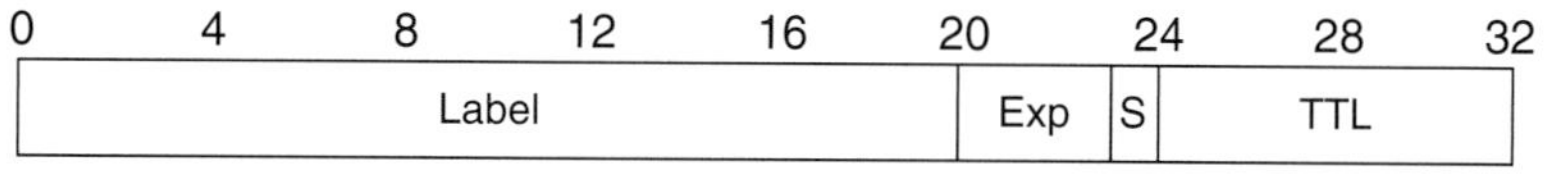

0–20	20–23	23–24	24–32
Label	Exp	S	TTL

Bit positions: 0 4 8 12 16 20 24 28 32

Figure 16-7: MPLS header structure

Label Insertion

MPLS is handled differently on an Ethernet network than with ATM or frame relay. MPLS is basically a virtual circuit system in which the path has to be set up before the data can flow and the label holds what in essence is the virtual circuit identifier. ATM and frame relay also use these concepts and have the structure to hold virtual circuit identifiers.

Ethernet

Ethernet, on the other hand, does not. The MPLS label is inserted after the data link header of Ethernet and before the network layer header. The MPLS label is also referred to as a "shim" and its insertion after the Ethernet header is illustrated in Figure 16-8.

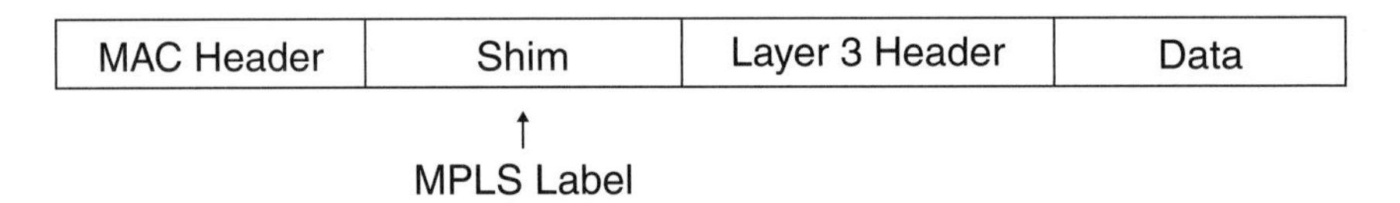

Figure 16-8: MPLS over Ethernet

The layer-2 EtherType 8847 immediately following the layer-2 source MAC address identifies the packet as containing the MPLS shim. Although the network layer is assumed to be IP, the "multi-protocol" in MPLS refers to the fact that any layer-3 protocol, such as IPX, could be used with MPLS.

ATM

Asynchronous transfer mode uses fixed 53 octet cells of which 48 octets are data and five octets are the header. The five-octet header contains the virtual circuit identifier (VCI) and virtual path identifier (VPI). The header is sufficiently large to hold the four-octet MPLS label. Therefore, when MPLS runs over ATM, the VCI/VPI fields hold the MPLS label. Notice that the ATM network in effect becomes an MPLS network. Most ATM switch vendors have upgraded their switches to provide MPLS capability. This allows the service providers to easily upgrade their networks. In fact, most ATM networks allow native ATM and MPLS to coexist side-by-side. This is often referred to as "ships in the night." Figure 16-9 illustrates MPLS over ATM.

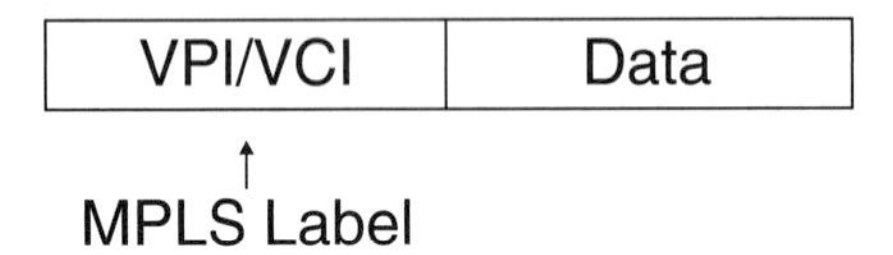

Figure 16-9: MPLS over ATM

Frame relay

Frame relay also uses virtual circuits. Its header contains the data link circuit identifier (DLCI). When MPLS runs over frame relay, the MPLS label is placed in the DLCI field. Figure 16-10 illustrates MPLS over frame relay.

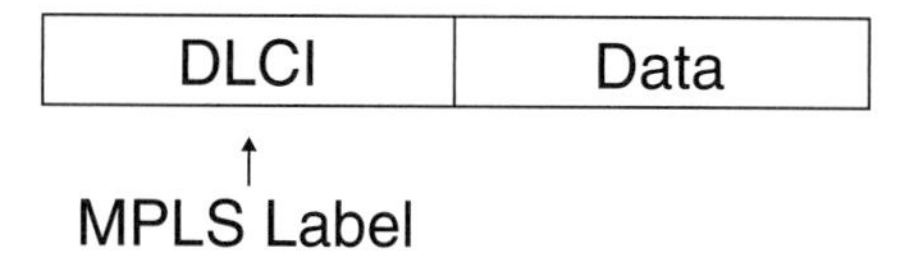

Figure 16-10: MPLS over frame relay

Now that we understand MPLS, we can understand why it is becoming a favorite of telecom carries. It provides QoS while still running over their in-place infrastructures of frame relay and ATM.

SUMMARY

Section 16.1: QoS Overview

This section provides an overview of QoS and the parameters that affect it including bandwidth, delay, jitter, lost packets, and echo.

Section 16.2: 802.1Q and 802.1p

This section discusses the 802.1Q/802.1p standard, which provides QoS for Ethernet.

Section 16.3: Type of Service

This section looks at the ToS field in the IP header and how it was originally intended to provide QoS for IP traffic.

Section 16.4: Differentiated Services

This section looks at differentiated services, which is the modern use for the ToS field in the IP header.

Section 16.5: Multiprotocol Label Switching

This section discusses MPLS, which is favored by the telecom carriers.

Hands-on Lab 16-1

Setting DiffServ Codepoint

In order to give precedence to a VoIP packet it needs to be marked for special treatment. In this exercise you will configure X-Lite to mark the voice packets with a DiffServ codepoint.

1. Open up the X-Lite softphone.
2. Place your mouse cursor anywhere on the X-Lite softphone and right-click. From the menu, select **Options...**
3. On the far left of the Options screen, at the bottom of the windows, click on the **Advanced** button. Because of its position on the screen, this button is easy to overlook.
4. Select **Quality of Service.**
5. Under **Audio QoS**, select **Use DSCP / ToS value.** Type in a decimal value from Table 16-1. The value can actually be within the range of 0 to 63, although 0 is the default.

   For illustration purposes for the rest of the exercise, assume that you use decimal 36 (CS4, AF42) as the code point.

   Save your configuration change by pressing the **OK** button.
6. Start Wireshark.

   Start capturing packets in Wireshark.

   Make a telephone call to another student.

   Talk for a short while and then terminate the call.

   Stop the packet capture in Wireshark.

   In the Filter box, type in RTP. You should now have a list of the RTP packets that capture your telephone conversation.
7. Select any RTP packet. In the decode section of Wireshark, expand the Internet Protocol section.

   How does Wireshark decode the Differentiated Service field? Does it match the configuration change that you made in X-Lite?

   If you had used 36 for your code point, then you would see DSCP 0x24 (24 hex = 36 decimal) and Class Selector 4.

8. Wireshark recognizes the differences between Diffserv and IP ToS. You can configure Wireshark to decode the IP type of service field either way. In order to change this parameter, use the following path in Wireshark.

   Edit > Preferences. . . > Expand **Protocols** > Scroll down to **IP**

   Find the option: **Decode IPv4 TOS field as DiffServ field:**

   This option should have a check mark. Leave it.

9. Exit Wireshark and X-Lite.

Review Questions Implementing QoS

1. VoIP devices can mark packets, even different kinds of packets, in order to apply a Per Hop Behavior (PHB). What is the advantage of PHBs?

   a) It allows voice packets to travel between service providers.
   b) All packets that need to be treated the same are grouped together.
   c) Voice packets can be distinguished from data packets.
   d) They identify packets that can be dropped when a link is oversubscribed.

2. Under what circumstances would using Virtual LANs be beneficial for VoIP?

   a) Your WAN link is oversubscribed and you want to increase bandwidth.
   b) You want to upgrade your Ethernet connections for your VoIP phones to 100 Mbps and use gigabit Ethernet for connections between switches.
   c) You want to improve the delay and jitter characteristics of your VoIP system.
   d) You want to segregate voice traffic from data traffic for performance and security reasons.

3. The IP ToS field was originally used to mark IP packets for QoS, but has largely been replaced by DiffServ. What factor contributed to the loss of popularity of ToS?

   a) ToS was not popular during the early Internet years and routers didn't bother to support it.
   b) ToS was designed for the use of the military and the military no longer uses the Internet.
   c) ToS supports only six precedence levels and this isn't considered sufficient for today.
   d) ToS is considered inefficient because one bit is always zero and is wasted.

4. You capture a voice packet with a protocol analyzer and examine the DiffServ / ToS field. The value is 184 decimal. What DSCP value is this?

   a) Class Selector 3 with medium drop probability
   b) Class Selector 1 with low drop probability
   c) Class Selector 5
   d) Expedited forwarding

Review Questions: continued

5. Which one of the following statements regarding MPLS is the most accurate?

   a) An MPLS system uses normal routing to select a path to the destination, but then uses MPLS to forward the packets.
   b) MPLS changes the structure of an ATM cell and frame relay frame.
   c) The TTL field in the MPLS label is always initialized at 0.
   d) MPLS uses the existing fields in the Ethernet header to carry information.

CHAPTER

17

H.323

Objectives

Upon completion of this section, the reader will:

- *Understand that H.323 was designed for VoIP, multimedia, and data transmission.*
- *Appreciate that H.323 uses many protocols, including some from other standards.*
- *Know how call signaling works and how connections are made.*

INTRODUCTION

H.323 defines how audio and video control information is formatted and packaged for transmission over the network. H.323 is a family of protocols that supports audio, video, and conferencing. This module, however, concentrates on the H.323 functions as they relate to VoIP.

17.1 H.323 OVERVIEW

H.323 is a family of protocols used to control VoIP. When discussing H.323, the reader quickly runs into information overload simply because H.323 was not designed just to handle VoIP; it is a very broad set of standards that provides for video and data conferencing as well as audio.

H.323 is defined by the International Telecommunication Union (ITU) and is an umbrella for other protocols. H.323 is not a single protocol that stands alone. Instead, it sensibly uses a wide variety of other protocols that are perfectly capable, even if they are defined by other standards setting bodies. This in fact is one of the strengths of H.323. Because the ITU did not have to design all of the components of H.323 from the ground up, it was implemented very quickly.

A call model, similar to the ISDN call model, eases the introduction of IP telephony into existing networks of ISDN-based PBX systems. A smooth migration toward IP-based PBX systems is a particular strength of H.323. Within the context of H.323, an IP-based PBX is, simply speaking, a gatekeeper plus supplementary services.

H.323 covers the following areas as diagrammed in Figure 17-1:

- Control and call signals
  H.245 — used to negotiate channel usage and capabilities
  H.225/Q.931 — used for call signaling and call setup
  H.225 RAS — Registration/Admission/Status is a protocol used to communicate with a gatekeeper
- Audio codecs — G.711, G.722, G.723, G.728, G.729
- Video codecs — H.261, H.263
- RTP/RTCP — used for sequencing audio and video packets

Notice that there is no individual protocol called H.323. If you capture the exchange of setup packets for a VoIP call using a protocol analyzer, you will search in vain for a packet labeled H.323.

H.323 elements

H.323 defines four major components for a network-based communications system: terminals, gateways, gatekeepers, and multipoint control units.

- Terminals on a VoIP system are telephones and PCs equipped with VoIP capabilities. These are the end-point devices of the H.323 system.
- Gateways are the bridge to the PSTN.

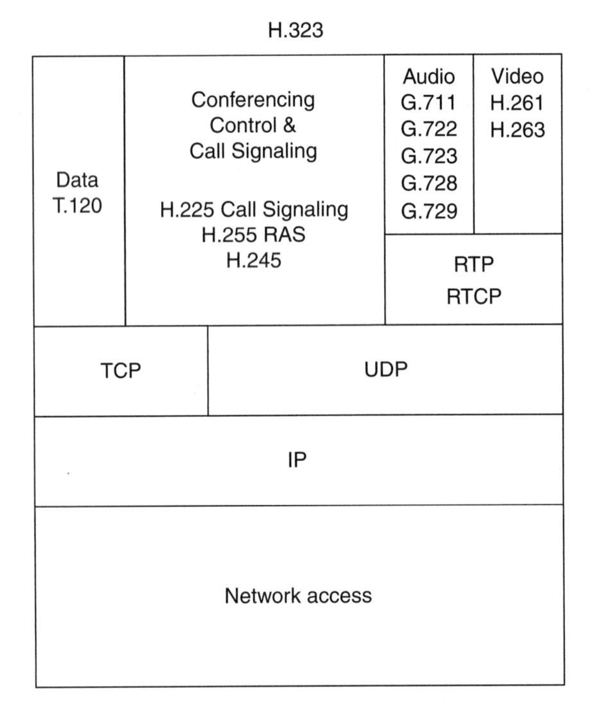

Figure 17-1: The H.323 protocol suite

- Gatekeepers act as the central point for all calls within the zone and provide call control services to registered endpoints. In a previous discussion, the H.323 gatekeeper was called the call manager. To review, the functions of the gatekeeper include address resolution and mapping, call initiation and establishment, admissions control, bandwidth control, and zone management.
- Multipoint control unit (MCU) supports conferences between three or more endpoints. Conferencing can be audio, video, or data.

17.2 CALL SIGNALING AND SETUP

The core of H.323 is setting up the telephone call. H.323 uses H.225 RAS, H.225 call signaling, and H.245 for call signaling and setup. All three protocols are used to set up and tear down a call, but they have different jobs to do. Let us examine the three protocols from the perspective of the role they play in the system. The interaction among the protocols is diagrammed in Figure 17-2.

H.225.0 has two parts, RAS (registration, admission, status) and call signaling.

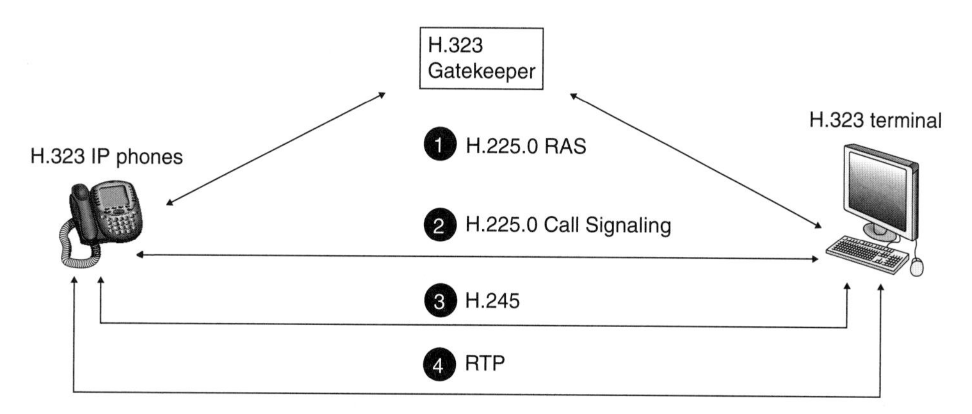

Figure 17-2: H.323 call setup

H.225.0 RAS

H.225.0 RAS (registration, admission, status) messages define communications between endpoints and a gatekeeper. H.225.0 RAS is only needed when a gatekeeper exists. It should be noted that an end-point to end-point H.323 call can be made that does not require a gatekeeper. In this case, the end user will need to know and be able to input the address of the other end-point directly.

Unlike H.225.0 call signaling and H.245, H.225.0 RAS uses unreliable transport for delivery. In an IP network, H.225.0 RAS uses UDP over ports 1718 and 1719.

H.225.0 RAS communications include:

Gatekeeper discovery: Gatekeeper discovery is used by end-points to find out their gatekeeper. An end-point that needs to find the transport address of its gatekeeper(s) will multicast a gatekeeper request (GRQ) message. One or more gatekeepers may reply with a gatekeeper confirm (GCF) message containing the gatekeeper transport address.

Endpoint registration: Once a gatekeeper exists, all endpoints must be registered with it. This is necessary because gatekeepers need to know the aliases and transport addresses of all endpoints in its zone to route calls.

Endpoint location: Gatekeepers use this message to locate endpoints with a specific transport address. This process is required, for example, when the gatekeeper updates its alias-transport address database.

Other communications: A gatekeeper performs many other management and control duties such as admission control, status determination, and bandwidth management, which are all handled through H.225.0 RAS messages. Table 17-1 lists RAS messages.

RegistrationRequest (RRQ)	This message is sent to a gatekeeper by an endpoint to request that the endpoint be registered and become under the control of that gatekeeper.
RegistrationConfirm (RCF)	A positive reply from a gatekeeper to a RegistrationRequest message.
RegistrationReject (RRJ)	A negative reply to a RegistrationRequest message.
AdmissionRequest (ARQ)	A request from an endpoint to a gatekeeper for permission to participate in a call. The gatekeeper formulates its reply based on several factors, including the availability of bandwidth in the network.
AdmissionConfirm (ACF)	A positive response from a gatekeeper to an AdmissionRequest message.
AdmissionReject (ARJ)	A negative response to an AdmissionRequest message.
LocationRequest (LRQ)	This message is sent to a gatekeeper to request its address translation service.
InfoRequest (IRQ)	This message is sent from a gatekeeper to request status information of its recipient.

Table 17-1: H225.0 RAS messages

H.225.0 call signaling (including Q.931)

H.225.0 RAS is used only when a gatekeeper is available on the system, but H.225.0 call signaling is always used. The major portion of H.225.0 call signaling is another protocol called Q.931. Q.931 is the call setup protocol of ISDN. By using Q.931, H.323 can easily integrate with ISDN and voice PBXs.

Q.931 is used to set up and tear down calls through an ISDN network. Q.931 is comparable to TCP in at least one respect. They are both responsible for the setup and tear-down of sessions on their respective networks. H.225 is transmitted through a network using TCP, usually to port 1720. Table 17-2 lists the H.225.0 call signaling messages.

Setup-UUIE	This message is used to initially request that a call be set up. It is the equivalent of dialing a number on a normal telephone.
CallProceeding-UUIE	This message is sent to the calling party to indicate that the call is currently being processed by the called terminal.
Alerting-UUIE	This indicates to the calling party that the called terminal is ringing.
Connect-UUIE	This is sent from the called terminal back to the calling terminal. It is the equivalent of answering the call. After a connect has been received, the actual transmission channels need to be set up using H.245.
ReleaseComplete-UUIE	This message can be sent by either party participating in an active call. It is an indication that the sender wishes to end the call. An acknowledgment is not required, although it is sometimes sent.

Table 17-2: H.225.0 call signaling (Q.931) messages

H.245 — Media control

The flexibility of H.323 requires that end-points negotiate to determine compatible settings before audio, video, and/or data communication links can be established. H.245 uses control messages and commands that are exchanged during the call to inform and instruct. The implementation of H.245 control is mandatory in all end-points.

H.245 provides the following media control functions:

Capability exchange: H.323 allows end-points to have different receive and send capabilities. Each end-point records its receiving and sending capabilities (e.g., media types, codecs, bit rates, etc.) in a message and sends it to the other end-point(s).

Opening and closing of logical channels: H.323 audio and video logical channels are unidirectional end-to-end links (or multipoint links in the case of multipoint conferencing). Data channels are bidirectional. A separate channel is needed for audio, video, and data communication. H.245 messages control the opening and closing of such channels. H.245 control messages use logical channel 0, which is always open.

Flow control messages: These messages provide feedback to the end-points when communication problems are encountered.

Other commands and messages: Several other commands and messages may be used during a call such as a command to set the codec at the receiving end-point when the sending end-point switches its codec.

H.245 control messages may also be routed through a gatekeeper if one exists.

SUMMARY

Section 17.1: H.323 Overview

This section looks at the protocols and components of the H.323 system.

Section 17-2: Call Signaling and Setup

This section looks at the roles that H.225.0 RAS, H.225.0 call signaling, and H.245 have in the call signaling process of H.323.

Hands-on Lab 17-1

Capturing H.323 Packets

In this exercise, you will install SJphone, another softphone. SJphone is used in this exercise because it supports H.323 calls. After you install SJphone, you will register with your SIP server and prove it works. Next you will use it as an H.323 phone and capture the packets as you make the telephone call. Because you do not have an H.323 gatekeeper, you will need to contact your partner directly by using the IP address of her computer.

Install the SJphone VoIP phone

1. Before you start this exercise, you must close down and exit from any softphone, such as X-Lite, that you have running.
2. The instructor has made the SJphone phone software installation file available or you have downloaded it from the Internet.
3. Start the installation. Use all defaults. Start the program.
4. You will be asked to provide an account name and password to initialize the service profile. Use the account that you have on the SIP server.
5. In order to register with the SIP server, you must modify a profile. Place your mouse cursor in the gray information box at the bottom of the SJphone screen and right click. Select **Options...**
6. Select the **Profiles** tab and select **SIP server.** Click the **Edit** button. Select the **SIP Proxy** tab. In the **Proxy (URI)** field, type in sip: the *IP address of the SIP* server (CSX SIP server). Click the **OK** button. Click the **Initialize** button. Provide the account name and password when asked. The account is the extension number, and the password must be accurate. SJphone should now initialize.
7. After you have registered SJphone, return to the 3CX management console. Click on **Line Status.** The extension that you registered shows the status of On Hook. You can now make a telephone call to someone in the class.

Change to the H.323 protocol

In order to use the H.323 protocol, you need to create and use a new profile.

8. Select **Options...** as before. Select the **Profiles** tab and click on the **New** button. Figure 17-3 shows the **Create New Profile** window that appears.

   The profile type shows the flexibility of SJphone. You can make a direct station-to-station SIP or H.323 call as well as use an H.323 gateway or gatekeeper.

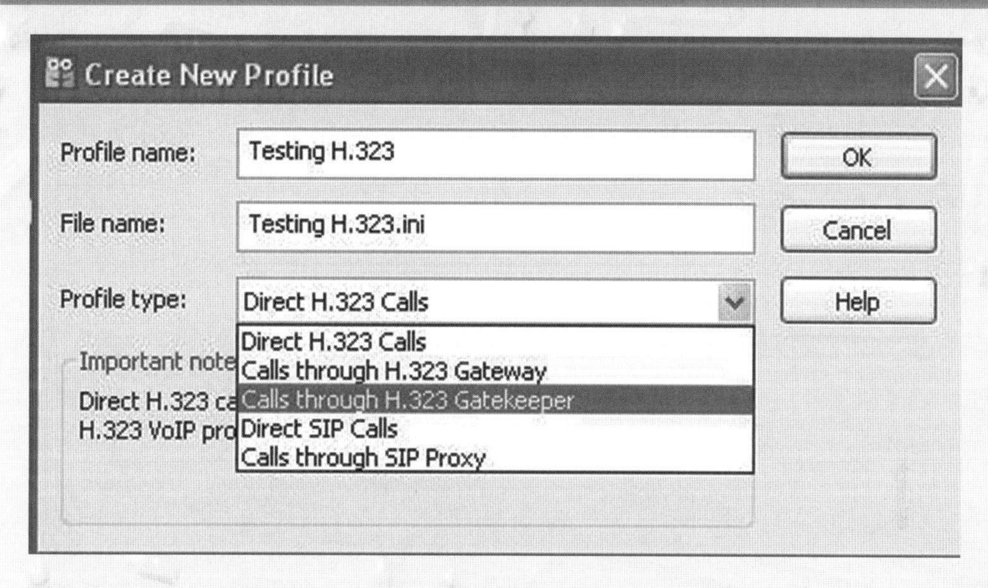

Figure 17-3: SJphone profiles

9. Type in a profile name. This will appear on the list of profiles. The file name will be filled in automatically. Select **Direct H.323** as the Profile type. Click **OK** twice. On the **Profiles** tab, your new profile should show its status as "in use." Click **OK.** The message in the status box of SJphone's front screen may take a while to change but eventually it should show the new profile name.
10. You are now ready to make a phone call to another person who has also configured SJphone for H.323 service. However, because this is a direct call, you must type in the IP address of the other person's machine.
11. Start your protocol analyzer. Make a call to another student and capture your call.
12. Stop the capture, exam it, and see if you can identify the H.225 and H.245 packets used to set up the call.

Hint: You cannot filter on H323. Instead filter on H225 or H245.

Review Questions H.323

1. Which protocol is used by H.323 to define communications between endpoints and gatekeepers?

    a) H.225.0 RAS
    b) H.245
    c) Q.931
    d) H.225.0 call signaling

2. Why is H.323 easy to integrate with ISDN and PBXs?

    a) Because it is controlled by the ITU.
    b) Because registration, admission, and status are controlled by H.225.0.
    c) Because it uses Q.931 as part of call signaling.
    d) Because the PBX manufacturers make a line card for H.323.

3. In an H.323 system, the handshake that negotiates the codec is provided by which protocol?

    a) RTP
    b) H.225.0 call signaling
    c) H.225.0 RAS
    d) H.245

4. If you use a protocol analyzer to capture the packets that provide the call setup for an H.323 telephone call, why can't you find any H.323 packets?

    a) The H.323 information is carried by the H.245 packet.
    b) H.323 is not a type of packet.
    c) You need to filter on Q931.
    d) Your protocol analyzer does not have the decode for H.323.

CHAPTER

18

SIP in Detail

Objectives

Upon completion of this section, the reader will:

- *Understand the role of the SIP protocol in the setup of telephone conversations.*
- *Have been exposed to the SIP architecture, which includes user agents, registration servers, and proxy servers.*
- *Appreciate how a session setup is made under SIP.*
- *Know the SIP message structure.*
- *Distinguish the SDP protocol from SIP and how it is used in conjunction with SIP to negotiate the codec used by the VoIP call.*

INTRODUCTION

Making a telephone call using VoIP technology requires two phases, the call setup and the conversation itself. SIP is one of two technologies used for the call setup phase (the other being H.323). Although SIP provides session setup, it doesn't negotiate which codec will be used by the call. This aspect of the setup is provided by the session description protocol (SDP). This chapter looks at the SIP and SDP protocols in detail.

18.1 SESSION INITIATION PROTOCOL (SIP)

Session initiation protocol (SIP) is an IETF standard (RFC 3261) for setting up sessions between one or more clients. It is one of the two important signaling protocols for voice over IP, gradually replacing H.323 in this role.

SIP's purpose

Although used for VoIP, it is important to appreciate that SIP was not designed specifically for VoIP. Its purpose is to facilitate the setting up of sessions between two parties for communication purposes and, as such, it can also be used for instant messaging, Internet conferencing, and presence and event notification.

Request-response protocol

SIP is a request-response protocol that closely resembles two other Internet protocols, HTTP and SMTP (the protocols that power the World Wide Web and e-mail); consequently, SIP sits comfortably alongside Internet applications. Using SIP, telephony becomes another Web application and integrates easily into other Internet services.

The concept of request-response simply means that SIP has only two message types: a request for a service or a response to that request.

SIP functions

SIP was designed to provide the following functions:

- SIP allows for the establishment of user location, i.e., translating from a user's name to the user's current network address.
- SIP provides for feature negotiation so that all of the participants in a session can agree on the features to be supported among them.
- SIP is a mechanism for call management—for example adding, dropping, or transferring participants.
- SIP allows for changing features of a session while it is in progress.

Any additional requirements for a telephone call are handled by other protocols and services.

Works with other protocols

The designers of SIP did not want to reinvent the wheel. They appreciated that other protocols were already available to fill certain functions of making a telephone call. SIP was designed to cooperate with these other protocols to build a complete multimedia architecture.

Examples of these other protocols include real-time transport protocol (RTP) for transporting real-time data, the real-time control protocol (RTCP) for providing QoS feedback, the real-time streaming protocol (RTSP) for controlling delivery of streaming media, the media gateway control protocol (MEGACO) for controlling gateways to the public switched telephone network (PSTN), and the session description protocol (SDP) for describing multimedia sessions. However, the basic functionality and operation of SIP does not depend on any of these protocols.

Session not content

It is important to distinguish between setting up a session and controlling the content of a session. The content includes the type and format of the data stream between the two parties. The content will be different for a telephone call and for video conferencing. SIP does not concern itself with the content, but can carry another protocol that does negotiate the content. The session description protocol (SDP) is the most common protocol used for this purpose. SIP will carry SDP as the payload.

18.2 SIP ARCHITECTURE

SIP uses the following components:

User agent: The UA is the user end-point device such as a VoIP telephone, PDA, cell phone, or a user personal computer with a soft phone. The user agent client initiates the telephone call with a SIP session.

Registration server: The registration server authenticates the user and adds the mapping between URL and network address to the location server's database. When the user agent starts up, the first message it sends is a REGISTRATION.

Location database: The location database maintains the database of name to location (IP address usually) mappings. The information in the database is usually acquired from the user agent registrations, but may be acquired in other ways as well, such as DNS. The database may be queried in various ways, although LDAP is the most common. When a user agent wants to connect to a remote SIP end-point, it queries the location database in the location server for the contact information.

Proxy server: Proxy servers, as their name suggests, act on behalf of user agents, routing SIP messages to correct destinations.

Redirect server: A redirect server differs from a proxy server in that it does not forward messages but simply does a location look-up and returns one (or more) addresses for the destination and leaves it up to the original user agent to contact the destination at these addresses directly.

A SIP server will include some or all of the preceding functions or the functions can be split between multiple machines.

18.3 SESSION SETUP

In order to set up a session, an exchange of informational packets between the user agents (the end-points) is required. In SIP terminology, these packets are called messages. Figure 18-1 is a simple example diagramming the exchange of messages during a basic conversation.

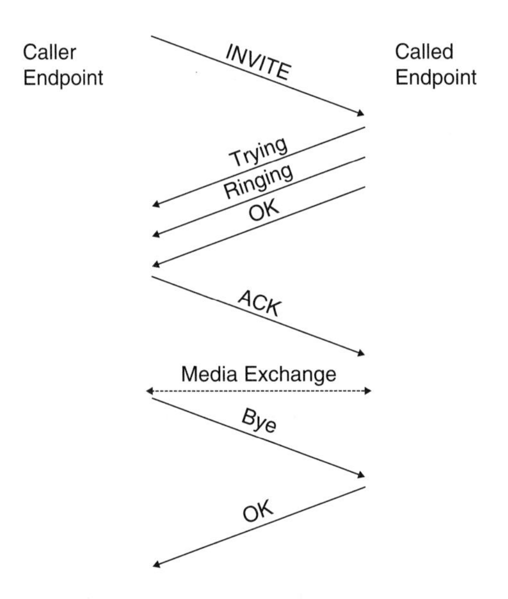

Figure 18-1: Exchange of SIP messages for a basic conversation

Figure 18-1 is a simple depiction of the session initiation process. At its core, the session starts with a caller picking up the telephone and sending a SIP INVITE message to the called. The called device responds with a Trying response indicating that the request is pending, and a Ringing response indicating that the device is sounding its bell but has not yet been picked up. When the telephone is answered, an OK response is sent to the caller. At this point, the session has been set up and the caller sends an ACK message to confirm. The caller and called now carry on a conversation, the *media exchange*. At the end of it, the telephones are replaced on their receivers; the session is closed when a BYE message is sent.

Before looking at a more complex scenario involving proxy servers, we need to examine the message types. The two message types are requests, called *methods*, and the *responses*, which are represented by codes.

Methods

The different types of requests are known as methods. Example methods from the previous simple setup scenario were INVITE and ACK. Table 18-1 lists the six basic methods and the seven method extensions. As development work progressed on SIP, additional methods were added in successive RFCs.

Basic Method Types	
INVITE	Initiate the call. This method includes information about the caller and called user and the type of media that is to be exchanged.
ACK	Sent by the user who initiates the INVITE. ACK is sent to confirm that the session is established.
BYE	Terminate (and transfer) call. This method can be sent by either user.
CANCEL	Cancel a pending request, including searches and ringing. After a session is established, a BYE is used to terminate the session.
OPTIONS	Features support by other side. Queries the other server or device for its capabilities. It can be used to check media capabilities before issuing an invite.
REGISTER	Used by a client to login and register its address with a location service.
SIP Method Extensions	
SUBSCRIBE	Enables a user to subscribe to events. There needs to be a method to alert the user when an event occurs.
NOTIFY	Notify users that a subscribed event has occurred.
MESSAGE	Used for instant messaging. A user sends an instant message to another user by sending a request that includes the MESSAGE method. This request sends the actual text in a body of a SIP packet.
INFO	Used for transferring information during a session, such as user activity.
SERVICE	The SERVICE method can carry a Simple Object Access Protocol (SOAP) message as its payload. This method is also used to search for contacts in a SIP domain.
NEGOTIATE	The NEGOTIATE method is used to negotiate certain kinds of parameters between the two entities. These may include security parameters, algorithms and compression.
REFER	Ask recipient to issue SIP request (call transfer).

Table 18-1: SIP methods

SIP response

Remember that SIP is a request-response protocol. A response to a request is in the form of a code. The code is a three-digit number that indicates the outcome of the request. The response also includes the outcome as a text phrase, which can be read in the packet header. The code can be used by the client software, whereas the text phrase can be flashed on to the screen for the user's benefit.

The response codes have numerical values between 100 and 699 with the first digit representing the response class. The following are the different classes of response codes.

Class Name	Description
1xx: Provisional	Request received, continuing to process the request.
2xx: Success	Action was successfully received, understood, and accepted.
3xx: Redirection	Further action needs to be taken to complete the request.
4xx: Client Error	Request contains bad syntax or server cannot fulfill request.
5xx: Server Error	Server failed to fulfill a valid request.
6xx: Global Failure	Request cannot be filled at any server.

The complete list of responses and their codes is found in Table 18-2. SIP extends the HTTP response codes. Where a response code is common between the two, they have the same meaning; however, not all of the HTTP response codes are relevant to SIP and vice versa. The 6xx class of codes was introduced with SIP and is not found in HTTP.

SIP Response Codes			
1xx Provisional	Request received, continuing to process the request	2xx Success	Action was successfully received, understood, and accepted
100	Trying	200	OK
180	Ringing	202	Accepted
181	Call is being forwarded		
182	Queued		
183	Session progress		
3xx Redirection	Further action needs to be taken to complete the request	4xx Request failure	Request contains bad syntax or server cannot fulfill request
300	Multiple Choices	400	Bad Request
301	Moved Permanently	401	Unauthorized
302	Moved Temporarily	402	Payment Required
305	Use Proxy	403	Forbidden
380	Alternative Service	404	Not Found

Table 18-2: Complete table of SIP response codes

		405	Method not allowed
		406	Not Acceptable
		407	Proxy Authentication Required
		408	Request Timeout
		410	Gone
		413	Request Entity Too Large
		414	Request-URI Too Long
		415	Unsupported Media Type
		416	Unsupported URI Scheme
		420	Bad Extension
		421	Extension Required
		423	Interval Too Brief
		480	Temporarily Unavailable
		481	Call/Transaction Does Not Exist
		482	Loop Detected
		483	Too Many Hops
		484	Address Incomplete
		485	Ambiguous
		486	Busy Here
		487	Request Terminated
		488	Not Acceptable Here
		491	Request Pending
		493	Undecipherable
5xx Server Error	Server failed to fulfill a valid request	6xx Global Failure	Request cannot be fulfilled at any server
500	Server Internal Error	600	Busy Everywhere
501	Not Implemented	603	Decline
502	Bad Gateway	604	Does Not Exist Anywhere
503	Service Unavailable	606	Not Acceptable
504	Server Time-out		
505	Version Not Supported		
513	Message Too Large		

Table 18-2: *Continued*

Message flow

Figure 18-2 illustrates the typical message flow between the caller and the called when two proxies are involved. Because of the shape of the two SIP telephones and two SIP proxies in the diagram and because this configuration is so common, it is referred to as the SIP trapezoid.

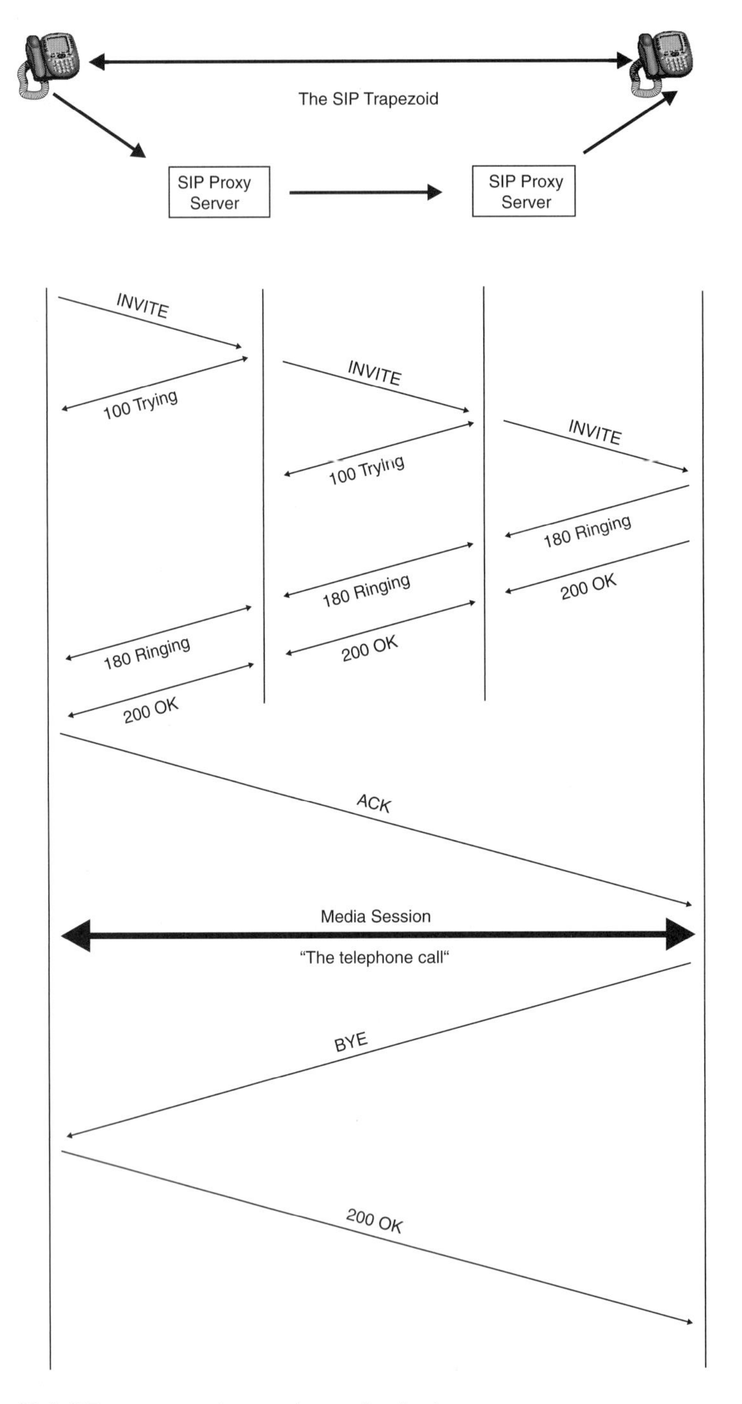

Figure 18-2: SIP messages when proxies are involved

The telephone of the caller will contact the SIP proxy that it has been assigned to with an INVITE request message. The caller SIP proxy will contact the recipient's SIP proxy also with an INVITE. The second proxy then contacts the called telephone, also with an INVITE request message.

The two proxy servers respond with a 100 Trying response to indicate that the setup of the session is still in progress. The called telephone sends a 180 Ringing response back and then a 200 OK response when the handset is picked up. Notice that these responses are sent back by the same path through the intervening proxy servers.

Once the session has been established, the caller's telephone knows the address of the called telephone and an ACK message can be sent directly. The telephone call can also proceed directly between the two telephones.

When one party hangs up, a BYE message is sent and the other telephone finishes with a 200 OK response.

This diagram illustrates only one of the many different message flows that SIP can support. For example, a proxy server may not be required if the destination address is directly known. Another possibility is that the called is directly on the caller's network and can be reached via an extension number. Again, a proxy server may not be required. Another common possibility is that the proxy servers continue to be advised of the state of the telephone call. This allows them to provide a busy signal to anyone else trying to reach the telephones while a call is in progress.

18.4 SIP MESSAGE STRUCTURE

Because SIP messages traverse multiple intermediary proxy servers, the message must contain all the information needed to make contact, including the paths and routing of the message.

Note that SIP doesn't deal with details of the conversation itself. These details, such as the type of media, codec, or sampling rate, must be dealt with by another protocol, such as session description protocol (SDP) or MIME.

There are only two types of SIP messages, request and response. Their structure is basically the same and they are diagrammed in Figure 18-3. The first line indicates the message type. This is followed by some header fields and finished off with the payload, usually SDP.

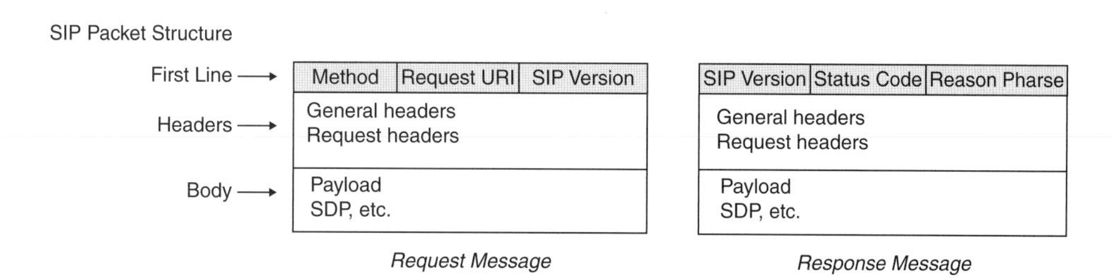

Figure 18-3: SIP request and response message structure

The sample messages shown in Figures 18-4 and 18-5 will give a better idea of how these work. The first message is a request (an INVITE) from Alice to Bob. The second is the response.

```
INVITE sip:bob@lrgnetworks.com SIP/2.0
Via: SIP/2.0/UDP pc33.lwc.com; branch=z9hG4bK776asdhds
Max-Forwards: 70
To: Bob <sip:bob@lrgnetworks.com>
From: Alice <sip:alice@lwc.com>;tag=1928301774
Call-ID: a84b4c76e66710@pc33.lwc.com
CSeq: 314159 INVITE
Contact: <sip:alice@pc33.lwc.com>
Content-Type: application/sdp
Content-Length: 142

(Alice's SDP not shown)
```

Figure 18-4: The SIP request message

The SIP request message in Figure 18-4 produces the SIP response message in Figure 18-5.

```
SIP/2.0 200 OK
Via: SIP/2.0/UDP
server10.lrgnetworks.com;branch=z9hG4bKnashds8;received=192.0.2.3
Via: SIP/2.0/UDP bigbox3.site3.lwc.com
        ;branch=z9hG4bK77ef4c2312983.1;received=192.0.2.2
Via: SIP/2.0/UDP pc33.lwc.com;branch=z9hG4bK776asdhds
        ;received=192.0.2.1
To: Bob <sip:bob@lrgnetworks.com>;tag=a6c85cf
From: Alice <sip:alice@lwc.com>;tag=1928301774
Call-ID: a84b4c76e66710@pc33.lwc.com
CSeq: 314159 INVITE
Contact: <sip:bob@192.0.2.4>
Content-Type: application/sdp
Content-Length: 131
(Bob's SDP not shown)
```

Figure 18-5: The SIP response message

First line

The request message starts with the required first line:

INVITE sip:bob@lrgnetworks.com SIP/2.0

This is correct because it follows the required syntax of:
Method: INVITE in this case
Request URI: the person who is being called
URI=Uniform Resource Identifier
SIP Version: SIP 2 in this case

The response message starts with the following first line:
SIP/2.0 200 OK

This is also correct for a response. The syntax is as follows:
SIP Version: SIP 2 in this case
Status code: 200
Reason phrase: OK is the text description of the code 200

Headers

Header fields are named attributes that provide additional information about a message. Headers for both requests and responses are treated together in this section. However, note the following:

- Although some headers can appear in both request and response, others are found in only one of these.
- There are many headers defined but only a few are actually found in any request or response.
- Some headers in the response are identical to those found in the corresponding request because they are simply copied into the response before being sent back.
- In theory, the order of the headers makes no difference to the operation of the protocol. However, if routing headers, particularly the VIA header, are placed close to the top, performance will be enhanced.

Some Header fields are mandatory and are found in all messages. These are: To, From, Via, Max-Forwards, CSeq, and Call-ID.

The Via header field

This field indicates the path taken by the request so far and indicates the path that should be followed in routing responses. The branch ID parameter in the Via header field serves as a transaction identifier, and is used by proxies to detect loops. The branch parameter is mandatory in the Via header field. For the request, Via contains the address (pc33.lwc.com) at which Alice is expecting to receive responses to this request. Note that for the response, there are three VIA fields listing both the proxy servers as well as Alice's PC.

The To header field

This field contains a display name (Bob) and a SIP or SIPS URI (sip:bob@lrgnetworks.com) toward which the request was originally directed. SIPS is the secure version of SIP.

The From header field

From also contains a display name (Alice) and a SIP or SIPS URI (sip:alice@lwc.com) that indicate the originator of the request. This header field also has a tag parameter containing a random string (1928301774) that was added to the URI by the softphone. It is used for identification purposes.

The Call-ID header field

This field contains a globally unique identifier for this call, generated by the combination of a random string and the softphone's host name or IP address.

The CSeq header field

The CSeq or Command Sequence contains an integer and a method name. The CSeq number is incremented for each new request within a dialog and is a traditional sequence number.

Why are the To, From, Call-ID, and CSeq header fields identical in the request and response messages? Because when the called telephone created the response, it merely copied these fields from the INVITE request message.

The Contact header field

This field contains a SIP or SIPS URI that represents a direct route to the sender, usually composed of a username at a fully qualified domain name (FQDN). While an FQDN is preferred, many end systems do not have registered domain names, so IP addresses are permitted. While the Via header field tells other elements where to send the response, the Contact header field tells other elements where to send future requests. Notice that in the request message (INVITE) the contact was Alice, but in the response message (OK), the contact was Bob.

The Max-Forwards header field

This field limits the number of hops a request can make on the way to its destination. It consists of an integer that is decreased by one at each hop.

The Content-Type header field

The Content-Type contains a description of the message body. In this case the body is made up of an SDP payload and is identified as application/SDP.

The Content-Length header field

This contains an octet (byte) count of the message body. The message bodies of the request and response in our example have different lengths, 142 and 131 bytes respectively.

SIP URI – Every endpoint on the VoIP system has a SIP URI for identification. The URI for Bob Jones at ABC.com might be:

sip:bobj@abc.com

Notice how much this URI resembles an e-mail address. This is not an accident. If placed on a Web page, clicking the URI will instigate a phone call to that endpoint.

If calling a telephone number on the PSTN, the URI will look like:

SIP:5551212@gateway, where gateway is the name of the machine that acts as the gateway to the PSTN.

18.5 SESSION DESCRIPTION PROTOCOL

The session description protocol (SDP) works in partnership with SIP to set up VoIP calls. SIP's role is to set up the session, while SDP's role is to describe it. SDP is not restricted to VoIP calls or even designed for it. SDP was actually designed for describing multimedia sessions for the purposes of session announcement, session invitation, and other uses. Curiously, RFC 4566, which describes SDP, also specifically mentions that negotiating media encodings is outside the scope of SDP. Nevertheless, SDP is used to negotiate the codecs used by VoIP calls.

SDP is a general purpose description protocol and not specifically tied to SIP. Other transports can also carry it, such as session announcement protocol, real-time streaming protocol, electronic mail using the MIME extensions, or HTTP. In the same vein, SIP is not tied to using SDP; it could use other methods to describe parameters, such as XML.

How codecs are negotiated with SDP

When the SIP INVITE is sent to the recipient, SDP is carried as the payload. The SDP message includes a list of codecs that the sending device can support. The list is ordered from the most preferred codec to the least. When the recipient device responds with the SIP 200 OK message, it carries SDP including its own list of codecs in the order of preference. Only codecs that are common to both lists can be considered and the codec with the highest preference between the two lists is used. This is known as the offer/answer model.

Figure 18-6 displays the decode of a sample SDP message. The highlighted packet shows a SIP INVITE message with SDP as the payload. The media description field lists five codecs that the sending device is willing to use, including PCM A-law, PCM u-law, and GSM.

SDP message structure

SDP is a text-based protocol, not binary. This means that you can read the information from the packet directly, as Figure 18-6 illustrates. Some of the information that

SDP must convey include session name and purpose, time that the session is active, the media comprising the session, and other information needed to receive the media such as addresses, ports, formats, and so on.

The description itself is a list of phrases with the structure <type>=<value>. In an unusual naming scheme, the <type> is referred to by a single unique letter, for example v= indicates the protocol version, while o= indicates the originator and session identifier.

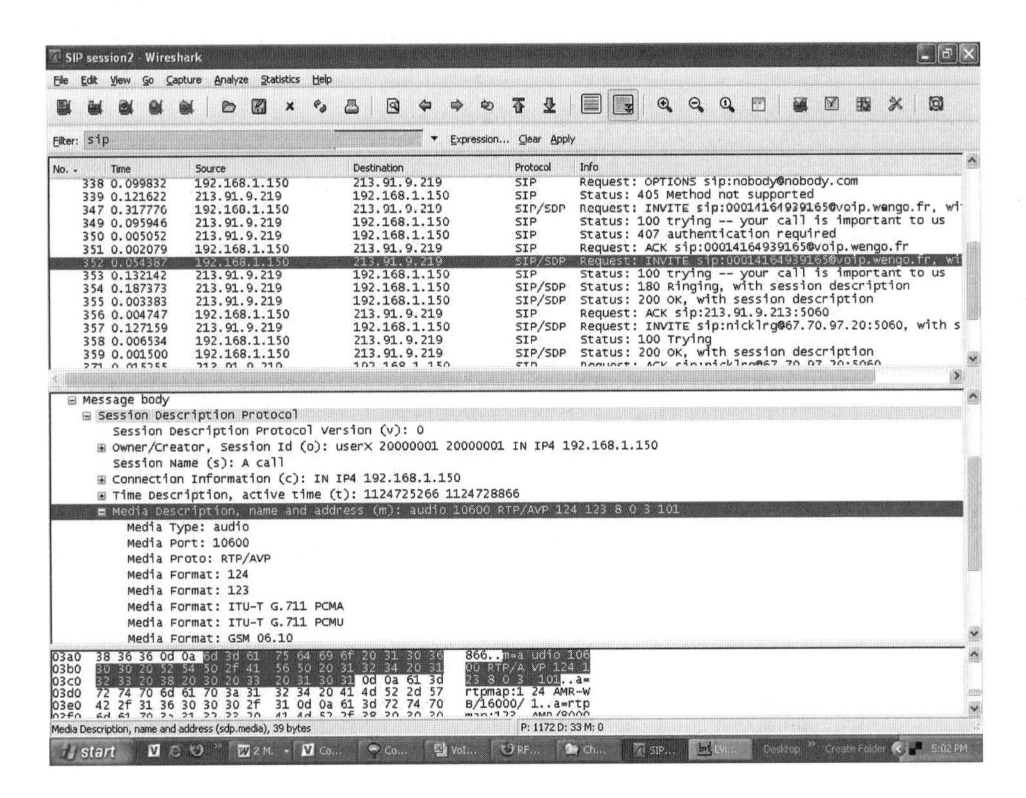

Figure 18-6: Decode of the SDP packet listing available codecs

Table 18-3 lists the SDP description fields. Note that four are mandatory and the rest are optional. If you examine Figure 18-6, you should be able to identify the v=, o=, s=, c=, t=, and m= fields.

v=	protocol version	required
o=	originator and session identifier	required
s=	session name	required
i=	session information	optional

Table 18-3: SDP description fields

u=	URI of description	optional
e=	e-mail address	optional
p=	phone number	optional
c=	connection information – not required if included in all media or at session level	optional
b=	zero or more bandwidth information lines	optional
t=	time the session is active	required
r=	zero or more repeat times	optional
z=	time zone adjustments	optional
k=	encryption key	optional
m=	media name and transport address	required
a=	zero or more session/media attribute lines	optional
i=	media title	optional

Table 18-3: *Continued*

SUMMARY

Section 18.1: Session Initiation Protocol (SIP)

This section introduced the idea that SIP is used to set up sessions for VoIP and looked at the full list of functions that it provides. It also discussed the fact that SIP has a resemblance to two other popular protocols of the Internet, HTTP and SMTP.

Section 18.2: SIP Architecture

The components used by the SIP architecture were introduced in this section. The components included the user agent, registration server, proxy server, and location database.

Section 18.3: Session Setup

The conversation between SIP clients for setting up and tearing down a session were discussed in this section.

Section 18.4: SIP Message Structure

A detailed look at the SIP message structure was discussed in this section. This is useful for anyone trying to troubleshoot a SIP session.

Section 18.5: Session Description Protocol

This section looked at the session description protocol and how it works in conjunction with SIP to negotiate the codec used by a VoIP call.

Hands-on Lab 18-1

Examining a SIP Connection with a Protocol Analyzer

1. Start Wireshark on your computer and start capturing packets.
2. Establish a call with another person in the classroom using X-Lite. Make some small talk and then hang up.
3. Stop capturing packets in Wireshark.
4. Type SIP in the filter box to list only SIP packets. Confirm that you captured the setup of your telephone call.
5. Confirm that you have the standard SIP handshake between the two devices. At a minimum, this should include the INVITE method, 100, 180, and 200 responses and the ACK method.
6. Wireshark can provide a graphical representation of the SIP handshake. Follow these steps:

   Statistics > **VoIP Calls** > select the VoIP call from the list > **Graph**

You see a graph similar to the one in Figure 18-7.

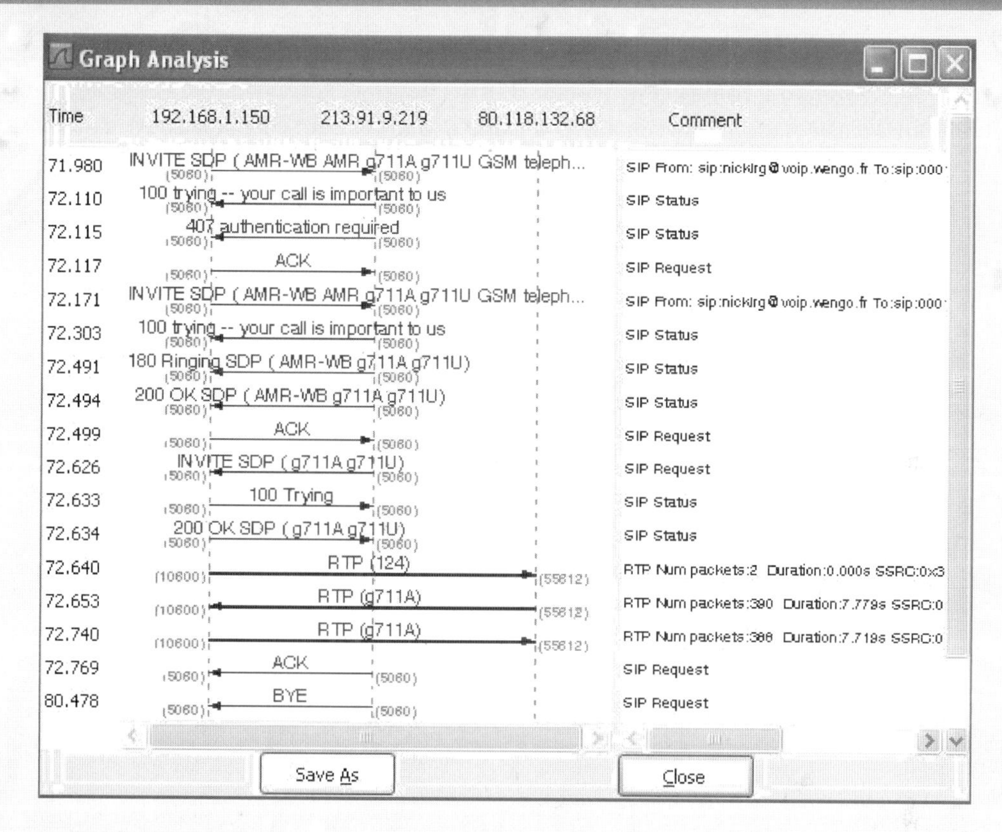

Figure 18-7: Graph analysis of a SIP session in Wireshark

Review Questions SIP in Detail

1. Which of the following functions does SIP provide?

   a) Negotiation of the content type
   b) An association between a user's name and his network location
   c) Feedback on quality of service
   d) Management of streaming media

2. SIP is similar to HTTP because

   a) a SIP URI can be placed on a Web page.
   b) they both use TCP as their transport protocol.
   c) they are both supported by Web browsers.
   d) it is a request response protocol.

3. When making a VoIP telephone call using SIP to someone on a remote network, the user agent will first contact

   a) the remote user.
   b) the local proxy server.
   c) the registration server.
   d) the remote proxy server.

4. You want to contact Sally using VoIP and type in SIP:sally@abc.com at your softphone. What did you just type in?

   a) Uniform resource identifier
   b) Fully qualified domain name
   c) Universal resource locator
   d) E-mail address

5. You are trying to reach a contact using VoIP and SIP. The other person is already talking on the telephone. What happens now?

   a) The remote telephone will ring while the person is talking.
   b) The SIP proxy will provide a busy signal.
   c) Your telephone will continue to ring the other telephone.
   d) The other person's telephone will provide a busy signal.

CHAPTER

19

Voice Gateways

Objectives

Upon completion of this section, the reader will:

- *Understand the functions of a voice gateway and its role as the interface between the VoIP system and the PSTN.*
- *Know how to integrate a voice gateway into a VoIP system in order to acquire access to the PSTN.*
- *Appreciate what FXO and FXS ports are.*
- *Understand the gateway architecture known as Megaco/H.248 and how it supports PSTN access for large organizations, service providers, and the telephone companies.*

INTRODUCTION

A voice gateway interfaces between a VoIP system and the traditional PSTN system. Since the PSTN still provides the majority of communications today and VoIP can't be expected to take over until well into the future, voice gateways are a standard fixture in today's communication systems. This chapter looks at monolithic gateways in which all of the gateway functions are in one device. It also looks at decomposed gateways in which the gateway functions are spread among several devices. In this latter case, the MGCP/Megaco protocol is used to communicate between these gateway components.

19.1 VOICE GATEWAYS

A gateway by definition connects two dissimilar systems. The two systems that need connecting are the VoIP system and the public switched telephone system (PSTN).

Functions of a gateway

The functions required in the gateway include (see Figure 19-1):

- PSTN interfaces — This function includes the PSTN signaling interface that terminates signaling protocols such as ISDN Q.931, and the PSTN media interface that terminates media streams such as pulse code modulation (PCM) voice streams.
- VoIP interfaces — This function includes the VoIP signaling interface that terminates H.323 (including RAS, Q.931, and H.245) or SIP, and the packet media interface that handles RTP.
- Signaling conversion — This function typically translates between ISDN Q.931 signaling and H.323/SIP signaling for call control.
- Media transformation — The media gateway function typically translates between the 64 Kbps PCM streams and RTP streams of various speeds.
- Connection management — This function must internally coordinate between signaling flows and media transformations. This involves creating, modifying, and deleting the association between the PSTN and Internet flows during the lifetime of a call.

Gatekeeper is not a gateway

Don't confuse a gatekeeper with a gateway. An H.323 gatekeeper serves as the central point for all calls within its zone of control. Gatekeepers perform address translation, admissions control, bandwidth control, and zone management functions. In addition, gatekeepers can perform call control, call authorization, bandwidth management, and call management functions.

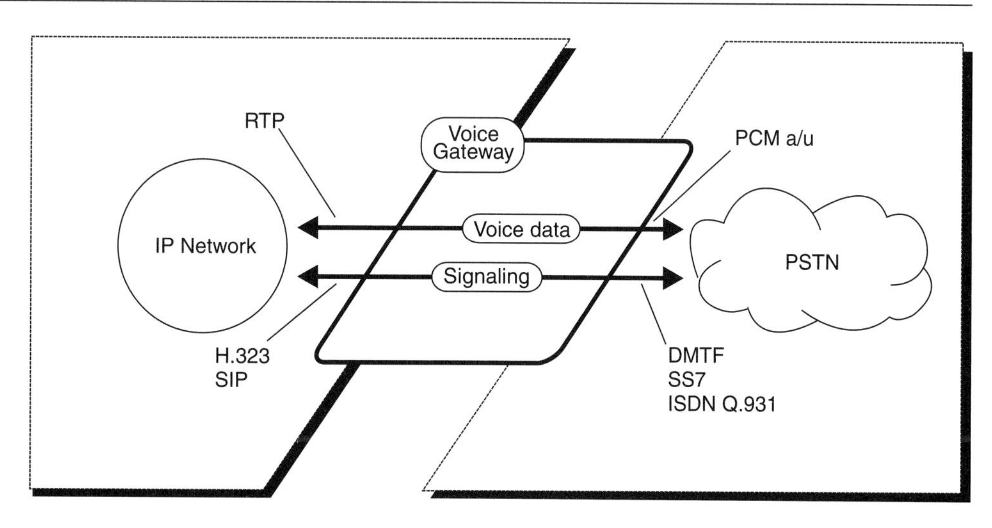

Figure 19-1: Functions of a voice gateway

19.2 MONOLITHIC GATEWAYS

A gateway typically sits between two different voice systems and translates signaling and the media stream. Normally these functions are performed by one device, so why do we call attention to that fact by calling it a monolithic gateway? The fact is that these functions can be split between devices; this is called *decomposing* the gateway, and that topic is covered in the next section on gateway protocols.

Which device?

VoIP gateways can be purchased as one of the following devices. Remember that when you are reading the sales brochures, a VoIP gateway is often referred to as a media gateway.

- Dedicated VoIP gateway device
- PBX with a gateway function
- Router with a gateway function
- IP-PBX
- Software that runs in a personal computer. The open source software called Asterisk is an example.

When choosing the gateway, the most important criteria is that it supports the signaling and media of your choice on the IP and PSTN sides.

For the IP side,

- The signaling is H.323 or SIP, although in rare cases it could be IAX2 or Cisco's SKINNY protocol.
- The media is RTP.

On the PSTN side,

- The signaling is SS7, DTMF (dual tone multifrequency) for analog, or ISDN's Q.931.
- The media is an analog telephone line or digital T1 or ISDN line.

The analog ports are called FXO, FXS, or possibly E&M ports, which are covered in the next section.

Ports and connectors

Figure 19-2 illustrates the possible ports and connections on a standalone VoIP gateway. Any device you choose must have the correct complement of ports for the services you are trying to support. Don't expect the gateway that you purchase to have all of the ports depicted in the graphic. Here is a summary of these ports.

Ethernet port

This is your IP LAN and connects to the Ethernet switch.

PRI/BRI port

This is the digital data port. It is used for a T1 line or ISDN service. You have to contract for this type of service from a service provider or telephone company. A separate digital line has to be run into your premises in order to take advantage of the service. Digital services will range from 64 Kbps for a fractional T1 (DS0) up to 1.544 Mbps for a full T1 (DS1) line. Remember that each 64 Kbps of bandwidth can support one telephone conversation.

FXO port

FXO is an acronym for Foreign eXchange Office. This is the connection to the public switched telephone network. Run a cable between this connection and the wall plate into which a telephone is normally plugged. Appreciate that only one telephone call can take place over one connection. If this is not acceptable, purchase a voice gateway with multiple FXO ports and provision multiple telephone lines.

FXS port

FXS is an acronym for Foreign eXchange Subscriber. This port has several minor uses. It can connect an analog PBX into your VoIP system. It can also be used to attach an analog telephone to your VoIP system. If you wish to implement fax over VoIP, you could attach a fax machine to this port. However, sending a fax signal over a voice codec is generally considered to be poor quality and unreliable and therefore not recommended.

E&M port

The E&M (ear and mouth) port is used to connect an analog PBX to your system. The FXS port can also be used to connect a PBX. The difference between E&M and FXS is in their signaling methods and the wiring. The requirements of the PBX are what determine which method you need to use.

RS-232 port

Although depicted in the diagram as an RJ11 connection, this could also be a DB-9 or DB-25 connector. This port is used for out-of-band management. By attaching a modem and dialing in or by using an attached terminal, the VoIP gateway can be managed and you can troubleshoot when the network is down.

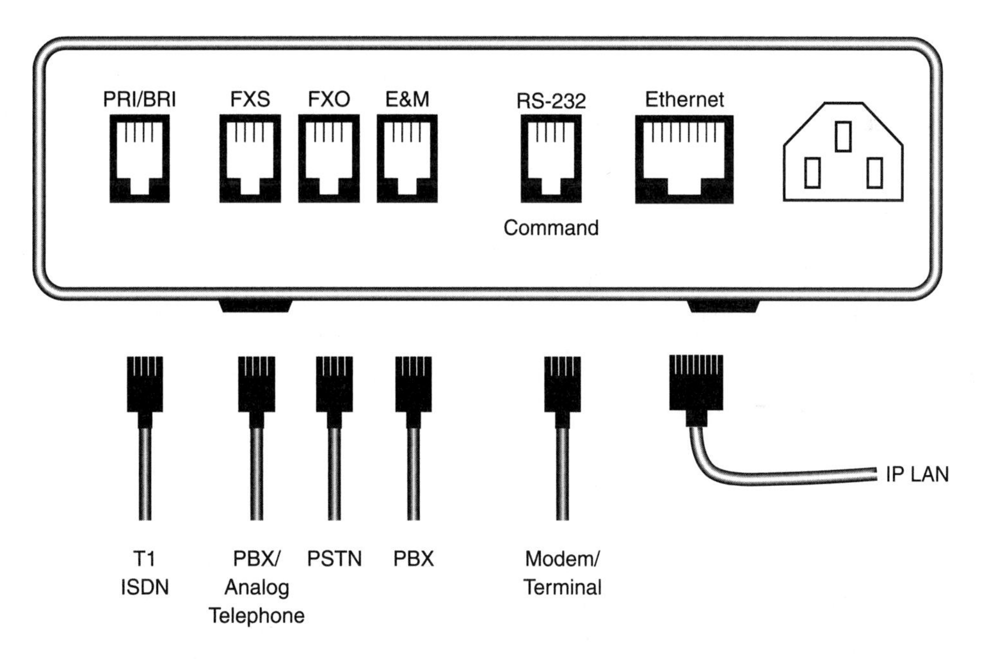

Figure 19-2: Voice gateway connections

19.3 FXO AND FXS

FXO and FXS ports are used to make analog connections to the PSTN or other analog devices. If you want your VoIP gateway to interface to the public telephone system, you have to understand how these work. These ports are illustrated in Figure 19-3.

The FXS interface delivers the telephone service from the telephone company's central office (CO). Think of it as the telephone plug on the wall. It must be connected to "subscriber equipment" such as a telephone, fax, or modem. The FXS interface provides the

- dial tone
- battery power
- ring tone

Remember that the FXS points to the *subscriber*, at least the subscriber's equipment.

The FXO is the complement to the FXS. It is the analog telephone equipment, the telephone, fax, or modem. It receives the service from the PSTN system. The FXO interface provides the following service:

- on-hook/off-hook indication (loop closure)

Remember that the FXO points to the telephone company office.

Because the FXS port generates a ring, it must be connected to a device that can detect the ring, which, of course, is the FXO.

Additional devices

If you add a PBX to our simple scenario, it becomes more complex. Assuming that you have analog telephones attached to the PBX, then the PBX will have both FXS and FXO ports. The telephones are FXO devices and will attach to the FXS interface on the PBX. The FXO interface on the PBX will attach to the PSTN. See Figure 19-3.

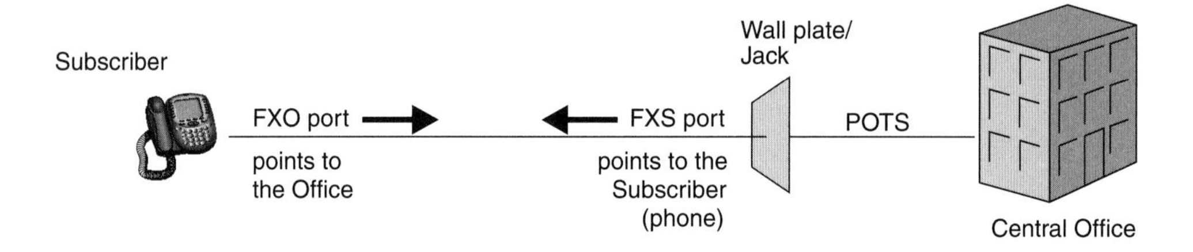

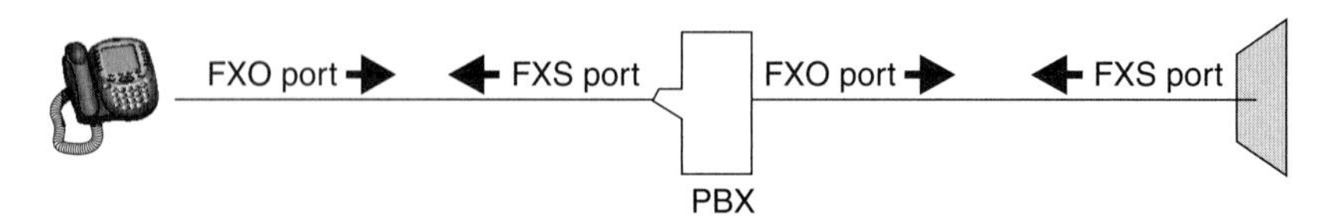

Figure 19-3: FXO and FXS ports

E&M interface

Occasionally, you may run across an E&M port in the voice gateway. The E&M interface connects trunk lines, usually called tie lines. E&M is a signaling technique used mainly between PBXs or other network-to-network telephony switches to avoid problems such as glare (when both ends "pick up" at the same time). E&M interfaces use four or six wires, as opposed to the two wires that FXS/FXO interfaces use. The additional wires are the E and M leads, and refer to the direction of the signaling (an E lead at one end is connected to the M lead at the other, so when the M(outh) lead signals a call, the E(ar) lead sees the signal and can "seize" the trunk so the call can be processed).

The term "ear and mouth interface" is sometimes used as a synonym for a telephone handset itself, or for a headset-and-microphone combination that allows hands-free operation.

Analog telephone adapter (ATA)

An analog telephone adapter has one purpose, to attach an analog telephone to a VoIP network. If you don't have or don't want to purchase an IP telephone, an ATA will solve your problem. The ATA has an Ethernet jack for the IP network and an RJ11 jack for a telephone. Of course, the RJ11 jack is a FXS port. If your ATA was supplied by your telecommunications company that is providing your home or small-business VoIP service, then the ATA is acting as a residential gateway to your service provider's network.

Figure 19-4 finishes this section by diagramming a summary of the VoIP gateway connections.

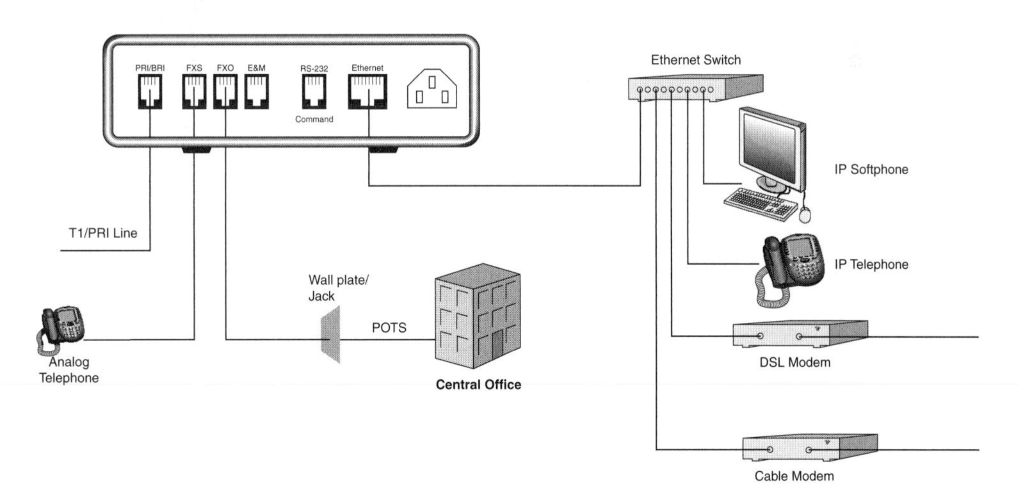

Figure 19-4: Summary of the VoIP gateway connections

19.4 MEGACO

Decomposing the voice gateway

A monolithic gateway works well between a small or midsize VoIP system and the PSTN and is cost effective because all components are integrated into a single device.

However, a monolithic gateway is not suitable for large systems. The reasons for this are varied but include the following:

- Not all components need the same capabilities. There will be a need for more of some devices than of others. It is more cost effective to design in just the capabilities that are needed in a device. Specialization is key. Devices dedicated to one task are more efficient.
- Components need to be distributed in a large system, some inside, some at the edge, and some external to the system. Economies of scale can be achieved only by the separation of functions.
- Control may need to be centralized. Telephone companies and service providers need to control the networks that they manage, even the networks that they manage for their customers. Large organizations will centralize the management of their telephone system into one department where the expertise can be concentrated.

The telecom industry appears to have reached broad consensus that the best answer lies in separating the call processing function from the physical switching function and connecting the two via a standard protocol. In softswitch terminology, the physical switching function is performed by a media gateway (MG), while the call processing logic resides in a media gateway controller (MGC). The protocol used to communicate between the MG and the MGC is MGCP or Megaco/H.248. See Figure 19-5 for the relationship between the MG and the MGC.

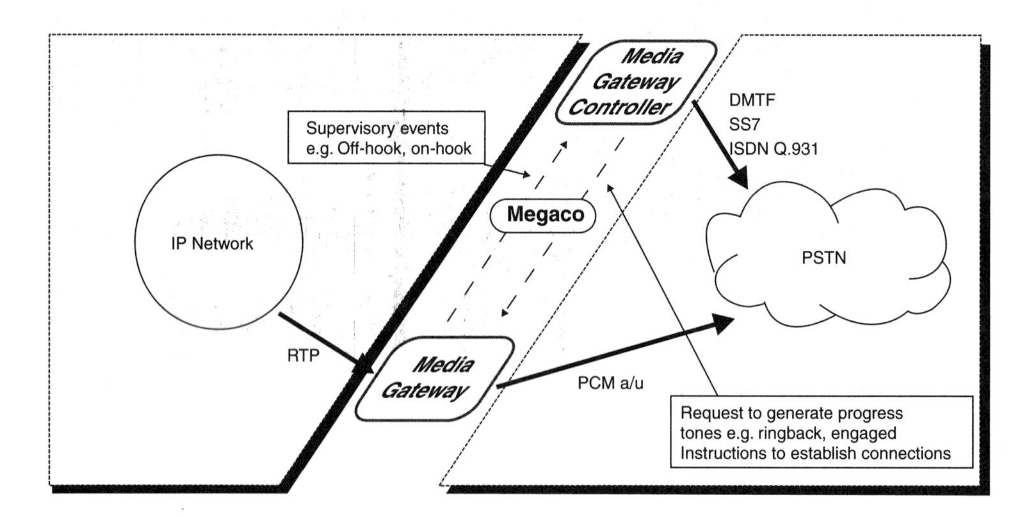

Figure 19-5: Megaco communicates between MGs and MGCs

Only large systems need apply

The division of duties and the proliferation of devices make sense only on large networks. This approach can be used by large organizations, service providers, and telephone companies. In fact, it is a necessary step in the evolution of the switched public telephone system to one based on VoIP.

If you administer a small to midsize VoIP network, you will probably never need to worry about gateway protocols.

Gateway protocols

The gateway protocols are used to communicate between the devices that perform the functions in the decomposed voice gateway. The two protocols that the following discussion is based on are the media gateway control protocol (MGCP) and MEdia GAteway COntroller (Megaco) protocol. The relationship between these two needs explaining.

MGCP was based on previous protocols (SGCP and IPCD). Although RFC 2705 describing MGCP was published, it was never ratified by the IETF as a standard. One of its weaknesses is that it has only limited support for networks other than the PSTN. MGCP had early deployment and is a reality in many networks today. Nevertheless it is not the direction that the industry is moving and is not truly an open standard.

Megaco evolved from the basic MGCP components and the two have many concepts in common. The drive to develop Megaco was the need to provide for requirements that MGCP did not address properly. The implementation is different and they are not directly compatible, but they may both be supported in any device. Megaco addresses the same type of applications as MGCP but in a more generic and elegant way. For the purposes of our discussions, MGCP and Megaco can be considered together.

Megaco and H.248

Megaco was developed by the IETF and the ITU standards setting bodies. The ITU standard for the Megaco protocol is called H.248 and therefore it should be referred to as Megaco/H.248.

Although the standards are unified and are the same, it is interesting to see how the two bodies differed in their viewpoints. The IETF was interested in a text-based protocol in which the values inside the protocol header are text, the approach used by SIP and HTTP. The ITU was interested in a binary-based protocol, similar to H.323 as well as IP, TCP, and UDP. Leaving aside the merits of each approach, the IETF view prevailed and support for text coding in Megaco is mandatory and support for binary is optional. Table 19-1 lists the differences between Megaco and MGCP.

Megaco/H.248	MGCP
A call is represented by terminations within a call context.	A call is represented by endpoints within connections.
Call types include any combination of multimedia and conferencing.	Call types include point-to-point and multipoint.
Syntax is text or binary.	Syntax is text.
Transport layer is TCP or UDP.	Transport layer is UDP.
Defined formally by the IETF and ITU.	Managed by the industry.

Table 19-1: Differences between Megaco/H.248 and MGCP

Component devices

MGCP/Megaco/H.248 is designed for very large VoIP systems in which specialization in devices is the rule, not the exception. Although we have already mentioned media gateways and media gateway controllers, the following is a more complete listing and description of these components. Refer to Figure 19-6 for a graphical representation of the relationship between them.

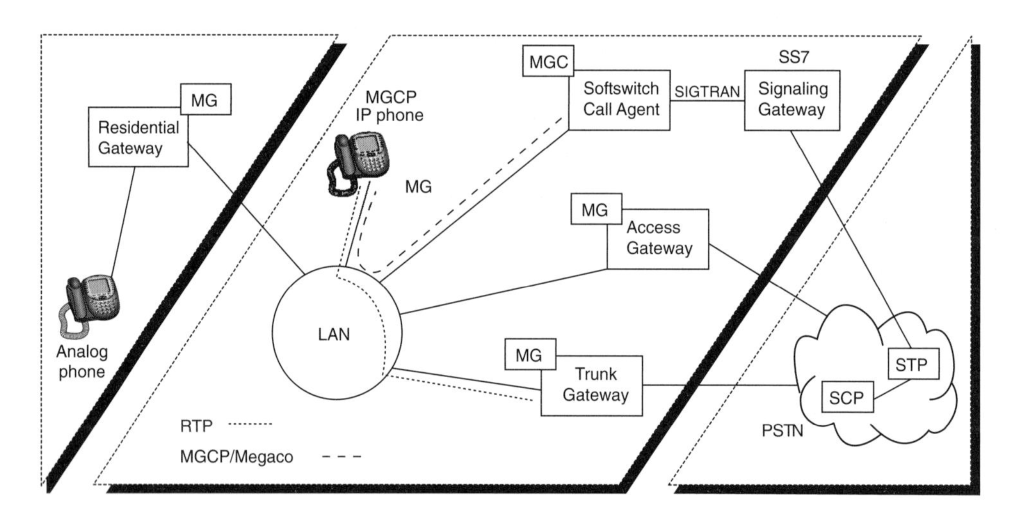

Figure 19-6: Components of the MGCP/Megaco system

Media gateway controller (MGC)

A media gateway controller handles the registration and management of resources at the media gateway(s). A media gateway controller exchanges messages with central office switches via a signaling gateway (described below). Because vendors of media gateway controllers often use off-the-shelf computer platforms, a media gateway controller is sometimes called a softswitch or a call agent.

Signaling gateway

A signaling gateway provides transparent translations of signaling between switched circuit and IP networks. The signaling gateway may terminate SS7 signaling or translate and relay messages over an IP network to a media gateway controller or another signaling gateway. Because of its critical role in integrated voice networks, signaling gateways are often deployed in groups of two or more to ensure high availability.

There is a special protocol called SIGTRAN, which is used to transport the SS7 signals in IP packets. The signaling gateway would use SIGTRAN when communicating with the MGC or another signaling gateway.

Media gateway (MG)

A media gateway terminates voice calls from the public switched telephone network, compresses and packetizes the voice data, and delivers compressed voice packets to the IP network. For voice calls originating in an IP network, the media gateway performs these functions in reverse order.

The following examples are specialized media gateways that are dedicated to a particular type of interface.

Trunking gateways

A trunking gateway acts as an interface between the telephone network and a VoIP network. Such gateways typically manage a large number of digital circuits.

Voice over ATM gateways

These gateways operate much the same way as VoIP trunking gateways, except that they interface to an ATM network.

Residential gateways

Residential gateways provide a traditional analog interface to a VoIP network. Services such as Vonage use residential gateways.

Access gateways

Access gateways provide a traditional analog or digital PBX interface to a VoIP network.

MGCP or Megaco telephones

Why is a telephone considered a media "gateway"? Because it converts signals, audio to digital in this case, just as any gateway does. MGCP/Megaco telephones are actually rare in comparison to SIP or H.323 telephones. Examples include the CISCO 7940G and 7960G and Nortel i2400 VoIP telephones.

The design intent of the Megaco IP Phone is to keep it determinedly simple while providing required support for fully featured business telephone functions and the flexibility to allow for a very wide range of telephone configurations. The approach used to achieve this goal is to provide a very simple and direct master/slave control model in which very little feature intelligence is required in the end device. This design intent matches the Megaco/H.248 protocol approach well.

Master/slave

There are two approaches to communication protocols in widespread use, peer-to-peer and master/slave. The main difference in these two approaches is how intelligence is distributed in network edge devices and communication servers. System and component cost, flexibility, ease of manageability, and centralized control are all affected by the approach taken. However, if properly designed, both approaches can coexist in a network and the network can take advantage of the strengths of each.

The peer-to-peer approach, exemplified by SIP and H.323, handles high level operations and features such as "make a call" and "hold." Peers generally negotiate with other peers to implement call features and other services. Although in theory peers have enough intelligence to "go it alone," they typically avail themselves of the services of central machines, such as registration servers and gatekeepers, to complete their tasks. All peer systems have problems with complexity and scalability and all peers have to be upgraded for new features to be introduced.

The master/slave approach is exemplified by MGCP and Megaco/H.248. This approach allows the gateway function to be split between intelligent (master) controllers and nonintelligent slaves. A master controller can control and drive multiple slaves, which are optimized for their interface function and devoid of superfluous functions, while at the same time being less expensive, less complex, and without the need for upgrades when any new feature is introduced. Under Megaco, the media gateway controller/call agent/softswitch is the master, while the many media gateways and IP telephones are the slaves.

Megaco/H.248 command set

Megaco/H.248 is a simple protocol in the sense that it has a small verb (instruction) set. Table 19-2 lists the Megaco/H.248 command set.

Command	Direction	Description
ServiceChange	MGC <->MG	Notify the responder of the new service state
Add	MGC ->MG	Add a connection
Modify	MGC ->MG	Change a connection characteristic
Subtract	MGC ->MG	Tear down a connection
Move	MGC ->MG	Move an endpoint from one connection to another connection (call waiting)
AuditValue	MGC ->MG	Determine the characteristics of an end-point
AuditCapabilities	MGC ->MG	Determine the capabilities of an end-point
	MGC <->MG	Notify the responder of an event (on-hook) Registration Impending or completed restart

Table 19-2: Megaco/H.248 command set

SUMMARY

Section 19.1: Voice Gateways

This section gives an overview of what a voice gateway does, particularly the need to continue the control signals from one system to another and the need to re-code the voice content itself.

Section 19.2: Monolithic Gateways

This section discusses a self-contained gateway that holds all of the gateway functions in one device. This is the type of device that a small or medium-sized organization or a residential user would use.

Section 19.3: FXO and FXS

This section covers the analog ports on a voice gateway that are used to connect to the PSTN. The FXO, FXS, and E&M ports are discussed.

Section 19.4: Megaco

This section covers the MGCP and Megaco/H.248 protocols. These are used by large organizations, service providers, and telephone companies when the functions of the voice gateway must be distributed over multiple devices. These systems are prized for their centralized control and management.

Review Questions Voice Gateways

1. The difference between a monolithic and decomposed VoIP gateway is that

   a) A monolithic gateway connects a VoIP system to the POTS whereas a decomposed gateway can connect the VoIP system to T1 or ATM lines.
   b) The monolithic gateway contains all the gateway functions in one device, whereas the decomposed gateway distributes the functions among multiple devices.
   c) A decomposed gateway can also act as a gatekeeper to an H.323 system, whereas a monolithic gateway can't.
   d) Decomposed gateways can perform multiple functions, whereas a monolithic gateway can only perform one.

2. When connecting a VoIP gateway to a POTS line, the port on the gateway will be

   a) an FXO port.
   b) an E&M port.
   c) a BRI port.
   d) an FXS port.

3. Under which condition would you use an E&M port on a gateway?

   a) If there is no FXO or FXS port.
   b) When the other device is a PBX.
   c) When the other device requires it.
   d) If you are connecting to a digital line.

4. The MGCP protocol is losing popularity in favor of Megaco because

   a) it is controlled by the IETF.
   b) it requires licensing fees.
   c) it is a peer-to-peer protocol.
   d) it has good support only for the PSTN.

5. Megaco is the same as

   a) H.323.
   b) H.248.
   c) MGCP.
   d) MGC.

CHAPTER

20

Setting up a Modern VoIP System

Objectives

On completion of this section, the reader will:

- *Have reviewed the connectivity options that can be made between VoIP and other systems.*
- *Appreciate the services that direct inward dialing (DID) can provide, including the ability to give toll free calls without using an 800 number.*
- *Understand that a well-thought-out dial plan is a prerequisite for a successful implementation of a VoIP system.*
- *Know what the ENUM service can provide.*
- *Be able to troubleshoot connection problems when forwarding calls through NAT and firewalls.*

INTRODUCTION

This section gathers together topics that have not found a home in the book so far, but which nevertheless are important when implementing a VoIP system. The first section provides a wrap-up and comparison of outside connections for your VoIP system; you need to match the characteristics of your future connection with the quantity of voice traffic your organization generates. Next, a dialing plan must be designed before the VoIP system goes in, and until it is implemented the system can't go into production. DID and ENUM are services that add value to your system; integrating them into your VoIP plans at the beginning is important if you want to take advantage of them. The final topic deals with troubleshooting VoIP when it crosses a NAT and firewall combination. It too must be planned for if you want to avoid problems after the system is implemented.

20.1 OUTSIDE CONNECTIONS FOR YOUR VoIP SYSTEM

A VoIP system needs to be connected to the rest of the world in order to be useful. When planning your VoIP system, you have many connectivity options and many service providers to choose from. Not every service provider offers every service. This section provides a summary of your connectivity options and their characteristics.

Analog line to the PSTN

- This is a traditional telephone line.
- It may be suitable for a home or a small office.
- Only one call at a time on one line.
- Multiple simultaneous calls require multiple physical lines.
- Simple setup. Run the wire from the FXO port of the VoIP appliance to the wall.

Digital line to the PSTN

- Requires installation of a digital line to the central office of the telephone company.
- Will be either a T1, PRI, or BRI service.
- Multiple calls over one line; maximum calls determined by number of channels. The number of channels is as follows: T1=24, PRI=23, BRI=2.
- Requires a T1/PRI port on the media gateway.

Asynchronous high-speed service to the Internet

- Provided by local cable or telephone company.
- Asynchronous because it is higher speed down to the customer than up to the Internet. This is an issue because VoIP uses an equal stream of data in both directions while making a call.

- Will be shared with television if a cable service and analog telephone calls if a telephone service.
- Number of simultaneous VoIP calls will be restricted or quality will be degraded because of these factors.
- ATA is specified by the VoIP service and needs to be purchased by the consumer.
- Analog telephones are attached to the ATA.

Synchronous high-speed connection to the Internet

- Critical factor is that the connection is synchronous and provides equal capacity in both directions.
- Can be many types of services, including T1/PRI, frame relay, ATM, metropolitan Ethernet, and synchronous DSL. Generally, these are referred to as leased lines.
- Simultaneous VoIP calls will reflect the higher capacity of these links and quality will be good unless the link is saturated.
- Suitable for medium- and large-sized organizations.

20.2 DIRECT INWARD DIALING (DID)

Although you may have many telephones at home, you only have one telephone line. If you want to make a telephone call, but someone is in the middle of a call, you will have to wait or ask the person to finish her conversation. This is a situation familiar to every father who has a teenage daughter.

The PBX does DID

The situation for an organization is different. Many people do need to make calls outside the organization simultaneously. Although many lines are required, should a line be provisioned for every telephone? Clearly not! A feature traditionally implemented in PBXs solves this problem. The idea with DID is that one telephone number is published for the company. When someone calls the company number, he is asked to dial in an extension number. The telephone corresponding to that extension number rings and then the two parties can converse directly. The PBX handles the routing of the call. This system is illustrated in Figure 20-1.

The advantages to this system are obvious. Only one telephone number is assigned to an organization, which conserves telephone numbers. Only one telephone number need be published by the organization, which saves on confusion. Internally, the organization can have 10 or 1,000 telephones; it makes no difference. In addition, the organization will contract for the number of trunk lines that it needs to maintain acceptable service; this decision is independent of the number of telephones that it has internally.

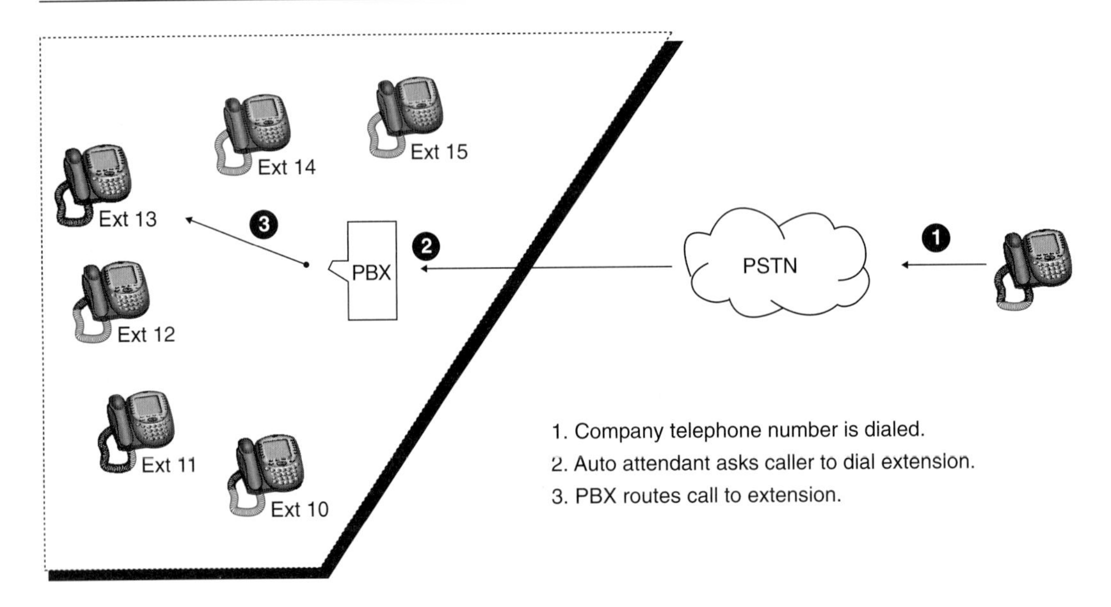

Figure 20-1: The PBX handles DID

DID extends VoIP's reach

The VoIP system has the same issues as the analog system and solves them the same way. In fact the DID concept is even more valuable for VoIP since it can be extended over the Internet. For VoIP, the call manager takes the place of the PBX and routes the call internally to the correct extension. For an incoming call from the PSTN, the system works exactly as described for the PBX.

However, for VoIP we can add one additional capability. We can apply to a DID provider and receive additional telephone numbers that need not be in our area code.

A DID service provider works as follows. He installs a VoIP server and media gateway in a distant city. This is known as a point of presence (POP). The VoIP server will support the protocols SIP, H.323, or IAX2, that his customers want. He receives a block of telephone numbers from the local telephone company. He provides connectivity between the VoIP servers in all his locations by using private leased lines or the public Internet. The DID provider can now provide you with a local telephone number in the other area code for which you wish to have a presence. Anyone in the distant area code can now dial your company thinking that you are local. There are no long distance charges involved although you will still need to pay service charges to the DID provider.

The advantages to this system are very valuable.

- The fees you pay are smaller than if you had used 800 numbers.
- The fees you pay are fixed and don't vary with distances.

- Your customers may not know where you are located; you appear local to them.
- It is easy to provision a DID number in many locations as business needs dictate; it is just as easy to give up a number when it is no longer needed.

Figure 20-2 illustrates how DID works. Assume a company in Seattle wants to provide a service to customers in the Chicago area. It finds a DID service provider who has a POP in Chicago and can therefore provide a telephone number in the 312 area code. The DID service provider has a SIP server at its head office in Los Angeles as well as at the POP in Chicago and in Chicago it has a media gateway that connects to the PSTN there. The company provides the Chicago local number to its customers. When a Chicago customer calls the local number, it is routed through the PSTN to the media gateway at the POP and a connection is made using the SIP protocol back through the SIP proxy server in Chicago through the SIP proxy server in Los Angeles and onward to the company office in Seattle. In Seattle, the company SIP server will accept the connection and forward the call to the extension that the caller dialed in.

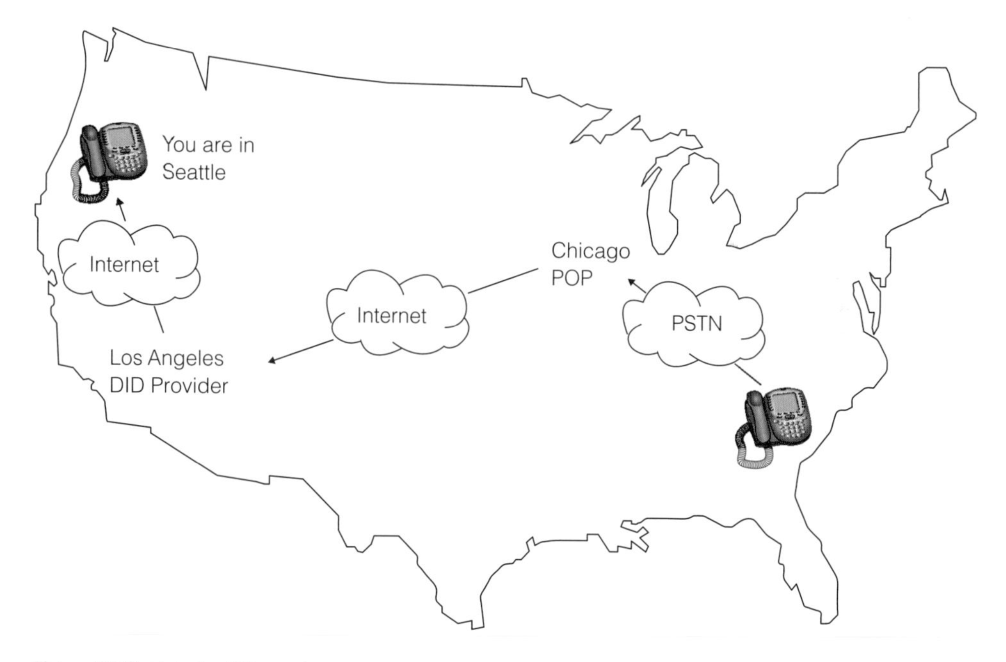

Figure 20-2: Using a DID number

The links in this scenario, between Chicago and Los Angeles and between Los Angeles and the company's premises in Seattle, can either be leased lines or the general Internet. Leased lines offer better quality of service, but the cost will be greater. The DID service providers offer a large selection of locations from which to choose local numbers. Only the largest could provision the equipment for the POPs in so many locations and even the largest couldn't manage POPs in overseas locations. The reason they can offer such a large selection is that they negotiate peering arrangements with DID service providers in other parts of the country and the world who manage their local POPs.

To explore the DID numbers and services available, search for DID service providers on the World Wide Web.

20.3 THE DIAL PLAN

When someone telephones your VoIP number, how is the call handled? Although the outcome is successful if your telephone rings, a complicated procedure called the dial plan has to be stepped through for this result. A lot of work has to be done in the background to make this system work. Consider the following:

- The dial plan has to manage both inbound as well as outbound calls. For example, how does the system distinguish between an outbound call destined for the PSTN, a call going to another VoIP system, or an extension within the same organization?
- How does the system handle emergency 911 calls and information 411 calls?
- How are long distance calls to the PSTN managed and who is allowed to make them?
- For inbound calls, you need to provide a menu system, call on hold, busy signals, and message box services.

Inbound and outbound rules

The dial plan is sometimes referred to as inbound and outbound rules. As this implies, the dial plan is a series of rules that are enforced when a particular condition is met. The following situations are handled by the dial plan:

- The called doesn't answer in a specified time; for example, if Bob doesn't pick up the phone in 25 seconds, then send the call to the voice message box.
- The call comes in at a specific time of day or a day of the week; for example, if the call comes in on the weekend, play an out-of-office message.
- The caller is placed on hold or is put in a queue to wait; the dial plan may specify that some action takes place. Typical actions might include playing music or a commercial, playing a message at regular intervals to reassure the caller of her importance and finally ringing an operator if the wait time is too great.

- The system needs to take action when the caller dials certain keys in response to voice prompts; for example, dial 0 for the operator, dial 5 for sales, and so on.
- The user wants to hear messages or manage his voice message box remotely.
- The user has left the office but the follow-me feature of the system will attempt to forward the call to his cell phone and if that fails, to his home phone.

For outbound calling, the dial plan needs to know how to handle the following types of calls:

- Interoffice
- Intraoffice to remote locations
- Local PSTN
- Long distance PSTN
- Toll-free long distance PSTN
- VoIP destinations
- Emergency 911
- Information 411

The administrator of the VoIP system will be responsible for the design of the dial plan, which must be implemented before the system can go online. Some specific tasks involved include:

- Determine the numbering system for the extensions, for example will 2, 3, or 4 digits be used for extension numbers?
- Assign prefixes for dialing between zones on the IP network or out to the PSTN. The typical example here is to prefix outbound calls to the PSTN with the number "9."

Configuration files and pattern matching

After the VoIP system designer creates a list of inbound and outbound rules, those rules have to be configured in the VoIP system. Although many modern VoIP systems allow the administrator to create these rules in a graphical user interface, the rules themselves are written to a configuration file. It is easier to troubleshoot the dial plan by examining the configuration file in a text editor and experienced administrators will create, edit, and troubleshoot the configuration file directly.

The following is a very generalized and simplified dial plan configuration file. It is not complete or specific to any VoIP system but the intent is to illustrate some of the incoming and outgoing rules found in the dial plan. The trunks section lists the channels the VoIP system can use to communicate with the outside world. The incoming and outgoing rules sections list their respective rules. Each rule is on a separate line and starts with what the rule applies to, in this case "exten" stands for "extension." The rule, which is on the right-hand side of the => is composed of three sections: the extension number dialed in, a step priority, and a command along with any parameters that the command needs. The first command (s,1,Answer) is used when the connection is made, but the caller has not dialed an extension as yet. The letter *s* (for "start") is used for the extension number. The first step is to answer the call. The second step for extension *s* is to play a welcome message. Thereafter, each command uses an extension number as its first parameter.

```
[Trunks]
; what trunks are available
trunks => SIP                              ;DID channel
trunks => FXO                              ;PSTN channel

[Incoming rules]
exten => s,1,Answer                        ;s means start, step 1, the answer command will
                                           answer the phone
exten => s,2,Playback (Welcome-msg)        ;step 2, still with start, playback command will play the
                                           welcome-msg
exten => 0,1,Dial(SIP/900)                 ;caller pressed 0 for operator, Dial command will try to
                                           connect to SIP extension 900
exten => 0,2,Playback (no-answer)          ;operator doesn't pick up, playback no-answer message
exten => 0,3,Hangup                        ;hang up
exten => 5,1,Dial (SIP/Sales)              ;caller pressed 5 for the Sales department, DIAL
                                           command will try to connect to the SIP extension
                                           registered as "Sales"
exten => 5,2,Message(Sales)                ;Sales doesn't pick up, go to the message box of Sales

[Outgoing rules]
exten => 911,1,Dial (FXO/911)              ;dial 911 on the PSTN
exten => _NXX,1,Dial (SIP/NXX)             ;any number between 200 and 999 will be forwarded to
                                           that extension using SIP
exten => _9NXXXXXX,1,Dial (FXO/NXXXXXX)    ;numbers with a 9 prefix are routed to the PSTN, the 9 is
                                           stripped from the number
exten => _011.,1,Playback (not authorized) ;international calls that require 011 are not authorized
```

It is not possible to list every possible outgoing number in the outbound rules that could be dialed. Pattern matching is used to represent many numbers with a pattern. Notice that in the preceding outgoing rules, patterns always start with the underscore (_) to indicate that they are patterns. The symbols that stand in for numbers are as follows:

N = any numeral between 2 and 9
Z = any numeral between 1 and 9
X = any numeral between 0 and 9
[146] = matches the numerals listed, namely 1, 4, and 6. This is not the number 146.
[3–8] = matches any number in the range, namely 3, 4, 5, 6, 7, and 8
. = matches anything, a wildcard

Here are some examples:

NXX = any number between 200 and 999
9NXXXXXX = local dialing prefaced with a 9
1NXXNXXXXXX = long distance pattern prefaced with a 1
011. = any number prefaced with a 011. Note the period after the 011. 011 is used to reach numbers outside the North American Dialing Plan.

Exercise 20-1

Creating a Dial Plan

You are the consultant who has been hired by ACME Enterprises to design and install its VoIP system. Your next step is to design the dial plan and translate the incoming and outgoing rules into commands you can type into the configuration file.

After consultation with ACME management, you learn the following information about the company and its corporate requirements for the VoIP system. The company has purchased a SIP server with media interfaces for the PSTN and a PRI service to a service provider. The server also has a high-speed connection to the Internet and the company has received DID numbers for cities in which it does business. The FXO interface connects to the PSTN and hence the local telephone company. It is expected that all local calls, including 911 and 411 calls, will be routed here. The company is located in Chicago and has a number in the 312 area code. For all outside calls, the user must dial 9. This includes local and long distance calls; however, you are worried that someone calling 911 may misdial. The PRI interface is connected to a leased line and 23 channels are available on it. The provider of the PRI service has offered much better long distance charges, such that management just couldn't turn it down. Therefore, all long distance calls must be placed through the PRI interface. Telephone calls outside of the North American dialing plan will not be allowed. All employees will use SIP telephones. There are 70 employees and extra telephones will be needed for reception, warehouse, and loading dock areas. All of these internal telephones will have three-digit extension numbers in the 300 to 399 range. Any internal user can reach any other internal user by just dialing the three-digit extension.

When a call is received, a message will be played by the auto attendant. If no number is dialed in for 10 seconds, the auto attendant will hang up. If the caller dials in 0 for operator, the call will be forwarded to the operator at extension 100. The caller can also dial 5 for the sales department and 6 for shipping. These two departments have been assigned extensions 105 and 106 respectively. The message box for these special accounts has names mapped to them: operator, sales, and shipping. If the caller dials the extension for a normal user, the call is routed to her extension. If the employee doesn't pick up the telephone in 10 seconds, the call will be routed to the message box where the caller can leave a message. All the message boxes for the employees are in the 700 range and derived from their extension number except 7 replaces 3. Therefore the message box for extension 314 is 714.

For this exercise you need to fill in the following chart. In the first column, fill in the description for the rule; in the second column, fill in the rule itself as you would type it into the configuration file. Some values have already been filled in for your guidance. When you are finished, you can compare your answers with those in Appendix 1.

Exercise 20-1 cont.

Incoming and outgoing rules description	Rules in configuration file
	[Trunks] trunks => SIP ;DID channel trunks => FXO ;PSTN channel trunks => PRI ;PRI channel
[Outgoing rules]	
911 routed to PSTN	exten => 911,1,Dial(FXO/911) exten => 9911,1,Dial(FXO/911)
411 routed to PSTN	
All area code 312 calls routed to PSTN	
All long distance calls routed through the PRI interface	
Dialing internal extension numbers	
Refuse any number dialed outside North America	
[Incoming Rules]	
Auto attendant answers call	exten => s,1,Answer
Auto attendant plays message	
Auto attendant waits 10 seconds	
Auto attendant hangs up if no number dialed in during the wait period	
Caller dialed 0	
Telephone rings for 10 seconds	
Operator doesn't pick up	
Accept message	
Hang up	
Caller dialed Sales	
Telephone rings for 10 seconds	
No one picks up	
Accept message	
Hang up	
Caller dialed Shipping	
Caller dialed extension	

20.4 ENUM

The ENUM service provides a link between PSTN telephone numbers and a reference for VoIP devices. Let's say that you know John Doe's PSTN telephone number but you don't know his VoIP reference, such as his SIP URI. DNS may provide the lookup if John Doe's information has been registered using the ENUM service. Although the word ENUM has its root in "enumerate", it is neither an acronym nor does it stand for anything.

E.164

It helps to appreciate that the telephone numbering systems on our planet have been organized by the International Telecommunication Union (ITU) into a system known as the E.164 numbering scheme. Because this system is universally recognized, you can make a telephone call to any destination in the world. The numbering plans themselves, however, are not standardized. The structure of telephone numbers is controlled by local authorities and they definitely do not agree on how these should be designed. In general, a telephone number is composed of at least three parts: a country code, a regional identification, and a subscriber identification.

Take the North American Numbering Plan (NANP) as an example because it is so familiar. It is used by the United States, Canada, and the Caribbean countries, 24 countries and territories in all. The country code for NANP is 1, which also doubles as the long distance prefix. The next three digits are the area code, which narrows the calling area to a maximum of 7.9 million numbers. The next three digits are the exchange code and usually identify the telephone company's central office node where the switching equipment is located. The final four digits identify the subscriber uniquely. The 11 digits in the number fit comfortably with the maximum 15 digits specified by E.164. Other countries use different number formats. For example a telephone number in the United Kingdom, complete with country code, might look like this: +44 (0)8704 20 30 96. In contrast, a telephone number in India might look like this: 91- 44 – 42103174. This disparity in telephone numbers must be accounted for when designing the ENUM system.

ENUM in DNS

As described in the chapter on DNS, DNS provides two types of lookups: forward and reverse. A forward lookup starts with a domain name and returns an IP address. A reverse lookup starts with an IP address and returns a domain name. The ENUM service is just a reverse lookup except that we start with an E.164 number instead of an IP address. Implementation is also a little tricky because unlike an IP address, which is always a fixed length (four octets for version IPv4), E.164 numbers can be varying lengths.

To provide a lookup if you know the telephone number, follow these steps.

1. Provide the PSTN telephone number, which technically is called an E.164 number.

   504 5551234

2. Turn it into an FQDN. The top level domain is ARPA and the second-level domain is E164. Note how the telephone number is reversed.

   4.3.2.1.5.5.5.4.0.5.e164.arpa

3. Submit the FQDN to DNS.
4. DNS replies with the URI: sip:john.doe@abc.com

ENUM uses the special naming authority pointer (NAPTR) service record in DNS.

The benefit to ENUM is that VoIP devices become more generally available. It is easier to input numbers than alpha characters on a keypad and everyone knows how to dial a telephone. The concept of the uniform resource identifier (URI) is a little confusing to many people. While the URI of sip:john.doe@abc.com is entirely legitimate, it is awkward to input on a keypad. ENUM makes all services that use SIP, such as instant messaging as well as VoIP, available from any device with a keypad.

To avoid confusion, a distinction needs to be drawn at this point. The device that you use to make the call is still a VoIP device. It is not a telephone on the PSTN. You are using an IP telephone and need to get in touch with John Doe. If you dial his E.164 number and prefix it with a 9, the dial plan will route your call over the PSTN and his telephone will ring. If, however, you dial his E.164 number without using the 9 prefix and John Doe is registered with the ENUM service, then his IP telephone will ring instead. There are many reasons why you would choose one route over another; long distance charges may be one reason. In any case the ENUM service just increases the flexibility and reach of VoIP.

If you wish to make use of ENUM, than you will need to find a service provider who has the ENUM service installed and the e164.arpa domain configured on his DNS server. Do a search on the Internet to find these service providers.

20.5 TRAVERSING NAT AND FIREWALLS

Forwarding a VoIP transmission across a network address translation (NAT) device can be a problem. First we will examine the difficulties involved and then some possible solutions. As a reminder, the mechanisms of NAT were examined in Chapter 10, "Addressing," for IP addressing.

NAT itself provides a simple service but because it is inevitably combined with a firewall and/or router, using it becomes complicated. Nevertheless services such as HTTP, FTP, or e-mail have no difficulty, so why is NAT such a problem for VoIP?

Two characteristics of VoIP are particularly troublesome. First, SIP, the VoIP signaling protocol, uses IP addresses found in the SIP header to route calls. Secondly, SIP and RTP, which carry the conversation, are not treated as a single data stream by the devices in between the endpoints. This can prevent the conversation taking place even after a connection is made.

SIP and IP addresses

In Chapter 18, "SIP in Detail," SIP was discussed at great length. Recall from Figure 18-2 that the INVITE message from the caller is passed through one or more proxy servers until it is forwarded to the called device. The called device sends various packets, such as RINGING and OK, back through the intermediate proxy servers. A final ACK is sent by the caller to the called, in order to finish the connection stage of the call. At this point RTP takes over and transmits the media packets directly between the end-points. How do the end-points know the IP addresses of each other since they have only communicated with the intermediate proxy servers until the final ACK? SIP helps this process along by placing the caller's IP address in the Via field of the SIP header.

As Figure 20-3 illustrates, the IP address of the originating SIP telephone is placed in the Via field of the SIP header. This is important because the IP address of the originating device has long since disappeared from the IP header. Refer back to Figure 18-6 if you need to refresh your memory about the SIP header and the VIA field. The problem here is that the original IP address is not routable on the Internet because it is a private IP address.

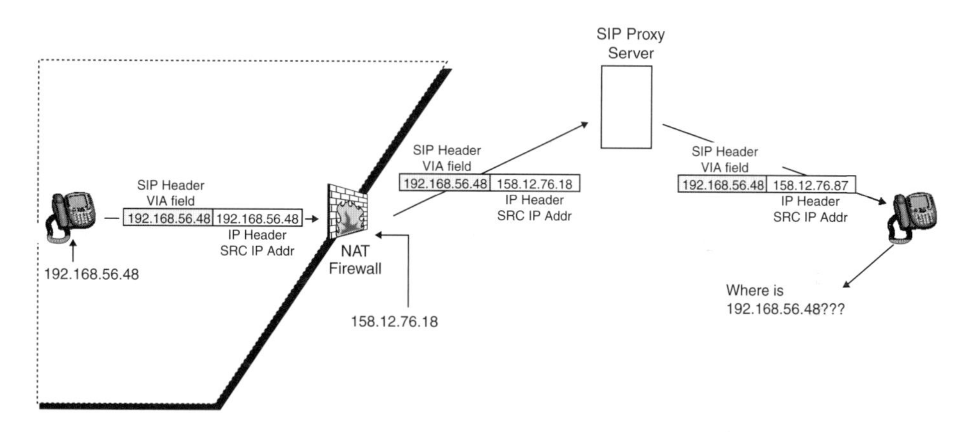

Figure 20-3: The IP address of the caller is placed in the SIP header

NAT and inbound connections

The firewall function of the NAT/firewall device protects the private network by refusing inbound packets if they are not a response to a request that originated inside the private network. This would clearly break a VoIP system if outbound calls were allowed but inbound calls were not. What good would a telephone system be if no one could call you?

This restriction can be relaxed if a port is opened on the firewall. To implement this, the firewall is configured to listen for a particular port number and forward the packet to a particular address. As an illustration, if you had a Web server with IP address 172.16.1.1 then you would configure the NAT/firewall to watch out for any packets from the Internet with TCP port 80, the HTTP port, in the TCP header and forward it through the firewall to the internal network to IP address 172.16.1.1.

This technique will work if you have a SIP proxy server on the inside of your private network. Set up a firewall rule that looks for the SIP port number 5060 and forward these packets to the SIP server inside your network. If you do not have a SIP server inside your private network but do have numerous SIP phones, this technique fails. The NAT/firewall has no way of knowing to which of the many internal IP telephones to forward the SIP packet.

RTP and the NAT/firewall

RTP has numerous issues with NAT and firewalls. It is not uncommon to find a situation in which SIP can make the connection between two VoIP telephones, but then no conversation can be heard, or heard in only one direction. Whenever you encounter this symptom, start by looking at RTP and the NAT/firewall.

RTP uses dynamic ports

RTP uses UDP and there is no fixed port defined for it. The ports are dynamic because they can be negotiated during the SIP connection phase. If the port can't be opened on the firewall because we don't know what the port is, the voice stream will fail. It also doesn't make sense to open a multitude of ports on the firewall because that would defeat the purpose of having a firewall. This issue can be easily dealt with, however. Decide on a port number for RTP, open it up on the firewall, and configure your devices to use that port number.

SIP and RTP are not associated for the firewall

Although SIP and SDP negotiate the parameters that RTP uses, the firewall doesn't know about their association. They are separate protocols, one using TCP and the other using UDP, with separate ports. If SIP is successful making a connection through the firewall, it doesn't help RTP. That is why the telephone may ring, but there is no one at the other end. RTP must be configured independently of SIP in the firewall.

Overcoming NAT/firewall problems

One obvious way to avoid NAT/firewall problems is to avoid NAT altogether. However, this requires giving each device a public IP address. This is not a realistic approach on the vast majority of networks today.

A very common approach is to use keep-alive packets from the inside of the private network to the proxy server on the public network. The IP address/port binding is kept in the NAT/firewall as long as there is activity through it. By sending keep-alive packets every 30 seconds or less, the binding is kept active. Therefore packets that originate on the public network can be passed through to VoIP devices on the private network. The REGISTER SIP packet sent at regular intervals from the SIP device to the proxy server acts very nicely as a keep-alive packet.

STUN server

Simple traversal of UDP through NATs (STUN) is a network protocol designed to alleviate the IP address problem with SIP. Recall that the problem is that the destination device uses the IP address found in the Via field of the SIP header. Because this is the original IP address of the caller's device, which is normally a private IP address and therefore not routable, the call will fail.

The VoIP device can use the STUN server as follows. The VoIP device first contacts the STUN server to learn what the public side IP address is for the NAT/Firewall. The STUN server simply reads this information from the Source IP address field in the IP packet that it just received. The STUN server replies to the VoIP device with this information. The VoIP device can now create a SIP REGISTER or INVITE packet and place this information into the Via header. The called VoIP device can now communicate with the caller by directing packets to the NAT/Firewall. This is illustrated in Figure 20-4.

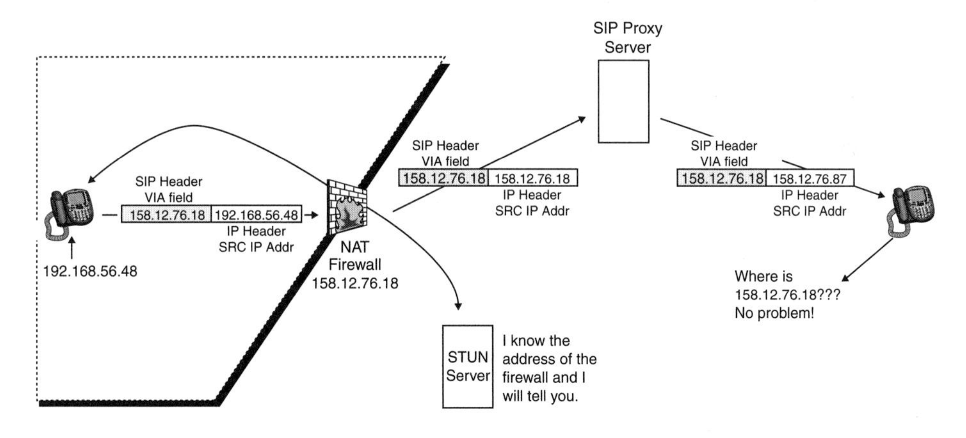

Figure 20-4: SIP uses a STUN server to provide NAT traversal

In order to implement STUN, you require a STUN server on the public network or Internet and you must configure your VoIP devices with the IP address of the STUN server.

SUMMARY

Section 20.1: Outside Connections for Your VoIP System

You have many options for connecting your VoIP system to the outside world, including analog lines to the PSTN, digital lines to the PSTN, asynchronous services, and synchronous services. This section provides a summary of these options.

Section 20.2: Direct Inward Dialing (DID)

Direct inward dialing (DID) allows a company to provide a single telephone number that connects to all extensions inside the company. This feature is implemented with traditional analog PBXs and is now available with IP-PBXs as well. DID is even more valuable for VoIP because an organization can provide local access to its clients from a long distance location.

Section 20.3: The Dial Plan

The dial plan is often called the inbound and outbound calling rules. The dial plan instructs the telephone system what to do whenever there is user action. For example, dialing a local number requires different action than dialing a long distance number. The dial plan must cover every contingency, both inbound and outbound, that could occur. An example was given of a user misdialing 911 by putting an extra 9 in front of it.

Section 20.4: ENUM

The ENUM service provides a link between PSTN (E.164) telephone numbers and a VoIP URI. The link is provided by a reverse lookup in DNS. ENUM provides a convenience. You can dial the other person's PSTN number and her VoIP telephone answers, not her analog telephone. The VoIP telephone needs to be able to distinguish between the two, typically because the 9 prefix is used for the PSTN.

Section 20.5: Traversing NAT and Firewalls

Forwarding a VoIP transmission across a network address translation (NAT) device can be a problem. NAT changes the source IP address inside a packet because it is not reachable from the public Internet. When the destination VoIP tries to respond, it fails because it can't find the originating device. Two possible solutions are described. VoIP devices behind the NAT/firewall can send keep-alive packets to the external proxy server, which keeps the IP address/protocol binding current in the NAT/firewall. Another technique is to use a STUN server.

Review Questions: Setting up a Modern VoIP System

1. A telephone conversation requires equal bandwidth in each direction. Because of this property, which type of connection is the least efficient for a VoIP call?

   a) PRI service
   b) Television cable service
   c) Frame relay
   d) Synchronous DSL

2. What is the basic problem with using SIP through NAT?

   a) RTP uses dynamic port numbers.
   b) The recipient VoIP device believes that the packet came from the proxy server.
   c) The IP address of the source VoIP device is embedded in the SIP header.
   d) Calls that originate from the public Internet are blocked by the NAT/firewall.

3. An important group of your customers are located in a distant city and the long distance charges they incur for telephone calls they make to you are very expensive. You use both the PSTN and a VoIP system for your business. How can you provide less expensive calls and keep your customers happy?

   a) Set up a point of presence in the remote city.
   b) Sign up for an 800 number with the telephone company.
   c) Ask them to install softphone software on their computers and use the Internet to make calls to you.
   d) Use direct inward dialing and acquire a local number from a DID provider.

Continues next page

Review Questions: continued

4. You are using a VoIP system with 200 users at your local location and 150 users at a remote location. Your dial plan requires that your users dial just the four-digit extension if they want to talk to someone at the same location as themselves. However, if they want to dial an internal extension at the remote location, they have to dial 8 first. All the extensions at one location are in the range 1000–3000; all the extensions at the other location are in the range 4000–6000. Which string matches the pattern for dialing an extension at the remote location?

   a) _8ZXXX
   b) NXXX
   c) 8ZXXX
   d) _8NXXX

5. The ENUM service uses which kind of DNS query?

   a) Request
   b) Forward
   c) Iterative
   d) Reverse

Appendix 1 Answer Key

Chapter 1:

Review Questions: 1. b, 2. a, 3. d, 4. c, 5. a

Chapter 2:

Review Questions: 1. b, 2. c, 3. d, 4. a, 5. a

Chapter 3:

Review Questions: 1. d, 2. a, 3. b, 4. c, 5. a

Chapter 4:

Review Questions: 1. d, 2. a, 3. c, 4. b, 5. a

Chapter 5:

Review Questions: 1. c, 2. a, 3. a, 4. d, 5. b

Chapter 7:

Review Questions: 1. b, 2. a, 3. d, 4. c, 5. c, 6. b, 7. d, 8. d, 9. a, 10. c

Chapter 8:

Exercise 8-2, Ethernet Decodes
Packet #1 E_II, IP (08 00)
Packet #2 E_II, IPv6 (86 dd)

Packet #3 E_II, ARP (08 06)
Packet #4 802.3, length 00 3f hex
Packet #5 E_II, 802.1Q (81 00)
Packet #6 E_II, MPLS (88 47)
Review Questions: 1. b, 2. d, 3. a, 4. a, 5. d

Chapter 9:

Review Questions IP: 1. a, 2. d, 3. b, 4. b, 5. c
Review Questions ICMP: 1. d, 2. b, 3. d, 4. d, 5. b

Chapter 10:

Exercise 10-1: 1. c, 2. b, 3. a, 4. d
Exercise 10-2:

1. Which of the following may be assigned to a host? If the address can't be assigned to a host, why not?

Address	Yes or why not
57.200.258.34	No, 3rd octet = 258
223.223.255.223	Yes
230.57.0.1	No, multicast
126.0.0.0	No, host ID all 0's
191.80.20.0	Yes
87.167.76.100	Yes
0.40.50.200	No, starts with 0
127.100.2.2	No, loopback address
255.255.255.255	No, all networks broadcast
56.0.0.201	Yes
199.21.133.255	No, local network broadcast

2. What class are the following IP addresses?

223.55.231.5	C
76.234.98.123	A
134.55.76.105	B
225.0.0.56	D
191.14.200.98	B
240.57.99.1	E
126.127.1.254	A

Exercise 10-3:

A. No problem
B. Default gateway address incorrect
C. Default gateway address all 0's
D. Wrong default gateway, duplicate IP address
E. Duplicate IP address
F. Wrong network ID
G. IP address is that of default gateway
H. IP and default gateway address are the same

Exercise 10-4:

1. a) 57, b) 191, c) 84
2. a) 1111 0000, b) 0110 0101, c) 1011 1011
3. A, B, C

Exercise 10-5:

a) subnet 128
b) subnet 192
c) subnet 203.208
d) subnet 160
e) subnet 92

Exercise 10-6:

Address #1	Address #2	Subnet mask	Yes or No
192.168.6.45	192.168.6.70	255.255.255.192	No
201.56.166.177	201.56.166.190	255.255.255.240	Yes
150.67.24.200	150.67.24.225	255.255.255.224	No
185.22.177.88	185.22.175.88	255.255.248.0	No
66.12.7.92	66.12.213.79	255.255.0.0	Yes
45.159.254.45	45.129.33.178	255.224.0.0	Yes

Exercise 10-7:

1. 6 subnets
2. 255.255.255.240
4. 197.245.12.45, 197.245.12.73

Review Questions: 1. a, 2. b, 3. d, 4. b, 5. c

Chapter 11:

Review Questions: 1. a, 2. a, 3. b, 4. c, 5. d

Chapter 12:

Exercise 12-1:

1. DHCP ACK
2. 192.168.1.109
3. 00:03:ff:d4:09:7a
4. 192.168.1.1
5. 1 day
6. 192.168.1.1
7. 206.47.244.104, 206.47.244.12

Review Questions: 1. b, 2. a, 3. a, 4. d, 5. a

Chapter 13:

Review Questions: 1. c, 2. d, 3. c, 4. a, 5. c

Chapter 14:

Exercise 14-1:

Q N DiffServ
S MGCP/Megaco
T RSVP
D Frame Relay
N IP
S SIP/SDP
T UDP
Q D 802.1p
Q N IP TOS
M RTP
D Ethernet
S H.323
D ATM
S RTSP
D/N MPLS (between D and N)
Q RTCP

Review Questions: 1. c, 2. a, 3. b, 4. d, 5. c

Chapter 15:

Review Questions: 1. b, 2. c, 3. a, 4. d, 5. c

Chapter 16:

Review Questions: 1. b, 2. d, 3. c, 4. d, 5. a

Chapter 17:

Review Questions: 1. a, 2. c, 3. d, 4. b

Chapter 18:

Review Questions: 1. b, 2. d, 3. b, 4. a, 5. b

Chapter 19:

Review Questions: 1. b, 2. a, 3. c, 4. d, 5. b

Chapter 20:

Exercise 20-1: Creating a Dial Plan

Incoming and outgoing rules description	Rules in configuration file
	[Trunks] trunks => SIP ;DID channel trunks => FXO ;PSTN channel trunks => PRI ;PRI channel
[Outgoing rules]	
911 routed to PSTN	exten => 911,1,Dial(FXO/911) exten => 9911,1,Dial(FXO/911)
411 routed to PSTN	exten => 411,1,Dial(FXO/411)
All area code 312 call routed to PSTN	exten => _9NXXXXXX,1,Dial(FXO/NXXXXXX)
All long distance calls routed through the PRI interface	exten => _1NXXNXXXXXX,1,Dial(PRI/1NXXNXXXXXX)
Dialing internal extension numbers	exten => _3XX,1,Dial(SIP/3XX)
Refuse any number dialed outside North America	Exten => _011.,1, Playback (not authorized)
[Incoming Rules]	
Auto attendant answers call	exten => s,1,Answer
Auto attendant plays message	exten => s,2,Playback (Welcome-msg)
Auto attendant waits 10 seconds	exten => s,3,Wait(10)
Auto attendant hangs up if no number dialed in during the wait period	exten => s,4,Hangup
Caller dialed 0	exten => 0,1,Dial(SIP/100)
Telephone rings for 10 seconds	exten => 0,2,Wait(10)
Operator doesn't pick up	exten => 0,3,Playback(noans_leave msg)
Accept message	exten => 0,4,Message(Operator)
Hang up	exten => 0,5,Hangup

[Incoming Rules]	
Caller dialed Sales	exten => 5,1,Dial(SIP/105)
Telephone rings for 10 seconds	exten => 5,2,Wait(10)
No one picks up	exten => 5,3,Playback(noans_leave msg)
Accept message	exten => 5,4,Message(Sales)
Hang up	exten => 5,5,Hangup
Caller dialed Shipping	exten => 6,1,Dial(SIP/106)
Telephone rings for 10 seconds	exten => 6,2,Wait(10)
No one picks up	exten => 6,3,Playback(noans_leave msg)
Accept message	exten => 6,4,Message(Shipping)
Hang up	exten => 6,5,Hangup
Caller dialed extension	exten => 3XX,1,Dial(SIP/3XX)
Telephone rings for 10 seconds	exten => 3XX,2,Wait(10)
No one picks up	exten => 3XX,3,Playback(noans_leave msg)
Accept message	exten => 3XX,4,Message(7XX)
Hang up	exten => 3XX,5,Hangup

Review Questions: 1. b, 2. c, 3. d, 4. a, 5. d

Appendix 2 Analyzing VoIP with Wireshark

WIRESHARK THE PROTOCOL ANALYZER

Wireshark is a protocol analyzer, or "packet sniffer" application, used for network troubleshooting, analysis, software and protocol development, as well as education. It has all of the standard features of a protocol analyzer. In addition, it has many extra features that work particularly well with VoIP. Wireshark was originally made available as a program called Ethereal. Ethereal is no longer under development and all future enhancements are made available in Wireshark. Wireshark is open source software and it is free.

WinPcap the capture driver

Wireshark doesn't have its own code to capture packets. Wireshark uses WinPcap to capture packets on the Windows platform. This is not an issue if you download Wireshark today since WinPcap will also be downloaded and installed. If for some reason this is not the case, you can download WinPcap from its own site. WinPcap is open source and available free. Not wanting to reinvent the wheel, many network programs use WinPcap as their packet capture driver. Besides Wireshark, some examples include Nmap, Snort, and WinDump.

Find the software

Wireshark can be downloaded from:

http://www.wireshark.org/

This should include WinPcap. If you need to download WinPcap independently, you can find it at:

http://www.winpcap.org

Installation

Double-click the Wireshark installation file. Accept all defaults and it will install normally.

Note: The default installation will also install WInPcap. If you already have WinPcap installed because another program installed it, don't install it again. Deselect this option when it is offered to you.

Filtering

Because the network can generate thousands and millions of packets in a very short time, this sheer volume can overwhelm the analyzer and relevant information goes undetected. It is crucial that the analyzer can filter the traffic and present a subset of the packets to us. Some important elements that you may filter on include station addresses, network addresses, types of protocol, and error conditions.

There are two types of filtering of interest: capture filters and display filters.

Capture filters

We can control which types of packets to capture, for example, by IP address or by protocol. By capturing only packets of interest, we simplify the capture display and reduce the size of our capture buffer.

Alternatively, we may capture only a portion of the packet. It is common to be interested in the headers of a packet and not the payload. By filtering on number of bytes, we can increase the number of packets that our buffer can hold.

Display filters

If we capture a broad range of packets, we may want to display only a subset of them. This makes it easier to concentrate on the packets of interest or highlight a pattern.

Filtering in Wireshark

Filtering in Wireshark is an art. Capture filters and display filters are handled separately and have different syntax. Naturally, this confuses many people. The capture filters are handled by WinPcap and therefore use WinPcap syntax. The display filters were written by the Wireshark development team and have a different syntax.

Applying capture filters

Here is the procedure:

1. Click the first icon (List Capture Interfaces).
2. Identify the network adapter that you want to use for your packet capture.

3. Click the Prepare button.
4. In the capture filter field, type in the filter.

See Figure Appendix 2-1.

Examples of capture filters include:
ether proto 0x0806 (for ARP)
tcp port 80
ether src not 00:0f:66:a9:4e:3f
host 192.168.1.1 or host 192.168.1.120

Table Appendix 2-1 gives a more complete list of WinPcap options.

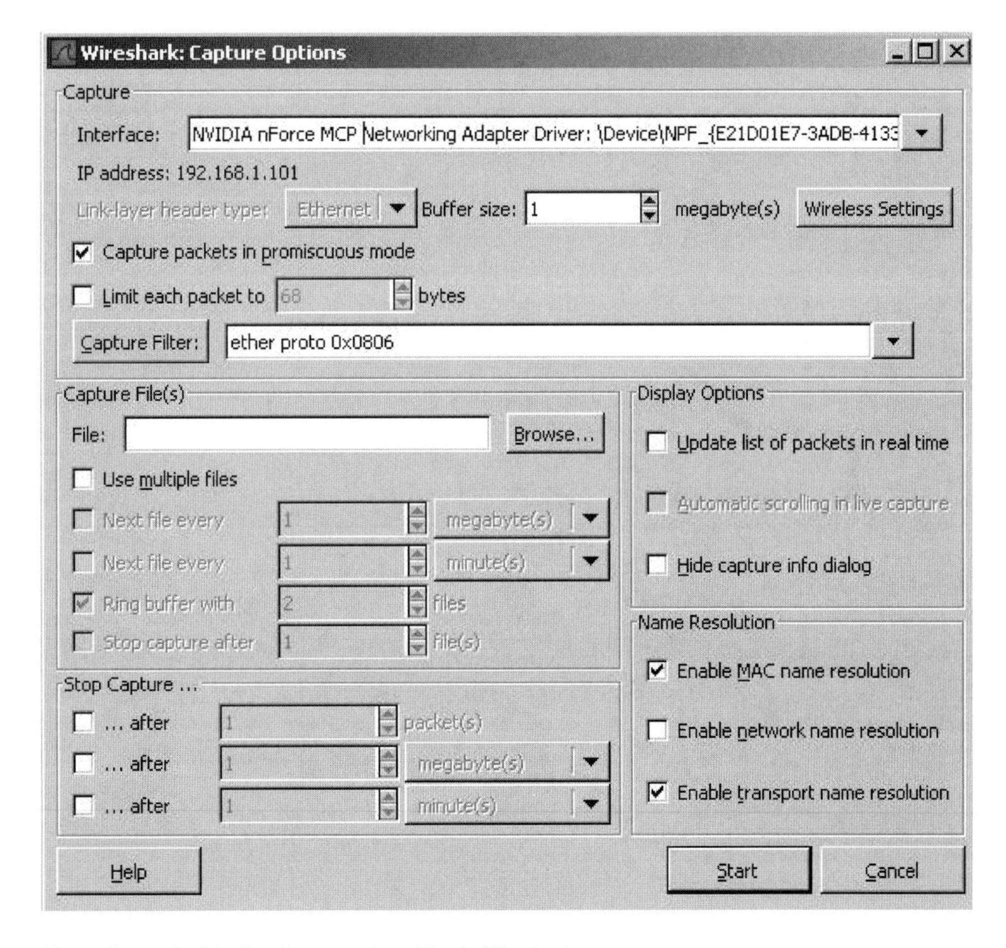

Figure Appendix 2-1: Creating a capture filter in Wireshark

Working with the Interface (if you are not capturing the traffic you think you should)		
Filter	Example	Comments
Filtering Packets		
host a.b.c.d host name	host 192.168.1.130 host denver	Filter on IP address Both destination and host
ether host a:b:c:d:e:f	ether host 00:0f:66:a9:4e:3f	Filter on Ethernet MAC address
port num port name	port 80 port http	Filter on port number or name Both destination and source
net a.b.c.d mask nmask net a.b.c.d/numbits	192.168.5.32 mask 255.255.255.240 192.168.5.32/28	Filter on a network, subnet or IP address range Both destination and source
ether broadcast ip broadcast	ether broadcast ip broadcast	Filter on Ethernet broadcasts Filter on IP broadcasts
ether multicast ip multicast	ether multicast ip multicast	Filter on Ethernet multicasts This is shorthand for ether[0]& 1!=0 Filter on IP multicasts
src . . . . . . src dst . . . . . . dst	src host 192.168.5.200 ether src 00:0f:66:a9:4e:3f dst port 80	Only match if the source or destination address equals. This can be prefixed to host, net, or port or appended to ether.
icmp, ip, tcp udp, arp	arp	Filter on these protocols by name
ip proto number	ip proto 1	Filter on IP protocols by number 1=icmp, 6=tcp, 17=udp, 47=gre, 50=ESP, 51=AH, 115=L2TP
. . . and . . ., . . . or . . ., . . . not . . ., (. . .)		Combine multiple criteria
- F file	-F filename.txt	Read the criteria expression from a file rather than the command line. Useful if you use the same criteria often or it is long and complex.
Complex arguments		
[m:n]	'ether[0] & 1 !=0'	Offset byte from the beginning of the header. Optionally, you can include the number of bytes. The example is looking at the first byte [0], offset 0, in the Ethernet header
Relationships >, <, >=, <=, =, != Binary operators +, –, *, / Binary AND (&) Negation ('!' or 'not') Concatenation ('&&' or "and") Alternation ('||' or 'or')		

Table Appendix 2-1: WinPcap command summary

Applying display filters

You need to have some packets in the capture buffer before you can see the effects of display filters.

Procedure for simple display filters:

1. Type your filter into the filter field.
   Simple display filters:
   The protocol name (IP, ARP, TCP, UDP, ICMP, RTP, SIP, BOOTP)
   ip.addr==192.168.1.200

Procedure for complex display filters:

2. Click on the Expression button beside the filter field.
3. Select the field name. In order to do this, find the protocol you are interested in and expand it with the plus (+) symbol.
4. Select the field you are interested in.
5. Select the relation. For all relations, except "Is present," you will need to type in a value.
6. Click the OK button.

Filter shortcut

If you select a packet in the capture buffer, you can create a filter by right-clicking to bring up a menu.

7. Select a packet in the capture buffer, place your cursor on either the source or destination address, right-click to bring up a menu.
8. Select Apply as filter and then Selected.
9. The filter will be created for you. Note that the filter will be for the source or destination address and that it will be for the IP address or the Ethernet MAC address, depending on which was used in the display.

USING WIRESHARK WITH VoIP

How can I filter for VoIP traffic?
Use SIP in the display filter for SIP messages.
Use H225 or H245 in the display filter for H.323 messages.
Use RTP in the display filter for actual voice transmission.

How can I see the SIP Methods and replies in my sample?
Statistics > SIP. . . > Create Stat.
You can create a filter, for example from a particular IP address

How can I see all VoIP calls in my sample?
Statistics > VoIP Calls

How can I see the SIP handshake for a particular call?
Statistics > VoIP Calls > Select the call > Graph

How can I see all of the packets that belong to a call, including the SIP and RTP packets?
Statistics > VoIP Calls > Select the call > Prepare Filter
Change back to the main screen.
A filter will have been created and the filter window should be green. Click Apply.

How can I listen to the voice stream?
Statistics > VoIP Calls > Select the call > Click the Player button > Modify the jitter buffer (a value of 20 ms works well) > Decode button > Select the media stream that you want to hear > Click the Play button.

Notes:

1. This feature was introduced with Wireshark version .99.5. If you have an earlier version, you need to download and upgrade your copy.
2. You can select multiple streams to listen to. This is required if you want to hear both sides of the conversation.
3. This feature works for only G.711 u-law and G.711 A-law streams. No other codecs are supported.

How can I look for problems in the voice stream?
Statistics > RTP > Show All Streams

Wireshark will display any RTP streams that it finds. Figure Appendix 2-2 is an example of this report. Note the following.

Lost: This is the number of packets lost during this capture. Too many lost packets will degrade the conversation.

Max Delta (ms): This is the latency or how long the packet takes to travel end to end. Latency of 50 ms is toll quality. Anything greater than 150 ms is considered unacceptable.

Mean Jitter (ms): Anything greater than 20 ms is considered unacceptable.

Pb?: This column indicates a problem. If there is an "x" beside an RTP stream, that stream has at least one problem.

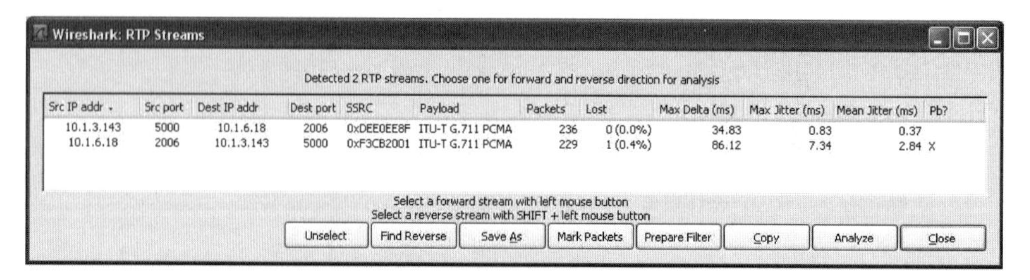

Figure Appendix 2-2: Wireshark displaying RTP streams

To see the RTP stream and any problems, click on the Analyze button. Figure Appendix 2-3 is the result and displays all of the packets in the stream. Scroll down until you find a packet with the Status as non-OK. In the example, a packet is marked as "wrong sequence nr(number)". This indicates that a packet is lost. If you look at the sequence numbers, you can see that one is missing. One missing packet is not a problem. The previous one will simply be repeated. However, many missing packets will result in noticeably degraded conversation.

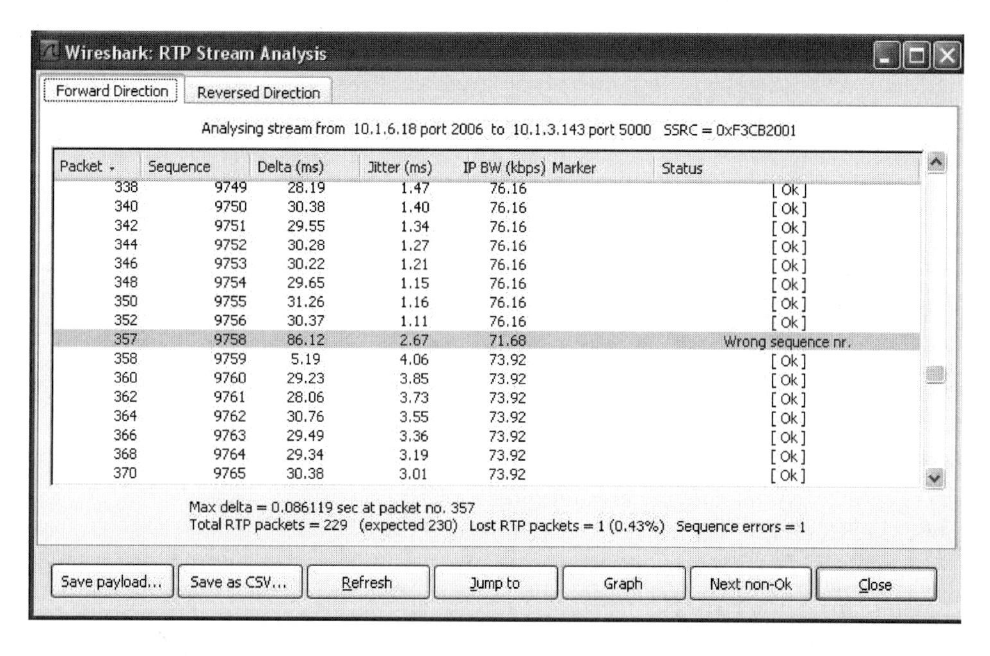

Packet	Sequence	Delta (ms)	Jitter (ms)	IP BW (kbps)	Marker	Status
338	9749	28.19	1.47	76.16		[Ok]
340	9750	30.38	1.40	76.16		[Ok]
342	9751	29.55	1.34	76.16		[Ok]
344	9752	30.28	1.27	76.16		[Ok]
346	9753	30.22	1.21	76.16		[Ok]
348	9754	29.65	1.15	76.16		[Ok]
350	9755	31.26	1.16	76.16		[Ok]
352	9756	30.37	1.11	76.16		[Ok]
357	9758	86.12	2.67	71.68		Wrong sequence nr.
358	9759	5.19	4.06	73.92		[Ok]
360	9760	29.23	3.85	73.92		[Ok]
362	9761	28.06	3.73	73.92		[Ok]
364	9762	30.76	3.55	73.92		[Ok]
366	9763	29.49	3.36	73.92		[Ok]
368	9764	29.34	3.19	73.92		[Ok]
370	9765	30.38	3.01	73.92		[Ok]

Figure Appendix 2-3: Analyzing an RTP stream

Appendix 3 VoIP Acronyms

ACELP	Algebraic-Code-Excited Linear-Prediction
ACK	Acknowledge
ADC	Analog/Digital Converter
ADPCM	Adaptive Differential Pulse-Code Modulation
AMR	Adaptive Multi Rate
ANSI	American National Standards Institute
APIPA	Automatic Private Internet Protocol Addressing
ARIN	American Registry for Internet Numbers
ARP	Address Resolution Protocol
ARPANET	Advanced Research Projects Agency NETwork
ASCII	American Standard Code for Information Interchange
ATA	Analog Telephone Adapter
ATM	Asynchronous Transfer Mode
AVVID	Architecture for Voice, Video and Integrated Data (Cisco)
BCM	Business Communications Manager (Nortel)
BOOTP	Boot Protocol
BPS	Bits Per Second
BRI	Basic Rate Interface
CBQ	Class-Based Queuing
CELP	Code Excited Linear Prediction
CNG	Comfort Noise Generation
CO	Central Office
CODEC	Code/DECode
CNAME	Canonical Name
CRC	Cyclic Redundancy Check

CS-ACELP	Conjugate-Structure Algebraic-Code-Excited Linear-Prediction
CSeq	Command Sequence
CSMA/CD	Carrier Sense Multiple Access/Collision Detection
CTI	Computer Telephony Integration
DHCP	Dynamic Host Configuration Protocol
DID	Direct Inward Dialing
DLCI	Data Link Circuit Identifier
DNS	Domain Name System
DoD	Department of Defense
DSL	Digital Subscriber Line
DSP	Digital Signal Processor
DSU/CSU	Digital (or Data) Service Unit/Channel Service Unit
DTMF	Dual Tone Multi Frequency
E&M	Ear and Mouth
ECLIPS	Enterprise Class Internet Protocol Solutions (Avaya)
ECN	Explicit Congestion Notification
ENUM	E.164 Number Mapping
FCS	Frame Check Sequence
FDDI	Fiber Distributed Data Interface
FDM	Frequency Division Multiplexing
FEC	Forward Equivalency Class
FIN	Finish
FQDN	Fully Qualified Domain Name
FTP	File Transfer Protocol
FXO	Foreign Exchange Office
FXS	Foreign Exchange Subscriber
GSM	Groupe Speciale Mobile
HTTP	Hyper Text Transport Protocol
IAX	Inter-Asterisk eXchange Protocol
ICANN	Internet Corporation for Assigned Names and Numbers
ICMP	Internet Communications Message Protocol
IEEE	Institute of Electrical and Electronics Engineers
IETF	Internet Engineering Task Force
IHL	IP Header Length
iLBC	internet Low Bitrate Codec
INTSERV	INTegrated SERVices
IP	Internet Protocol
IPG	Inter Packet Gap
IPX/SPX	Internetwork Packet Exchange/Sequenced Packet Exchange
ISDN	Integrated Services Digital Network
ISO	International Standards Organization
ISP	Internet Service Provider
ITSP	Internet Telephony Service Provider

ITU	International Telephone Union
IVR	Interactive Voice Response
KTS	Key Telephone System
LAN	Local Area Network
LD-CELP	Low Delay-Code Excited Linear Prediction
LER	Label Edge Router
LLC	Logical Link Control
LNP	Local Number Portability
LPC	Linear Predictive Coding
LSP	Label Switch Path
LSR	Label Switch Router
MAC	Media Access Control
MAN	Metropolitan Area Network
MCU	Multipoint Control Unit
Megaco	Media Gateway Controller Protocol
MG	Media Gateway
MGC	Media Gateway Controller
MGCP	Media Gateway Control Protocol
MIME	Multipurpose Internet Mail Extension
MIPS	Millions of Instructions Per Second
MODEM	MOdulate/DEModulate
MOS	Mean Opinion Score
MPLS	Multi Protocol Label Switching
MP-MLQ	Multipulse Maximum Likelihood Quantization
NANP	North American Numbering Plan
NAT	Network Address Translation
NIC	Network Interface Card
NTP	Network Time Protocol
OSIRM	Open Systems Interconnect Reference Model
OUI	Organizational Unique Identifier
PAM	Pulse Amplitude Modulation
PBX	Private Branch Exchange
PC	Personal Computer
PCM	Pulse Code Modulation
PCS	Personal Communication Services
PDA	Personal Digital Assistant
PDU	Protocol Data Unit
PHB	Per Hop Behavior
PING	Packet INternet Groper
PoE	Power over Ethernet
PoP	Point of Presence
POP	Post Office Protocol
POTS	Plain Old Telephone System

PRI	Primary Rate Interface
PSH	Push
PSTN	Public Switched Telephone Network
QoS	Quality of Service
RAS	Registration Admission Status
RED	Random Early Discard
RFC	Request for Comment
RJ	Registered Jack
RPE-LPC	Regular Pulse Excited - Linear Predictive Coder
RST	Reset
RSVP	Resource reSerVation Protocol
RTP	Real Time Protocol
RTCP	Real Time Control Protocol
SIGTRAN	Signaling Transport
SIP	Session Initiation Protocol
SCCP	Skinny Client Control Protocol
SDP	Session Description Protocol
SLA	Service Level Agreement
SMB	Small and Medium Size Business
SMTP	Simple Message Transport Protocol
SNR	Signal to Noise Ratio
SONET	Synchronous Optical Network
SS7	Signaling System 7
SCP	Signal Control Point
SSP	Signal Switching Point
STUN	Simple Traversal of UDP through NATs
STP	Signal Transfer Point
SYN	Synchronize
TCP/IP	Transmission Control Protocol/Internet Protocol
TDM	Time Domain Multiplexing
TFTP	Tiny (Trivial) File Transfer Protocol
TLD	Top Level Domain
ToS	Type of Service
TTL	Time to Live
USB	Universal Serial Bus
UDP	User Datagram Protocol
UPS	Uninterruptible Power Supply
URG	Urgent
URI	Universal Resource Identifier
URL	Universal Resource Locator
UTP	Unshielded Twisted Pair
VAD	Voice Activity Detection
VLAN	Virtual LAN

VoIP	Voice over Internet Protocol
VPI	Virtual Path Identifier
W3W	World Wide Web consortium
WAN	Wide Area Network
WFQ	Weighted Fair Queuing
WRED	Weighted Random Early Discard
WWW	World Wide Web

Appendix 4
VoIP Glossary

802.1Q/802.1p: 802.1Q is an IEEE standard used to manage virtual LANS. It provides VLAN identification by inserting a four octet tag inside an Ethernet header. 802.1p provides prioritization information within the 802.1Q tag.

Analog Telephone Adapter: The ATA or the analog telephone adapter is the hardware device that connects the conventional analog telephone to the Internet through a high-speed bandwidth line, provides the interface to convert the analog voice signals into IP packets, delivers dial tone, and manages the call setup.

Bandwidth: Bandwidth refers to the quantity of traffic that can flow over a communications link. It is measured in bits per second (bps). Bandwidth is often mistaken for a measure of speed because the greater the bandwidth, the faster the objects show up on the screen. The speed of electrons through a copper cable, called the propagation delay or velocity of propagation, is constant and depends on the type of cable. Bandwidth, however, varies for any number of reasons.

CELP: Code Excited Linear Prediction, CELP and pronounced "kelp," is a vocoder that is based on analysis-by-search procedures. The code in the name refers to a code book that holds indexes to important parameters of the signal and are transmitted to give a better quality. CELP requires intensive computation but it produces natural speech. CELP is not used in its native format but is the basis of many codecs used in VoIP such as LD-CELP, MP-MLQ/ACELP, and CS-ACELP.

Centrex: Centrex is a service provided by the telephone companies that duplicates the functions of a PBX, except that the equipment is located at the telephone company premises. Centrex alleviates the company from having to look after its own in-house equipment.

Codec: Codec is a contraction of code/decode and is the function that converts an analog signal into digital data. A codec is integral to a VoIP system but anytime a measurement of an analog property, for example temperature or the saltiness of water, is converted into digital data, a codec is used.

Companding: Companding is derived from the words compression and expansion. It describes how the linear discrete values generated by PCM are compressed during encoding. The ITU has defined two logarithmic companding schemes which are very similar to each other and are used extensively throughout the world's telephone systems. μ-law (Mu-law) is used in North America and Japan, while A-law is used in the rest of the world.

Dial plan: The dial plan is the set of rules that the telephone system uses to direct a call when a number is dialed.

DID: Direct Inward Dialing (DID) is a service provided by a PBX such that all the telephone extensions inside a company can be reached over the limited number of channels provided by the telephone company to the business.

DiffServ: Differentiated Services is a quality of service technique that marks packet so that routers can queue them based on their priority levels. DiffServ makes use of a field in the IP header called Type of Service (ToS).

E.164: The E.164 is the telephone numbering systems used among all of the countries of the world and is organized by the International Telecommunication Union (ITU).

E&M port: The Ear & Mouth (E&M) port on a gateway is used to connect an analog PBX to a VoIP system.

FXO port: FXO is an acronym for Foreign eXchange Office. On a VoIP gateway, this is the connection to the public switched telephone network.

FXS port: FXS is an acronym for Foreign eXchange Subscriber. This port has several minor uses. It can connect an analog PBX into a VoIP system. It can also be used to attach an analog telephone or a FAX machine to a VoIP system.

GSM: Groupe Speciale Mobile is a popular codec used by VoIP, which began as a European cellular telephone speech encoding standard.

H.323: H.323 is an ITU standard that provides the structure for controlling real-time voice and videoconferencing on the Internet. The H.323 standard supports voice, video, data, application sharing, and whiteboarding and defines media gateways for conversion to packets.

iLBC: iLBC (internet Low Bitrate Codec) is a free speech codec suitable for robust voice communication over IP. The codec is designed for narrowband speech and results in a payload bit rate of 13.33 kbps. Because iLBC's speech frames are compressed independently, the iLBC codec enables graceful speech quality degradation in the case of lost frames, which occurs in connection with lost or delayed IP packets. iLBC is being standardized by the IETF and is used by the popular VoIP service, Skype.

ISDN: Integrated Services Digital Network (ISDN) is a telephone system and a service provided by telephone companies to their subscribers. It is a digital service that integrates data, voice, and multimedia. Two varieties are available. Basic Rate Interface (BRI) provides two 64 Kbps data channels and a 16 Kbps control channel and Primary Rate Interface (PRI) provides 23 data channels and one control channel, each of which is 64 Kbps. Although developed in the 1980s, ISDN did not become popular in North America because of the slow implementation by the telephone companies. Nevertheless, PRI has now become a popular way to connect company VoIP systems to the PSTN.

ITU: The International Telephone Union (ITU) is headquartered in Geneva, Switzerland, and is an international organization within the United Nations System where governments and the private sector coordinate global telecom networks and services.

Jitter: Jitter is defined as variable delay. In other words, the packets are showing up with different intervals between them. Jitter degrades the voice quality, which becomes noticeable to the listener.

Mean Opinion Score: The mean opinion score (MOS) provides a numerical measure of the quality of human speech at the destination end of the circuit. The scheme uses subjective tests (opinionated scores) that are mathematically averaged to obtain a quantitative indicator of the system performance.

MGCP/Megaco: The media gateway control protocol (MGCP) and MEdia GAteway COntroller (Megaco) protocol are two very similar protocols that are used to communicate between a media gateway controller and media gateways. It provides centralized control of multiple gateways in large systems.

MPLS: Multiprotocol Label Switching (MPLS) is a technique used to forward packets in an efficient and fast way through a network by adding a simple label to each packet at an edge router.

Nyquist theorem: According to the Nyquist theorem, the discrete time sequence of a sampled continuous function contains enough information to reproduce the function exactly, provided that the sampling rate is at least twice that of the highest frequency contained in the original signal. This theorem is responsible for the practice of taking 8,000 samples per second when encoding a voice conversation.

PAM and PCM: Pulse Amplitude Modulation (PAM) assigns a value to the amplitude of a sound sample. Pulse Code Modulation (PCM) quantizes the value into a digital value.

PBX (Private Branch Exchange): A PBX is a small telephone switch owned by an organization. It provides telephony services to a large number of corporate users while making use of a limited number of outside lines.

Per Hop Behavior: Per Hob Behavior (PHB) is the treatment given to a packet by a router on the next leg of the journey. An example of a PHB is giving the class of VoIP packets precedence over data packets when they are forwarded by the router. The service provider is responsible for the PHB of a class of packets on his network.

Playout buffer: The playout buffer is memory in a VoIP device that receives the voice packets in the uneven intervals at which they arrive and then plays them out to the receiver at a steady pace. It compensates for jitter.

POTS/PSTN: Plain Old Telephone System (POTS) and Public System Telephone Network (PSTN) are two acronyms used to refer to the traditional telephone system. They are often used when the public telephone system is contrasted with VoIP.

Power over Ethernet: Power over Ethernet (PoE) is a technology that provides power to a device over the Ethernet cable. By providing both power and data over the same cable, an extra cable is eliminated, thereby simplifying the system. PoE is a particularly valuable feature for IP telephones.

Quantization: Quantization is the process by which one sample of voice is placed into one octet. Quantization is the process of constraining an infinite number of values into a set of finite values. In simple language, quantization assigns a numerical value to a sample.

RAS: Registration/Admission/Status is the H.225.0 RAS protocol that is used by a terminal to communicate with the gatekeeper within the H.323 family of protocols.

Real-time Transport Protocol: The Real-time Transport Protocol (RTP) is a protocol that provides sequencing information and payload identification to streaming data. RTP carries the telephone conversation during a VoIP call.

SDP: The Session Description Protocol (SDP) works in partnership with SIP to set up VoIP calls. SIP's role is to set up the session, while SDP's role is to describe it.

Signaling System 7: Signaling System 7 (SS7) is the control system used by telephone companies to set up telephone calls as well as provide other services such as caller ID, conference calling and toll free long distance calls. SS7 is a digital, packet-based network that is out-of-band, meaning that the signals do not travel over the same lines as the telephone call.

Silence suppression: When there is a pause in the conversation, no signal is generated. The silence suppression feature of modern codecs will prevent null information from being transmitted. This decreases the bandwidth requirements for a conversation from 33% to 50%.

SIP: Session Initiation Protocol (SIP) is an IETF standard for setting up sessions between one or more clients. It can be used by VoIP as well as Instant Messaging, Internet conferencing, and presence and event notification. SIP is a request-response protocol that closely resembles two other Internet protocols, HTTP and SMTP.

Skype: Skype is a telephone service using VoIP, which allows a Skype user to make free telephone calls to other Skype users over the Internet. Skype also allows users to make telephone calls to anyone on the PSTN, but charges a small fee for this. Skype uses a proprietary call control protocol and iLBC as its codec. Of the many services available on the Internet of this nature, Skype is by far the most famous. The company is now owned by eBay.

Softphone: A program that runs on a personal computer that provides similar functionality to a telephone, including making and receiving calls, putting a call on hold, transferring a call, and conferencing. A softphone requires that a headset with a microphone and speakers be attached to the computer if it is to duplicate the capabilities of a telephone.

Speex: The Speex project is an attempt to create a free software speech codec, unencumbered by patent restrictions. Unlike many other speech codecs, Speex is not targeted at cell phones but rather at VoIP and file-based compression. CELP is the basis for Speex and it is well suited to Internet applications.

STUN: Simple Traversal of UDP through NATs (STUN) is a network protocol designed to facilitate making VoIP calls through a firewall using Network Address Translation. A STUN server allows clients to find out their public address, the type of NAT they are behind and the Internet side port associated by the NAT with a particular local port.

Superclass softswitch: Superclass softswitch is a type of software that telephone companies add to their central office TDM switches to add VoIP capabilities. The resulting combination is often called a "hybrid."

Toll Bypass: Toll bypass is used by corporations to decrease their long distance costs by using leased lines to transmit telephone calls between branch offices.

Type of Service: Type of Service (ToS) is a one octet field in the IPv4 header that provides precedence and type of service information. It can be used by routers to give priority to packets. In modern networks, the use of this field has been superseded by the DiffServ protocol.

Virtual LANS: Virtual LANs allow similar devices to be grouped together such that they can communicate easily with each other, but not with devices belonging to other VLANs. This allows the traffic from devices that generate voice data to be segregated from the traffic of devices that generate computer data.

Voice coding: Voice coding is the opposite of waveform coding when it comes to encoding speech for a VoIP system. It attempts to recognize the different parts of speech and encodes them accordingly. The voice coder "vocoder" cannot be used for encoding music.

Voice over IP or VoIP: VoIP is the technology used to describe telephone services over a Transmission Control Protocol/Internet Protocol (TCP/IP) network. The key concept here is that the telephone call is being made over a data network and this can be a private company network, the Internet or the network of a telephone company. The second crucial concept is that the voice traffic is not merely digital but also that it is packetized and transmitted using TCP/IP.

Waveform coding: Waveform coding refers to transforming a wave pattern into a bit pattern. Waveform coding is used for both voice and music. PCM is the technology used by waveform coding. The alternative way to encode voice is a voice coder (vocoder).

Index

D

N

U

V